Cambridge Natural Science Manuals.

BIOLOGICAL SERIES.

GENERAL EDITOR :—ARTHUR E. SHIPLEY, M.A.

FELLOW AND TUTOR OF CHRIST'S COLLEGE, CAMBRIDGE.

THE

SOLUBLE FERMENTS

AND

FERMENTATION.

THE

SOLUBLE FERMENTS

AND

FERMENTATION

BY

J. REYNOLDS GREEN, Sc.D., F.R.S.

TRINITY COLLEGE, CAMBRIDGE ;
PROFESSOR OF BOTANY TO THE PHARMACEUTICAL SOCIETY OF GREAT BRITAIN
FORMERLY SENIOR DEMONSTRATOR IN PHYSIOLOGY IN
THE UNIVERSITY OF CAMBRIDGE.

SECOND EDITION.

CAMBRIDGE:
AT THE UNIVERSITY PRESS.
1901

CAMBRIDGE
UNIVERSITY PRESS

University Printing House, Cambridge CB2 8BS, United Kingdom

Published in the United States of America by Cambridge University Press, New York

Cambridge University Press is part of the University of Cambridge.

It furthers the University's mission by disseminating knowledge in the pursuit of education, learning and research at the highest international levels of excellence.

www.cambridge.org
Information on this title: www.cambridge.org/9781107673953

© Cambridge University Press 1899

First edition 1899
First published 1899
Second edition 1901
First paperback edition 2014

A catalogue record for this publication is available from the British Library

ISBN 978-1-107-67395-3 Paperback

PREFACE.

THE various problems connected with the phenomena of fermentation have received attention during the past ten years from so many investigators in different countries, and are occupying the minds of so many people to-day, that it has seemed desirable to put together, as far as possible, the results reached up to the present time, and to put forward a view of our present position with regard to the whole subject.

The recent discovery by Büchner of *zymase*, the enzyme which is capable of setting up alcoholic fermentation, has attracted renewed attention to the whole question, and has for the first time clearly shown what has long been suspected, that the production of alcohol must fall into line with the other processes of decomposition which have for some years now been included under the one name of fermentation.

It has appeared to the writer desirable for many reasons to lay stress on the relations existing between fermentation in the broad sense and the general metabolic phenomena of living organisms. Recent discoveries have shown more and more plainly what a prominent part is played by enzymes in intracellular metabolism, till it has become clear that the distinction drawn between organised and unorganised ferments is based upon an incomplete acquaintance with the metabolic processes in both higher and lower organisms, and must now be abandoned entirely in the light of fuller knowledge. The discovery of zymase, already alluded to, causes the disappearance of almost the last resting-place of the distinction.

As soon as we cease to associate fermentation solely with the lowly forms of living organisms, the vitalistic theory of the writers of the school of de Latour and Pasteur, which makes alcoholic fermentation especially the expression of that life under different conditions of nutrition or aeration, is seen to be no longer sufficient to embrace all the facts. The more recent work takes the subject beyond the stage at which it was left by Pasteur, showing us that precisely similar operations are incidental to the life of the higher organisms. It thus becomes necessary to enquire into the relationship of protoplasm to metabolism, and to the association of ferments or enzymes with the living substance, and so to establish the intimate relationship of fermentation to the ordinary metabolic processes. It becomes possible to go further than this, and to consider by what chemical or physical processes the observed changes or decompositions are effected by protoplasm or by its secretions.

The comparatively recent speculations of Emil Fischer upon the configuration of the enzymes and the various bodies they attack, based as they are upon careful and painstaking investigations, direct us towards a new chemical hypothesis of their action, which, while not contravening the views of the vitalistic school of the earlier observers, certainly extends, if it does not complete them. That the work of enzymes will be ultimately shown to be chemical rather than physical in nature is rendered probable by the able researches of Croft Hill, to which attention is called at the conclusion of this work.

J. REYNOLDS GREEN.

CAMBRIDGE,
April, 1899.

PREFACE TO THE SECOND EDITION.

IN the two years which have elapsed since the publication of this work there has been a large increase in the number of memoirs bearing upon its subject. I have endeavoured, without unduly extending the size of the volume, to incorporate into it the bulk of the new matter and to include the principal memoirs, numbering more than a hundred, into the bibliography at the end. Several new enzymes have been discovered, *hadromase, tannase, lotase, galactase* and *vesiculase*, besides several oxidases, and some which effect reduction instead of oxidation. These will be found treated of in their appropriate connexions.

In issuing this new edition I should like to express my appreciation of the kindness of my scientific friends in welcoming the work as they have done.

J. REYNOLDS GREEN.

CAMBRIDGE,
April, 1901.

CONTENTS.

CHAPTER I.

THE NATURE OF FERMENTATION AND ITS RELATION TO ENZYMES.

Early views of fermentation ; work of Becher, Leuwenhoek, Lavoisier ; discovery of the true nature of yeast by de Latour, Schwann, and Kützing ; views of Pasteur, Liebig, and Naegeli. Discovery of enzymes ; organised and unorganised ferments ; vital phenomena of protoplasm and its relation to fermentation. Characteristics of enzymes ; classification.

CHAPTER II.

DIASTASE. (*AMYLASE, PTYALIN.*)

Discovery of diastase ; work of Kirchhoff, Leuchs, Payen and Persoz ; different sources of diastase. Diastase of translocation ; its distribution ; occurrence in seeds, leaves, tubers, pollen, fungi. Diastase of secretion ; structure of the scutellum ; variations in secretion during germination ; the aleurone-layer of grasses and its relation to diastase ; differences between the two varieties of vegetable diastase ; diastase in *Bacteria*.

CHAPTER III.

ANIMAL DIASTASE.

Discovery of diastase in saliva and in pancreatic juice ; distribution in the animal body ; secretion of saliva ; antecedent of diastase ; diastase in the liver ; distribution of diastase in the Invertebrata.

CHAPTER IV.

PREPARATION OF DIASTASE AND ITS COURSE OF ACTION.

CHAPTER V.

CONDITIONS OF THE ACTION OF DIASTASE.

CHAPTER VI.

INULASE.

CHAPTER VII.

CYTASE, AND OTHER CELLULOSE-DISSOLVING ENZYMES.

CHAPTER VIII.

SUGAR-SPLITTING ENZYMES.

CHAPTER IX.

SUGAR-SPLITTING ENZYMES (*continued*).

CHAPTER X.

GLUCOSIDE-SPLITTING ENZYMES.

CHAPTER XI.

PROTEOLYTIC ENZYMES. PROTEOLYSIS.

CHAPTER XII.

PROTEOLYTIC ENZYMES (*continued*).

CHAPTER XIII.

PROTEOLYTIC ENZYMES (*continued*). *VEGETABLE TRYPSINS.*

CHAPTER XIV.

FAT-SPLITTING ENZYMES. LIPASE. (*PIALYN, STEAPSIN.*)

CHAPTER XV.

THE CLOTTING ENZYMES. RENNET.

CHAPTER XVI.

THE CLOTTING ENZYMES (*continued*).

THROMBASE, (*THROMBIN*), THE FIBRIN-FERMENT.

Phenomena of the coagulation of the blood; early researches; the work of Schmidt and Hammarsten. Discovery of thrombase; its preparation from serum and from fibrin; nature of thrombase. Conditions affecting coagulation; influence of calcium salts; zymogen of thrombase; inhibition of coagulation, and conditions of action of thrombase; its possible nucleo-proteid nature; intravascular clotting. Crustacean fibrin-ferment. Myosin-ferment; comparison of myosin and fibrin; preparation of the enzyme and peculiarities of its action. Vesiculase. pp. 266—286.

CHAPTER XVII.

THE CLOTTING ENZYMES (*continued*). PECTASE.

The formation of vegetable jellies; early work of Fremy; discovery of pectase. Researches of Mangin on the pectic bodies in plants. Work of Bertrand and Mallèvre on pectase. Preparation of the enzyme and its relation to calcium salts; conditions of the action of pectase; its distribution in the vegetable kingdom; differences in the composition of the cell-wall; formation of the middle lamella. pp. 287—300.

CHAPTER XVIII.

AMMONIACAL FERMENTATION. *UREASE*.

Changes which take place in urine on standing. Organisms causing ammoniacal fermentation. Decomposition of urea and of hippuric acid. Urease; its detection; experiments of Musculus; of Lea; of Miguel. Conditions of activity of urease. Histozyme. pp. 301—307.

CHAPTER XIX.

OXIDASES, OR OXIDISING ENZYMES.

CHAPTER XX.

ALCOHOLIC FERMENTATION.

CHAPTER XXI.

THE FERMENTATIVE POWER OF PROTOPLASM.

CHAPTER XXII.

THE SECRETION OF ENZYMES.

Conditions antecedent to secretion; nature of stimuli causing its occurrence; process of secretion in animal cells; sequence of changes which may be observed; process in vegetable tissues. Zymogens. Part played by the nucleus in the act of secretion; work of Macallum; formation of enzyme a gradual process; relation between zymogens and enzymes. Zymogens of pepsin, trypsin, rennet, diastase. Vegetable zymogens. The zymogen of thrombase. Differences in behaviour between zymogens and enzymes.

CHAPTER XXIII.

THE CONSTITUTION OF ENZYMES.

Reactions of enzymes; fallacious character of colour tests; association of enzymes with various proteids; relation existing between the proteid and the enzyme. Are enzymes proteids? Difficulties in the way of accepting this view. Study of invertase by O'Sullivan and Tompson. Theory of the nucleo-proteid constitution of enzymes; work of Pekelharing, Macallum, Spitzer.

CHAPTER XXIV.

THE MODE OF ACTION OF ENZYMES.
THEORIES OF FERMENTATION.

Early theories of fermentation. Views of Valentinus, Libavius, Lemery, Willis and Stahl, Gay-Lussac. Discoveries of de Latour and his contemporaries; the vitalistic theory. Controversy between the supporters of the physical and the vitalistic theories. Views of Liebig, Pasteur, and Naegeli. Discovery of enzymes. Differences supposed to exist between their action and that of organised ferments. Views of Naegeli and Sachs. Fallacies underlying such supposed differences. Relation of fermentation to metabolism. Modern theories of fermentation. Researches of Fischer and the configuration hypothesis; theory of electric hydrolysis; theory of de Jager and Maurice Arthus; chemical hypothesis of Bunsen and Hüfner; researches of Croft Hill and their support of this theory. Destruction of enzymes by heat.

ERRATA.

On p. 149, line 21 *for* 4 cm.   *read* excess

p. 212, „ 26 }
p. 302, „ 25 }   „ amides   „ amido-acids

p. 233, „ 35 }
p. 341, „ 23 }   „ casein   „ caseinogen
p. 342, „ 16 }

p. 245, „ 33   „ two years later   „ in 1891

p. 268, „ 30   „ tyrein   „ casein

p. 427, „ 12   „ page 177   „ page 187

CHAPTER I.

LIKE so many of the phenomena which are associated with the presence of life, fermentation has been the subject of careful study extending over centuries of time. It has been the object of experiment at the hands of many investigators, and its nature has been the theme of vigorous controversy. Hypothesis after hypothesis has been advanced to explain the various phenomena which it presents, each in turn marking an advance in knowledge, but none of them sufficing to account satisfactorily for the whole of the facts observed. Though within the past two centuries very substantial advances have been made, much still remains undiscovered.

The term fermentation was first applied to the process which leads to the formation of alcohol, the knowledge of which goes back to very remote antiquity. The name probably arose from the copious evolution of gas which accompanies the production of the spirit, and which gives the liquid in which it is taking place the appearance of a gentle ebullition.

It is not surprising that in early times very mistaken ideas arose as to the nature of the process, based upon this appearance. Wherever a disengagement or evolution of a gas was noticed it was thought to be due to similar causes. Hence we find some of the older writers classing with fermentation the effervescence which ensues when an acid is brought into contact with chalk. The evolution of gas in the animal intestine was held more correctly to be a kindred phenomenon.

Though many of the reactions thus grouped together were possessed of very little in common, one process which was of very early date was associated correctly with fermentation. This was the action of leaven in the preparation of bread. The evolution of gas was observed in connection with the raising of the dough, though no further resemblance to the alcoholic fermentation was recognised. Indeed the speculations concerning leaven and its action were of the wildest order, some writers comparing it to the hypothetical philosopher's stone, and attributing to it the power of transforming the dough into something resembling itself.

One fact of importance came out amidst all the mass of confusion, though its interpretation left much to be desired. This was the discovery that a very small quantity of leaven was capable of fermenting or transforming an almost indefinite amount of dough. The latter was however considered to be converted into leaven.

Another point of resemblance between these two fermentative processes was known at a comparatively early date. Just as the aeration of dough was associated with the presence of leaven, alcoholic fermentation was found to be accompanied by the formation of a deposit in the fermenting liquid, which took the form sometimes of a sediment, sometimes of a scum floating on the surface. By many writers considerable importance was attached to this deposit, and, as in the case of the leaven, some special occult force was attributed to it, by virtue of which it set up the changes which could be observed in the liquid. It was consequently called a *ferment*.

The true nature of this deposit, like that of leaven, for a very long period was not investigated, nor the part it plays at all understood.

About the end of the sixteenth century the process of putrefaction was associated with the two fermentations so far mentioned, and the three shown to have much in common. The confusion however which led to the association of effervescence with them was maintained for some considerable time. It was not till 1659 that de la Boë pointed out that they differ considerably from effervescence, the chief reaction of the latter being one of combination, while in the cases of fermentation it is one of decomposition.

The first clear pronouncement upon alcoholic fermentation was made by Becher in 1682, and it marks an epoch in the development of our knowledge of the subject. This author ascertained that only saccharine liquids are capable of undergoing alcoholic fermentation, and he showed that the alcohol does not exist as such in the original "must" of wine, as had been supposed, but is formed during the operation of fermentation. Becher considered its formation to be due to a kind of combustion of the sugar, as he ascertained that air is needed to set up the phenomenon.

About the same time the deposit or scum which had been observed as an invariable accompaniment of fermentation was examined microscopically by Leuwenhoek, who showed it to be composed of little ovoid or spherical globules, but he was unable to determine their nature.

Though the subject was somewhat vehemently discussed and various theories were advanced to explain the observed phenomena, but little real advance was made during the next century. The acetic fermentation was discovered and the similarity of putrefaction and fermentation in general insisted upon. But the nature of Leuwenhoek's globules remained undetermined, many observers, especially Fabroni, holding them to be of animal origin. The work of Lavoisier towards the close of the eighteenth century threw a flood of light upon the process. The great chemist studied quantitatively the relations of the sugar to the derivatives formed from it during the fermentation, and came to the conclusion that the operation consisted of a splitting up of the sugar into two parts, one of which became oxygenated to form carbonic dioxide, while the other was converted into alcohol. He said that if it were possible to recombine these two bodies, sugar would again be formed.

About the year 1815 analyses by Gay-Lussac, Thénard and de Saussure fixed definitely the composition of sugar and alcohol. During the early years of the present century indeed the views of Lavoisier were the basis of much research from the chemical side, and the work resulted in determining the changes which take place in the sugar.

During the same period inquiries were made by many observers into the nature of the globules and their relation to the fermentation. Astier in 1813 and Desmaziéres in 1826 adhered to Fabroni's view that they were of animal origin, Astier holding that they could only live at the expense of the sugar which they decomposed.

Shortly afterwards Cagniard de Latour, repeating Leuwenhoek's experiments, saw that the globules consisted of a definite organism, capable of reproduction by budding, and belonging apparently to the vegetable kingdom. He concluded that probably they disengaged the carbonic dioxide and fermented the liquid by some effect of their vegetation.

This discovery, which is really the basis of the present views of the subject, was also made almost simultaneously by Schwann at Jena, and by Kützing at Berlin, who were confirmed by Quevenne, Turpin and Mitscherlich. The organism was referred by some to the Fungi and by others to the Algae, but its true systematic position was first ascertained by Meyen, who pronounced it a fungus and placed it in a new genus, to which he gave the name *Saccharomyces*.

These researches laid the foundation for the more complete and satisfactory views of Pasteur, whose investigations have thrown so much light upon the whole process of fermentation. The association of a definite organism with the decomposition of the sugar and the idea that the latter was in some manner connected with the manifestation of its vital processes, removed the question from the realm of chemistry in the narrower sense and gave it a place among the problems of physiology. Pasteur in studying the subject from the latter point of view made the very important discovery that the production of alcohol is accompanied by the coincident formation of glycerine and succinic acid, and he determined by quantitative methods that about 4 per cent. of the sugar which disappears during the progress of a fermentation gives rise to these two new derivatives.

Pasteur came to the conclusion that the exercise of the

fermentative power by the Saccharomycete, or yeast, was connected with nutrition in the absence of free oxygen, and that it was really the expression of the effort of the organism to obtain oxygen in the absence of a free supply. In this connection some investigations of Lechartier and Bellamy may be recalled. These observers made a number of experiments with ripe succulent fruits, which they kept for several months in an atmosphere devoid of oxygen. Under these conditions the fruits gave off continuously a certain quantity of carbon dioxide, and at the end of the time of observation the pulp contained a measurable amount of alcohol. Microscopic observation showed that this fermentation took place in the absence of yeast cells, and was in fact carried out by the living substance of the cells of the pulp. Pasteur confirmed these observations and regarded them as strengthening his hypothesis.

Schützenberger has argued with some force against the view that the decomposition of the sugar is related to the respiratory process of the yeast, preferring to regard it as ministering to its nutrition. A discussion of this point must however be left to a subsequent chapter.

Though the views of the Pasteur school did not at once meet with universal acceptance, being opposed strongly by Liebig and by Naegeli, both of whom supported a theory of molecular vibration as explaining the decomposition of the sugar, no doubt is now entertained that the living protoplasm of the vegetable cell is the ultimate cause of the fermentative process, and that the latter is the expression of some activity connected with the maintenance of its life. That it is not a specific peculiarity of the protoplasm of the yeast cell, but is shared by that of much higher plants, is equally clearly shown by the work of Lechartier and Bellamy already alluded to.

The influence of living cells of yeast in the fermentation or leavening of dough, and that of other vegetable organisms in the processes of putrefaction established during the present century show how essentially similar these processes are to the alcoholic fermentation. Micro-organisms have been found capable of setting up also many other decompositions comparable with all these.

Whilst Pasteur's views were gradually making themselves felt, work in another direction was progressing which was destined to materially enlarge our views of the whole subject. Quite in the early part of the present century (1814) Kirchhoff observed that germinating barley contained something which was capable of liquefying starch-paste and that in the process it formed a kind of sugar. Kirchhoff associated this power with the albuminoid material or gluten of the grain. In 1823 the observation was repeated by Dubrunfaut, and in 1833 Payen and Persoz extracted from germinating barley the substance which effected the decomposition. They steeped the grain in water for some time and filtered off the extract. On adding a large excess of alcohol to the filtrate they obtained a white flocculent precipitate, which when dried and dissolved in water was found capable of converting starch-paste into sugar. The change induced by this substance, which they called *diastase*, was held to be a kind of fermentation. It appeared to act in the same way as the globules of Leuwenhoek, the nature of which was at that time undetermined, as already explained. At the same time it was clearly an unorganised substance, though very little was definitely known about it. On this account it was distinguished as an *unorganised ferment*.

Two years earlier Leuchs had noticed that saliva possessed the same property as the germinating barley in that it was capable of converting starch-paste into sugar. In 1845 Mialhe showed that a preparation of diastase could be prepared from this animal secretion by the same process as Payen and Persoz had adopted for their barley extract.

This discovery was followed by others. In 1836 Schwann demonstrated the existence of *pepsin* in gastric juice, and showed that it decomposed indiffusible albuminous bodies into others that were capable of passing through membranes. Berthelot found that a watery extract of yeast, quite free from the cells of the plant, was capable of converting cane-sugar into two other sugars of simpler composition. Liebig and Woehler discovered a similar body in the seeds of the almond, which decomposed amygdalin with the formation of sugar and other products, and they were again struck by the similarity of the action to that of yeast in alcoholic fermentation. About 1860 Brücke prepared from the mucous membrane of the stomach the ferment first observed by Schwann which split up insoluble proteids, causing the formation of soluble peptone.

All these various ferments were recognized as associated in some way with the living substance of either animal or vegetable cells, and they were soon held to play an important part in the life of the organism from which they were extracted.

Two classes of ferments appeared to be thus indicated, the one a living organism, working only during its own processes of growth and multiplication; the other consisting of substances which could be extracted by solvents from the cells in which they were formed, and capable of setting up decompositions apart from the life of such cells. The two categories were consequently called *organised* and *unorganised* or *soluble* ferments. The term *enzyme* is now generally used to indicate the latter class.

In studying these various bodies and the ultimate relations of the one class to the other it is evident that attention must be given to the vital phenomena of protoplasm, and to the changes which take place in its substance during the manifestation of its life. The work of recent years has thrown much light upon the various operations that take place in cells, and we now know that these are for the most part, if not entirely, regulated by the behaviour of the living substance.

Various views have been put forward as to the arrangement of the living protoplasm. Though this is a matter which can by no means be regarded as finally established, there is some probability that it is disposed in the form of a network, the meshes of which are occupied by a material which is similar in composition but which is not living. The great characteristic of protoplasm is its instability; it is continually undergoing decomposition and reconstruction. Some of the residues of its breaking down are capable of being built up again into its substance; others are thrown off from it. Of these some are eliminated entirely from the organism, others are retained within it to carry on some of the more subordinate processes of its life or its nutrition.

An animal or vegetable cell is hence the centre of very vigorous activity; work is going on within it in the direction of incorporating material for the growth of the living substance, or of preparing material brought to it, so that it may be capable of such incorporation. Again some of its substance may be undergoing decomposition with a view to supplying the energy which it needs for the maintenance of its vital processes.

The chemical changes involved may be of three kinds. The decompositions may involve the incorporation of material into the actual substance of the protoplasm and the subsequent splitting off of various residues from the latter. Such appear to attend the formation of the various enzymes such as diastase; also the formation of fat, starch, and other compounds which can be seen in various cells. Other changes may take place without the establishment of such an intimate relationship with the protoplasm. They may be carried out by the protoplasm outside its own substance, the materials affected not being incorporated in it while the change is taking place. Such decompositions have been alluded to by various writers as caused by the fermentative action of protoplasm itself.

A third class of reaction may take place in the cell without the actual intervention of the protoplasm at all. It is probable that processes of oxidation and reduction are taking place among the substances which occupy the meshes of the protoplasmic network, and that quiescent as the cell appears it is the seat of many chemical reactions of this kind. Thus the formation of sugar in the cells of leaves under the influence of chlorophyll, which probably involves the polymerisation of some form of aldehyde, need not necessarily involve the action of the living substance, as such polymerisation is very frequent among aldehyde bodies.

Some of the decompositions of the latter class may be distinguished from the others by the fact that though protoplasm is not immediately concerned in bringing them about, it prepares from its own substance an enzyme by which the transformation is effected. The secretion or formation of this new factor belongs to the first class of reactions mentioned, but the material once secreted is endowed with the power of setting up and maintaining the decompositions in question. The ultimate purpose of the secretion is usually the digestion of some form of food material to prepare it for incorporation into the living substance itself. But recently reason has been found to believe that the processes of oxidation and reduction are carried out by similar agencies. We can consequently recognise in many organisms both digestive and respiratory enzymes. So many of these bodies are now known that it does not seem very unreasonable to put forward the view that all decompositions of this kind will ultimately be found to be carried out by such a mechanism. The number of cases in which the direct rather than the indirect intervention of protoplasm appears to be involved is continually growing smaller as further work upon metabolism proceeds. A very important addition to our knowledge in this direction has been made during the past two years by Buchner, who has ascertained that even the earliest known fermentative process, the formation of alcohol from sugar, is carried out by a soluble ferment which can be extracted from the yeast cell.

The constructive processes which take place in the cell have not been so clearly shown to be carried out by such secretions. In most cases in which a building up of complex from relatively simple substances takes place in the cell, it is carried out by the direct intervention of the protoplasm. Hence the study of fermentation is mainly directed to processes of degradation whereby complex substances are broken down into more simple ones. To this point we shall return later.

We may consequently for the present define fermentation to be *the decomposition of complex organic material into substances of simpler composition by the agency either of protoplasm itself or of a secretion prepared by it.*

We find instances of both these methods in unicellular and in multicellular organisms, as well in the animal as in the vegetable kingdom. By far the greater number of instances of the fermentative activity of the protoplasm as apart from secreted enzymes may be found in the so-called *organised* ferments, the yeasts and the great group of micro-organisms or *Schizophyta*. So prominent indeed are these plants in their power of exciting various decompositions that they were origin-

ally regarded as the only "ferments," the fact that the "fermentation" was merely incident to their own biological process
passing almost unnoticed. The theory that this power or
property puts them into a class distinct biologically from other
and higher plants is however quite disproved by the discovery
that many of them provoke the decomposition associated with
them by means of enzymes, which can with a little care be
prepared and separated from them with almost as much ease as
from the higher plants themselves. Moreover the enzymes they
secrete are identical with those which are prepared by the latter.

It follows from this that those processes of fermentation in
which the protoplasm is directly involved are intracellular. It is
only comparatively lately that this has been realised in the case
of many vegetable organisms. The presence of a cell-wall
clothing the living substance would render impossible the contact necessary to produce externally the decompositions observed.
This point still needs emphasising in the case of many of the
bacterial fermentations. Not only in unicellular but in multicellular organisms also this can be observed. Instances are
afforded by the alcoholic fermentation of ripe fruits noticed by
Lechartier and Bellamy, and more recently studied by Gerber;
by the transformation of glycogen into sugar in the muscles
and possibly the liver of the higher animals, though it is not
certain that the latter is not the seat of a fermentation by
a variety of diastase. The power of forming acids possessed
by various bacteria is shown also by the cells of the succulent
parenchyma of the higher plants. Though acetic acid is formed
by *Mycoderma aceti* from alcohol, and such parenchyma appears
to form the acids it contains from sugar, the protoplasm in
both cases is the active agent. The acids are more complex
in the latter case, but this is probably due to the character of
the metabolism of the two classes of cells respectively.

Turning to the processes of fermentation which are carried
out by means of enzymes we find more complexity. In the
simplest cases of unicellular plants and animals intracellular
fermentation is most general. The complex substance, which
is usually a food-material, is absorbed into the cell; an enzyme
is secreted there, and the work of transformation follows.
From various unicellular plants enzymes have been extracted

which do not escape during life, but which are responsible for various chemical reactions in the cells. Evidence of the existence of others in other cells has been obtained, though the enzymes have never been extracted. Thus in the digestive processes of certain unicellular animals, notably Amœba, Carchesium, and Actinosphærium, ingested food has been found, soon after its absorption, to be surrounded by a vacuole containing an acid liquid, and has been seen to dissolve gradually under the influence of the contents of the vacuole, in just the same way as similar food-material can be digested *in vitro* by artificial gastric juice.

But the process of intracellular digestion is not the only one which is carried on. Many cells after forming the enzyme excrete it from their substance into the surrounding medium, where it brings about decompositions the products of which are subsequently absorbed by the organism. Many bacteria have been found to behave in this way, but it is a phenomenon which is associated more generally with multicellular structures, both animal and vegetable. So long as a single cell or mass of protoplasm constitutes the whole organism, its metabolic and nutritive processes go on throughout its whole substance. But with the multiplication of cells in the body and the attendant differentiation of members and organs, special cells are found devoted to the preparation of enzymes. Others in turn are specially set apart for the absorption of nutritive materials. It is evident that in such cases the utilisation of the enzymes is only possible when they are excreted from the body and allowed to do their work in such positions that the products which they form can be brought easily into the absorptive region. Hence we get gradual differentiation of the complex glands, found for instance in the alimentary tract of the human body, which excrete their enzymes into a region which is external to the cells; indeed for many reasons the alimentary tract itself may be regarded as merely an invagination of the external surface of the body.

The distinction which was formerly drawn, as already mentioned, between organised and unorganised ferments is thus seen to be misleading. The fundamental difference which

appears when the two modes of activity are compared is found
to be a difference of differentiation of the organism. The
unicellular being appears in most cases to carry out its work
throughout its whole substance, and the magnitude of the sphere
of the changes it sets up being far greater than its own dimensions it appears only as something provoking large disturbances
in some organic compound, the fact that these are only
incidental to its own biological needs being thereby obscured.
The multicellular organism devoting only a definite part of its
substance, and that generally a small one, to the transformations, these are easily recognised as biological in their end and
aim.   Both unicellular and multicellular structures as we have
seen carry out the work by identical means, either the fermentative activity of their protoplasm or the secretion and
sometimes subsequent excretion of enzymes.

The study of fermentation thus resolves itself very largely
into an investigation of the destructive or decomposing power
of protoplasm, exerted either directly or by means of secreted
enzymes.   This may perhaps be narrowed still more, for as
already noted many of the cases in which protoplasm has been
supposed to act directly have been found to be additional
instances of enzymic powers.   The improved methods of preparation which have been discovered have in many cases led
to the isolation of enzymes where they have not been suspected.
The search for the enzyme which produces alcohol from sugar
has been carried on by many investigators, and as mentioned
above has now been crowned with success.   The importance
of this discovery can hardly be overestimated, as the yeast
cell has so long been held to occupy the most prominent
position as an organised ferment.   There still remain however
many cases in which the existence of an enzyme is only
hypothetical.

Turning to the consideration of the enzymes as a particular
group of substances we find that certain general features can
be ascribed to them in common, though they present great
differences among themselves.   A discussion of their possible
or probable composition must be deferred to a later chapter,
but for the present we may note that it is very difficult to

hazard more than a conjecture on the subject, as we have no criterion of their purity, nor indeed any satisfactory test for their existence except the demonstration of their powers.

Their activity is very largely dependant upon temperature, not being manifested at low temperatures approaching the freezing point of water; it increases gradually as the temperature rises, reaching its maximum at a point which varies somewhat for each enzyme, but which ranges between 30° and 50° C.; beyond this so-called *optimum* point it gradually diminishes, and finally disappears. By the action of a temperature approaching 100° C. they permanently lose their characteristic properties and are destroyed or decomposed. The deleterious effect of this high temperature is apparently dependant upon their being in contact with water, for if perfectly dry they will survive heating to a much higher point.

It is generally held that they do not themselves enter into the reactions which they set up, but provoke those changes without undergoing any alteration. Further, they are not destroyed by their own activity, the energy with which they work not apparently proceeding from any decomposition of their substance.

Certain facts which have been observed recently throw a certain amount of doubt on the accuracy of these generally accepted statements, but a discussion of these points must be deferred to a subsequent chapter.

The enzymes show a peculiar sensitiveness to the environment in which they find themselves. They are peculiarly influenced by the reaction of the solutions in which they are working, some being active only in acid, others in neutral, and others again in alkaline media. Nor is their activity alone affected, for slight alterations in the composition of a medium favourable to them will frequently result not only in the stoppage of their action but even in their own destruction. Their activity again is in all cases lessened and finally paralysed by the presence of an excess of the products which they form.

For convenience of discussion the enzymes have been classified according to the materials on which they work. The following groups are fairly well established.

(1) Those which transform insoluble carbohydrates of various kinds, ultimately producing soluble sugars. Of these we recognise the various forms of *diastase*, which attack starch and its allies; *inulase*, which decomposes inulin; *cytase* which hydrolyses cellulose.

(2) Those which transform sugars of the biose type into simpler sugars, usually hexoses. Of these we have instances in *invertase* which attacks cane-sugar; *glucase* which splits up maltose, and others decomposing other bioses.

(3) Those which decompose glucosides, giving rise to a form of sugar together with various aromatic bodies. These include *emulsin, myrosin,* and several others.

(4) The proteolytic group, whose members decompose various forms of insoluble proteids. Among them are *pepsin* and *trypsin,* which play so prominent a part in animal digestion.

(5) The clotting enzymes, which produce from various soluble bodies jelly-like substances which involve a coagulation of the liquid in which they are dissolved. Included in this group are *rennet,* which coagulates milk, *thrombase,* which plays a part in the coagulation of the blood, and *pectase* which is the leading factor in the formation of vegetable jellies.

(6) The enzyme which decomposes oils or fats, which has recently been named *lipase.*

(7) Oxidases, or oxidising enzymes, which, as their name implies, assist in the oxidation of various substances. Among them are *laccase* and *tyrosinase.*

This list is not a complete one, for a few other enzymes are known, though they appear to occupy isolated positions. Among them may be mentioned *urease,* which forms ammonium carbonate from urea, and the new *zymase* of Buchner, which is the alcohol-producing enzyme.

CHAPTER II.

DIASTASE. (*AMYLASE, PTYALIN.*)

OF the soluble ferments or enzymes diastase has been longest known. It was first observed in 1814 by Kirchhoff that germinating barley grains contain a principle which when extracted from them by water is capable of converting starch into sugar. The power at the time was considered to be due to some property of the gluten of the grain, and was not put down to a separate constituent of the barley. But little notice was taken of the discovery for some years, but in 1823 Dubrunfaut repeated the observation and slightly extended it, ascertaining that the transformation was most easily effected at a temperature of 65° C., and that the sugar produced was capable of alcoholic fermentation. In 1831 it was discovered by Leuchs that saliva also possessed the power of transforming starch into sugar.

These observations led the way to the work of Payen and Persoz, who were the first to prepare diastase from the extracts of germinated grain. In 1833 these observers steeped ground germinating barley for some time in water, and after filtration of the extract, precipitated from it, by the addition of strong alcohol, a white flocculent material which when dried and redissolved was found to have the power possessed by the original extract. Payen and Persoz extended their observations, subsequently discovering the same body in oats, wheat, maize, and rice during the process of germination and in the tubers of the potato while they were in course of growth.

The name *diastase* was given by its discoverers to the material precipitated by alcohol.

The observation of Leuchs in 1831 was followed up in 1845 by Mialhe, who applied to saliva the same treatment as that adopted by Payen and Persoz in the case of extracts of germinated grain. He added to saliva absolute alcohol in excess, and thereby produced a precipitate which like the vegetable diastase was found to possess a starch-transforming power.

He named this body *salivary diastase,* indicating its similarity to, if not identity with, the body prepared by Payen and Persoz. The name has fallen into disuse in works on physiology, the term *ptyalin* having been generally adopted in its place.

The power to saccharify starch was in the same year discovered to be present in another animal secretion, that of the mammalian pancreas. The discoverers were Bouchardat and Sandras, and their work extended so far as to enable them to claim that a variety of diastase can be extracted from that organ.

Thus at comparatively early dates there were established three sources of diastase, two of animal and one of vegetable origin. For a long time these were regarded as the special situations of the enzyme. Within the past three decades however a far more widespread distribution has been established both in the animal and the vegetable organism.

Dealing in the first place with investigations carried on in the region of plant life, we find the subject taken up again in 1874 by von Gorup-Besanez, who found the enzyme in several other varieties of germinating seeds. He was followed by Kosmann in 1877, Baranetzky in 1878, and Krauch in 1879. Owing to the investigations of this group of workers the presence of diastase was established in the ordinary vegetative members of the plant and in other resting bodies besides seeds. Thus Kosmann and Krauch recognized its presence in the leaves and shoots of the higher plants, and also in certain algae, lichens, mosses and fungi, while Baranetzky found it both in buds and potato-tubers. These extended observations led the latter writer to suggest that it is universally present in vegetable cells, so long as the latter are living.

In 1884 a very important research upon this subject was

carried out by Brasse. He used the leaves of the potato, dahlia, Jerusalem artichoke, maize, beet, tobacco, and castor-oil plant, of all of which he made watery extracts. On the addition of alcohol to these extracts a flocculent precipitate fell, which he collected and dried. On adding some of this powder to starch-paste the latter became liquefied and converted into sugar. The amount of sugar formed was then ascertained by observing the weight of cuprous oxide which it formed when it was boiled with Fehling's solution.

Brasse guarded his experiments from error due to the possible presence of micro-organisms by carrying out his digestions in the presence of traces of chloroform, which prevents the growth of such organisms.

The investigations of many writers during more recent years have gone far to establish the truth of Baranetzky's suggestion, while they have extended our knowledge of diastase so far as to show that at least two varieties of it exist in the vegetable organism.

The first of these is the more widespread; it apparently plays the principal part in the transport of starch from place to place about the plant and is most commonly met with in the vegetative organs. The other is almost if not entirely confined to germinating seeds, particularly those of the Gramineæ, and is probably the body first prepared by Payen and Persoz. These two varieties have been called for purposes of distinction, diastase of *translocation* and of *secretion* respectively.

Diastase of translocation.

This enzyme is the more widely distributed of the two, not only occurring in the vegetative organs but being present in the seed during the development of the embryo. Its action can be studied upon grains of starch in the vegetable tissues themselves, or it can be extracted from the tissues by water or by glycerine, and its behaviour tested by mixing such an extract with either a thin starch-paste or a solution of soluble starch. In the former case the diastasic solution gradually dissolves the starch grain, sometimes from the outside only, sometimes from the interior, but it never gives rise to any pitting or corrosion, so that the shape and transparency of the grain remain unaffected almost to the time of its disappearance. When the diastase is allowed to work upon a thin starch-paste, made by mixing about 1 gramme of starch with 100 c.c. of boiling water, the opalescent liquid slowly becomes more and more transparent till it is quite limpid; it then gradually undergoes almost complete conversion into sugar. If a true solution of starch is substituted for the starch-paste the transformation into sugar is much more rapid. The details of the action will be discussed in a subsequent chapter.

The action of this form of diastase takes place most energetically at a temperature of 45°—50° C.

Diastase of translocation has been described as occurring in leaves, shoots, and certain reserve-organs. In 1889 it was found by Kjeldahl to be present in the ungerminated grains of barley, a discovery confirmed by Brown and Morris the next year. Kjeldahl showed that while the extract of ungerminated barley was able to convert solutions of soluble starch rapidly into sugar its activity on thin starch-paste was but slight. Brown and Morris studied the formation and distribution of the enzyme during the process of ripening or maturation of the barley grain. The chief place of its occurrence is the bulk of the endosperm, in which it may be found during its development. It is always most plentiful in the part of the endosperm which is nearest the young embryo, and it appears to prepare the material for the nutrition of the latter as it increases in size. It thus appears in the barley grain at a much earlier period than the other variety of diastase, to be described later, which is only found after maturation and indeed at the commencement of germination. The translocation-diastase makes its appearance at a very early period in the development of the endosperm, and gradually increases until that development is completed but the grain not ripened. Brown and Morris estimated carefully the amounts of diastase present in barley grains at three stages of their formation;—(i) when the endosperm was half developed; (ii) when the development had reached two-thirds of its total amount; (iii) when it was

complete. The amount of the diastase was ascertained by
finding what amount of soluble starch could be converted into
sugar by the ground-up substance of the same number of barley
grains acting for the same time and under the same conditions.
The relative quantities were found to be in the proportion of
4·4 : 7·8 : 9·7, showing that as the embryo grows the quantity
of diastase found in its vicinity increases *pari passu* with the
growth, confirming the idea that the enzyme makes its appearance with a view to the formation and nutrition of the embryo.

Not only does this variety of diastase appear in the seed
outside the embryo, but when the latter begins to grow during
germination the same diastase is found to be developed in its
cells, not so much in the cotyledon as in the plumule and
radicle, the parts namely in which growth is most vigorous.
Even at this early stage Baranetzky's idea of diastase as a
necessary constituent of every vegetable cell seems to be
confirmed. The function of this amount of diastase is not
connected with the absorption of sugar from the endosperm of
the parent but with translocations in the body of the embryo
itself.

The diastase of leaves and shoots has much in common with
that of ungerminated grain and appears to be the same variety.
The disappearance of starch from the leaves of plants during
the hours of darkness, after its formation and deposition in the
chloroplastids during sunlight, has always been associated with
the presence of the enzyme, though no detailed investigations
into the conditions of the disappearance were made till recently.
In 1890 the existence of diastase in foliage leaves was controverted by Wortmann, who, in consequence of failure to extract
it in many cases, asserted that the conversion of this starch
into sugar was effected by the immediate action of the protoplasm of the living cell itself, and not by the agency of an
enzyme. This statement led to the reinvestigation of the
subject of the starch-transformations in leaves by Vines and by
Brown and Morris, with the result of fully establishing the
accuracy of the observations of Brasse and other earlier writers.
At about the same time an elaborate paper by Krabbe was
published which showed that the visible mode of dissolution of

the starch-granule within the cell corresponds to that observed when it is treated with diastase outside the plant.

Vines carried on his work largely with the leaves of grasses, which he extracted with water. Some of his extracts were filtered, and others used while still turbid. The latter were found to be more active than the former, a fact which suggests that Wortmann lost a good deal of the diastase contained in the leaf by the operation of filtration, the enzyme for some reason not having fully passed into solution. Vines was able to prepare a solution of diastase from the grass leaves which exhibited a fair degree of activity. He suggests that though the amount of enzyme at any given moment may be small, there is a constant secretion or formation of it, so that the total amount produced is sufficient to cause the conversion of all the starch translocated during the night.

The work of Brown and Morris appeared in 1893 and is the fullest and most detailed that has at present been carried out on the diastase of foliage leaves.

At the outset of their investigations they were able to explain the apparent absence of diastase from many leaves which was supposed to be indicated by the very small diastasic power of filtered extracts. They dried leaves of various plants in air at a temperature of 40° to 50° C. and reduced them to a fine powder. An extract of 10 grammes of such powder, the leaf of *Helianthus tuberosus* being used in the experiment quoted, was made by steeping it for some time in water to which a trace of chloroform was added. Finally the extract was filtered. Two digestions of solutions of soluble starch were then carried out, the extract described being used in one of them, while 10 grammes of unextracted powdered leaves were employed in the other. The relative diastasic activity in the two cases was found to be in the proportion of 7 : 1 in favour of the powdered leaves, the estimation being made by observation of the cupric-reducing power of the products of the digestions.

The authors advance two reasons to explain the difference, the first being the great resistance which protoplasm often offers to the separation of the enzyme, and the second the action of the tannin so frequently present in leaves, which renders it impossible in many cases to extract the enzyme from the cells.

Brown and Morris investigated the leaves of 34 species of plants belonging to a great many different Natural Orders and found that all contained a measurable amount, though the quantities differed very considerably. The Leguminosæ were especially rich, the common Pea (*Pisum sativum*) being extremely conspicuous among them; its diastasic activity was nearly one-half that of malt.

Though exact quantitative determinations of the enzyme were hardly if at all possible, comparative estimations were found easy to make. The method of measuring the amount of starch conversion which an extract is capable of carrying out under certain standard conditions and in unit time has been shown by Kjeldahl to yield accurate results, and from it he formulated a "law of proportionality" which may be expressed thus[1]; "providing a starch-transformation is not allowed to fall below a cupric-reducing power of κ 25 to 30 for the mixed products of hydrolysis, then under identical conditions of time, temperature, and concentration, *the cupric-reducing power is proportional to the amount of diastase originally present.*"[2]

Brown and Morris further showed not only that diastase is present in foliage leaves, but that it is possible to prove that its amount varies considerably from time to time, under differences in their environment. This point will be referred to more fully in a subsequent chapter, but for the present it will suffice to say that there is a considerable formation during darkness and a conspicuous diminution during bright sunshine.

The work of Brown and Morris receives material support from some investigations published in the same year by St Jentys. He has found that the diastase is formed only in small quantities as it is required. When an aqueous extract of leaves is prepared, it usually contains tannin, which causes the precipitation of the starch. The latter is only acted on with

[1] Brown and Morris. "On the Chemistry and Physiology of Foliage Leaves," *Journ. Chem. Soc. Trans.* 1893, p. 637.

[2] κ represents the apparent percentage amount of dextrose which a substance contains as determined by its cupric reduction. Thus κ 25 means that the reducing power of the substance is one-quarter that of dextrose.

difficulty under such conditions. Moreover the tannin precipitates the diastase itself.

Besides occurring in the actively vegetative parts of the
plant this form of diastase is found also in the reservoirs of
starchy material, and is particularly conspicuous there at the
time when a call is made upon the reserve supply, as happens
for instance when growth recommences.

An investigation of the potato by Prunet has brought to
light several very interesting facts connected with the formation
and distribution of this variety of the enzyme in the tissue of
the tuber. On the onset of the germination of that organ it is
possible to notice that the buds near its apex develop more
rapidly than those situated lower down. Prunet separated the
different regions of the sprouting potato, and made watery
extracts of each. He ascertained the diastasic powers of the
several extracts thus prepared by precipitating the enzyme with
excess of alcohol, filtering, and dissolving the precipitate, and
then allowing it to act upon thin starch-paste. His results are
comparative merely and perhaps only approximately accurate,
as he used for an indication of the progress of the digestion
simply the colour obtained by the addition of iodine. As the
starch in such a digestion disappears, this colour, which is at
first blue, changes in the direction of purple, then becomes red,
and finally is not developed. The method is not so accurate as
the measurement of the sugar produced, but it yields reliable
comparative results. He found that while diastase is present
in all parts of the tuber during the progress of its germination
it is in greatest amount near the points where active growth is
proceeding. The commencement of the growth of the young
shoots is accompanied by a production of the enzyme in their
immediate neighbourhood, and as they become longer the
formation of the diastase can be traced further and further
back. Prunet estimated at the same time the sugar existing
in the same regions of the potato and found a correspondence
between the amount of the two bodies.

In 1893 the writer carried out a series of investigations on
the processes of germination of the pollen grains of various
plants, in the course of which the presence of this form of

diastase was recognized in those of several species. Starch was found to be present in many of the pollens examined and to occur also in the tissues of the styles down which the tubes penetrate in the process of the germination of the pollen grain. The diastase was looked for by similar methods to those already described and was found in most cases.

The same enzyme was discovered in the tissue of the style in some of the flowers examined.

The quantity of diastase which can be extracted from pollen varies considerably from time to time. At the onset of germination of the grain it is usually considerably increased, and it appears to accompany the tube as the latter elongates, which suggests a formation not only in the grain but in its tube also. As the pollen loses its power of germination with the increase of age, it loses at the same time the diastase which it contains. Thus, as in the potato, it is possible to trace clearly the influence which the enzyme exerts on the process of germination.

The statement of Kosmann that diastase can be extracted from fungi was for a long time unsupported. It seems somewhat unlikely at first sight that it should occur there, as these plants do not contain starch. It has however been shown that glycogen is very frequently to be met with in them, and as glycogen resembles starch so very closely this *a priori* difficulty disappears.

In 1883 some investigations were made by Duclaux into the behaviour of *Aspergillus niger* under different conditions, and he ascertained that this fungus has the power of saccharifying starch solutions in the same way as germinated barley. More extended observations were carried out by Bourquelot ten years later, and the latter botanist was successful in proving that diastase can be extracted from a culture of the fungus at the time when it is in full fructification. The estimation of the amount is difficult if not impossible, for as we shall see later *Aspergillus niger* contains other enzymes, one of which acts upon the sugar to which the starch gives rise, decomposing it during the progress of the experiment.

In 1895 Bourquelot and Hérissey demonstrated the presence of the same diastase in the tissue of *Polyporus sulphureus*. It

exists in this fungus, as it does in *Aspergillus*, side by side with
other enzymes, but can readily be extracted by expressing the
juice of the large thallus and precipitating it with twice its
volume of 95 % alcohol. The diastasic power can be demonstrated by using the juice without this treatment.

According to De Bary, diastase can be extracted from the
plasmodium of *Aethalium*, a Myxomycete, and Stone states that
it is present in *Taka*, a preparation of *Eurotium oryzae*.

Morris and Wells have shown that diastase can also be
extracted from yeast cells, and many observers, especially
Lauder Brunton and MacFadyen, and Wortmann, have found
it to be excreted by bacteria. Its presence in these organisms
seems not to be constant, but to depend upon the medium in
which they are cultivated.

Hansen has detected diastase in the latex of *Ficus elastica,*
Carica papaya, and *Papaver somniferum*. It can be separated
from the latex by precipitation with alcohol.

The existence of diastase in tissues containing starch or
glycogen is not surprising when we regard its importance in
the digestion of these carbohydrates. It occurs, however, in
other tissues in which they are very rarely present, if at all.
In the case of the sugar beet, in which the reserve materials
take chiefly the form of cane-sugar, Gonnermann states that he
prepared from crushed germinating roots an enzyme which
digested starch-paste and amylodextrin. Baranetzky detected
diastase also in the carrot and the turnip.

Diastase of secretion.

This variety of diastase is especially connected with the
process of germination and can be most favourably studied in
the grains of grasses. If the progress of the development of the
embryo in the germinating seed is studied with the aid of
the microscope, changes in the starch-containing cells of the
endosperm can be observed at a time when the primary radicle
has attained a length of about 2 mm. The walls of the cells
become broken down, and the starch grains at this period begin
to undergo a process of dissolution which is entirely different
from that which can be observed in other regions of the plant.

Instead of gradually dissolving without alteration of their
shape, they show small indentations or pittings on their
surfaces. These gradually become deeper and wider, and the
grain is soon broken up into irregular pieces, while the
separate laminæ tend to split off from each other. This process
of erosion is followed later by the gradual disappearance of the
several portions of the grain.

The formation and behaviour of the diastase which is the
instrument in effecting this change has been studied especially
by Brown and Morris and by Haberlandt, who carried out their
work chiefly on different cereals, principally barley and rye.
They agree in attributing its origin to a true process of secretion
carried out by the embryo, which thus prepares the way for its
own nutrition, pouring enzymes, of which diastase is one, into
the tissue of the endosperm by the side of which the embryo
rests.

The embryo in the grasses lies at one end of the grain and
the portion of it which is in contact with the tissue of the
endosperm is the so-called *scutellum*, a mass of parenchymatous
tissue whose morphological nature has been much disputed,
but which is probably a specially differentiated portion of the
cotyledon. Over the surface of the scutellum, abutting on the
cells of the endosperm, there is a well-marked external layer or
epithelium. This is composed of closely set columnar cells,
about three times as long as they are broad, which are arranged
with their long axes at right angles to the surface. The
cell-walls are delicate and thin, and made of unchanged cellulose.
Each cell contains a finely granular protoplasm and a fairly
large nucleus. Prior to the researches of Brown and Morris
the functions of this epithelium were thought to be confined to
the absorption of digested reserve-materials, passing from the
endosperm to the embryo. These observers have shown that
they are also the seat of the formation of one, if not two
enzymes, the first of which is the diastase under consideration,
and the other is an agent in the preliminary destruction of the
cell-walls. A few hours after placing the grains of barley under
conditions favourable for germination a marked change takes
place in the contents of the cells. The protoplasm becomes

visibly coarser in structure and the granules increase in size
and number, becoming so prominent in the cell that the nucleus
is almost hidden. These changes are complete in about 1—2
days, and the condition of the cells remains thenceforward
unaltered as long as germination proceeds. The tissue of the
endosperm above the scutellum becomes depleted of its starch,
while the coarse granularity of the epithelium persists, and the
tissues of the embryo itself, particularly the cells of the scutellum,
become more and more charged with newly formed starch
grains. When the reserve store of starch in the endosperm
has been thus absorbed, the epithelial cells again change in
appearance, losing their granularity and becoming transparent
as at first. There is, however, an important difference between
their first and their final condition, as in the latter the nucleus
is no longer to be found, indicating that the cells have finished
their work and are on the way to disintegration. The process
of germination thus shows that as it commences, active
metabolism goes on in the epithelium, synchronous with the
dissolution of the cell-walls and the transformation of starch; it
continues as long as starch is being digested, and ceases when
that process is over. The obvious conclusion to be drawn from
the observations is that these metabolic changes are concerned
in the elaboration of the secretion, which causes the disappearance of the reserves of cellulose and starch; that is, that
the epithelium is primarily a glandular structure.

    That this conclusion is not merely hypothetical has been
demonstrated by careful experiments, carried out by Brown and
Morris, upon the artificial nutrition of barley embryos removed
from germinating grains. Such embryos were carefully dissected out and their scutellar surfaces freed from adhering
matter by careful stroking by means of a fine camel-hair pencil
moistened with dilute sugar solution. They were then placed
upon a thin layer of moist barley starch, so that their scutella
rested in contact with the latter. Within a very short time
transitory starch grains could be detected in the parenchyma of
the scutella, and microscopic examination of the starch grains
in contact with the surface of the epithelium showed them
to be corroded and in process of disintegration. In further

experiments the excised embryos were partially embedded in a ·5 per cent. gelatin solution in which small quantities of barley starch had been suspended. The embedding was carried out just as the gelatin solution was on the point of solidifying, so that the curved surface of each scutellum was in close contact with the cultivation medium. After a short time small pieces of the gelatin cut out from the mass just below the growing embryo showed the starch grains affected as before. Besides barley starch, the diastase that passed out into the gelatin was found to be capable of corroding and dissolving the starches of wheat, rice, and maize, but to have no evident action on those of the potato and kidney-bean.

Further proof of the connection between the secretion of the diastase and the epithelium cells was afforded by paralysing the latter by exposing them to chloroform vapour. After such treatment they had no action on the starch.

Freshly dissected embryos of *resting* barley were subsequently rubbed down in a mortar with a little chloroform water and allowed to stand for 24 hours. The extract was then filtered and added to an equal volume of a 2·5 % solution of soluble starch and digested for an hour at a temperature of 30° C. It was found to have no diastasic action. At the same time some more embryos excised from grains of the same sample of barley were cultivated for four days upon appropriate culture media. They were then treated exactly as the first ones and the extract was found to have under similar conditions considerable power of converting the starch into sugar. By both microscopical and chemical methods Brown and Morris therefore showed the *formation* of diastase to take place *during germination*, and further, that when formed it slowly diffused out of the epithelium.

These observers advanced further proof of the formation of the diastase by the epithelium by carefully removing the latter from the rest of the scutellum by gentle scraping. They found that scutella so denuded of their covering were capable of absorbing soluble carbohydrates if placed upon them, but they had lost their power of corroding starch grains. Extremely thin sections of the epithelium, taken in a plane tangential to the surface of the scutellum, and placed upon the mixture of gelatin and starch, exercised a corrosive and solvent action on the starch grains.

This secretory action of the epithelium is only manifested by grains in the process of germination. The power of forming diastase seems only to be acquired when the barley is quite ripe. Barley taken fresh from immature ears shows the gradual acquirement of the structures of the grain as already described, but the scutella of such unripe grains have no action on starch. The diastase which we have already seen to be present in ungerminated grain is located in the body of the grain itself and not in the scutellum. Moreover by its general behaviour it is proved to be of the translocation variety.

The increase in the amount of secretion diastase during the progress of germination is very marked. Comparing embryos after four days' germination with others taken three days later, the diastasic activities of the two were found by Brown and Morris to be as 16 : 66. Petit says that no diastase is found before the fourth day of germination.

Haberlandt has put forward the view that the diastase of secretion is not elaborated only by the scutellum but that the so-called *aleurone-layer* of the endosperm is also concerned in its production. This layer is three or four rows of cells in depth, and the constituent cells have thick and somewhat cuticularised walls. It is situated in the peripheral portion of the grain, just underlying the wall, which is composed of the fused pericarp and seed-coat. The contents of the cells are chiefly aleurone grains with a little oil, embedded in the protoplasm. Each cell contains a well-defined nucleus. The aleurone-layer covers the whole of the endosperm but becomes much less conspicuous in the neighbourhood of the embryo, where it is reduced to a single layer of cells.

Haberlandt's experiments were carried out upon rye, from which he thinks histological and chemical evidence is forthcoming to prove that in germination secretion of diastase takes place in these cells as well as in those of the scutellum. He states that during germination corrosion of the starch grains first appears between the scutellum and the aleurone-layer on the sides of the rye grain, and that as the process goes on diastasic action upon the starch takes place earlier in the cells underlying the aleurone-layer than in those of the central portion of the endosperm. The cells of the aleurone-layer assume the peculiar granular appearance of secreting cells, and project in papilla fashion into the interior of the endosperm. He removed pieces of the integument, which included the aleurone-layer, from rye grains which had germinated to such a point that the connecting tissue between the latter and the endosperm had been broken down, leaving the aleurone-layer in close connection with the seed-coat and pericarp. He then washed them with a brush wetted with a weak sugar solution, and placed them on moist filter-paper with the aleurone-layer upwards. Upon such preparations a small quantity of starch suspended in water was then carefully placed, and the whole kept at a temperature of 18—20° C. Corrosion of the starch grains took place within 24 hours.

Like the cells of the epithelium of the scutellum, those of the aleurone-layer exhibited no power of corroding starch until the onset of germination.

In their paper published in 1890 Brown and Morris opposed the views of Haberlandt, and maintained the opinion that the aleurone-layer does not play the part he suggested. In a more recent investigation carried on during 1897 Brown and Escombe made a very exhaustive series of experiments upon this point, and after carefully eliminating the influence of the action of micro-organisms and of the possible presence of translocation diastase in the starch-containing cells, they came to the conclusion that this layer does possess a secretory function, and that it forms a certain amount of diastase, but from the mode of its action on starch grains they considered the latter to be the translocation variety of this enzyme. The pitting and erosion of the grains of starch, so characteristic of the action of scutellar diastase, were not seen, but the process was one of more regular solution, accompanied in some cases by the splitting of the granule without previous pitting.

The aleurone-layer according to these observers is the chief seat of the formation of another enzyme, *cytase*, which dissolves the cell-walls, and which will be treated of in a subsequent chapter. Like the scutellar epithelium therefore it secretes two enzymes, cytase and diastase. Comparing the two regions it appears that cytase is most prominent in the secretion of the aleurone-layer, diastase in that of the scutellum.

The results of Haberlandt, Brown and Morris, and Brown and Escombe, have been confirmed by the investigations of other writers, especially Grüss, Hansteen, and Puriewitsch. Grüss holds further that the depletion of the starch-containing cells below the aleurone-layer is partly due to residual diastase existing in them.

The differences between the two varieties of vegetable diastase have been studied by Lintner and Eckhardt, who have compared them as they exist in raw and in germinated barley. In their experiments they prepared the diastase of secretion from extract of malt, of which it is the characteristic enzyme. They made extracts of malt and of barley of such strengths that both solutions had equal power of converting starch into sugar and were therefore comparable in the other peculiarities of their action. Causing such extracts to act upon solutions of soluble starch at several different temperatures, they were able to construct a curve for each which showed how different temperatures modified their behaviour. The curves were drawn so that their abscissæ represented the temperatures, and they erected on the abscissæ ordinates proportional to the cupric-oxide-reducing power which was found to be possessed by the solution at the end of the experiment. The two sets of curves differed materially. Malt diastase showed itself to possess an optimum point, or point of greatest activity, at 50°—55° C.; barley diastase, which is the translocation variety, was most powerful at 45°—50° C., an average of about 5° C. lower. At a temperature of 35° C. they were equally active; the barley diastase at 4° C. had as much power of starch conversion as the malt diastase had at 14·5° C.

Kjeldahl also has published some comparative experiments upon the two varieties. He confirms Lintner's results as to the optimum point of the activity of secretion-diastase, but makes the point at which it is destroyed by heat a little higher.

The two vegetable diastases may be thus compared :—

(1) Translocation diastase. This dissolves starch grains without corrosion; has a very slow action on starch-paste, though it readily converts soluble starch into sugar; works best at a temperature of 45°—50° C.; is much more active at a low temperature than secretion diastase.

(2) Diastase of secretion. This corrodes starch grains and disintegrates them before solution; rapidly liquefies starch-paste: works most advantageously at a temperature of 50°—55° C. It will withstand heating to 70° C. without destruction.

Diastase is secreted by some of the lowliest plants as well as those of higher organisation. In the latter the presence of the first variety is connected with the transformation of starch with a view to the transport of the latter from one part of the plant to another, and the enzyme always remains within the cells. Secretion diastase on the other hand is formed by a special glandular tissue, and is excreted from this tissue into a region where it may prepare nutritive substances to be subsequently absorbed by the cells which in the first place prepared the enzyme. The barley embryo in fact digests and absorbs the endosperm on which it grows parasitically.

When we pass from the higher to the lower forms we can recognize similar behaviour on their part. Diastase has been observed to occur in several bacteria, from some of which it is possible to extract it. These organisms do not uniformly contain it, nor is it present at all times in those which do secrete it.

Lauder Brunton and MacFadyen isolated it from Klein's scurf-bacillus and the Welford bacillus, but these microbes only secreted it when cultivated upon starch-paste. If the cultivation-medium was a meat extract, no diastase was produced. Wortmann showed in 1882 that certain bacteria exert diastasic powers on starch through excreting an enzyme when starch grains are their only available food.

In *Bacillus mesentericus vulgatus* diastase has been found to exist side by side with four other enzymes. The cholera bacillus of Koch can liquefy starch-paste when cultivated upon it, and some of the starch is transformed into sugar. It apparently does this by the action of its own protoplasm, as Wood, who observed the phenomenon, says no enzyme is secreted, or can at any rate be extracted from the cells. On comparing these results with the others already quoted it seems probable that his methods of extraction were defective, as it is difficult to see how the protoplasm of the bacillus and the starch-paste could come into contact, on account of the cell-wall clothing the organism. Protoplasm no doubt possesses a starch-transforming power, but it must come into contact with the starch in order to exercise it. This property of protoplasm will be more fully considered in a subsequent chapter.

Pfeffer was able to determine the secretion of diastase by *Bacterium megatherium*, and found that the amount this organism prepares depends to a large extent upon the amount of cane-sugar present in the culture medium. He extended his researches in this direction to several of the mould fungi, and found the same influence was exerted by the environment. *Penicilium glaucum* ceased to form diastase at all in the presence of 10 per cent. of cane-sugar, and even when the latter only amounted to 1·5 per cent. the starch was only slightly attacked. *Aspergillus niger* behaved differently, producing diastase even in the presence of 30 per cent. of cane-sugar. Different sugars have different powers of checking the secretion of the enzyme, cane-sugar and dextrose being more effective in this direction than maltose. As the former are the sugars which have the greatest nutritive value for the organisms this result appears to indicate that the formation of diastase in these plants only takes place under the stimulus of a semi-starvation. This is rendered the more probable in that if the plants are excited to more vigorous growth by being supplied with peptone, a greater proportion of sugar is necessary to inhibit the production of diastase.

CHAPTER III.

ANIMAL DIASTASE.

THE existence of a diastasic property in animal cells was first indicated in 1831 by Leuchs, who discovered saliva to have the power of converting starch into sugar. He did not take the matter further, and it was not till 1845 that the enzyme was prepared from the secretion of the salivary gland. This was effected by Mialhe, who obtained it by precipitation of saliva with excess of absolute alcohol. It was prepared in a purer form by Cohnheim in 1863 by a method that will be described in a subsequent chapter (page 45).

In 1845 Bouchardat and Sandras prepared the same enzyme from the secretion of the pancreas.

Further observations, made chiefly within the present decade, show that diastase is as widely distributed in the animal as in the vegetable body. Indeed though two varieties have not been indicated so clearly as in the latter case, we find two modes of action much resembling those of the translocation and secretion varieties found in plants. Thus the secretions of the salivary gland and of the pancreas are poured out into special regions of the alimentary canal to convert into a diffusible form the starch and glycogen of the animal's food, while in the cases of tissues containing the latter of these carbohydrates, an enzyme is produced which acts only on the glycogen in the interior of the cell, and from the latter the diffusible products are removed by a species of translocation.

Diastase is stated by Röhmann to exist in small amount in the succus entericus or secretion of the small intestine, though

in much smaller amount than in either saliva or pancreatic
juice. Hamburger has made a similar observation. Comparing
the relative diastasic powers of saliva, pancreatic juice, and the
secretion of the small intestine, by observing their power of
forming cupric-oxide-reducing sugar from solutions of starch,
he finds that taking the reducing power of glucose as unity for
purposes of a standard, the products of the action of the three
secretions gave reducing powers of ·31, ·36, and ·26 respectively.
This amount of reducing-sugar was produced by saliva and by
pancreatic juice in an hour or less, while it was only arrived at
under the influence of intestinal juice in rather longer than a
day. The last-named secretion therefore contains but a very
small amount of diastase.

The existence of the enzyme in the intestinal juice of the
sheep was proved by Pregl in 1895.

The other variety which in function at least corresponds to
translocation diastase is found in almost all tissues and fluids
of the body, though it is often present in very small quantities
and is difficult to extract. Different animals however show
considerable differences in these respects.

The liver and the muscles are the chief storehouses of
carbohydrate reserve-material in the body, where it takes
the form of large quantities of glycogen, which vary considerably from time to time. Carbohydrates leaving the
liver are found to be in the form of sugar, most probably
glucose, and the conversion of glycogen into the latter must
therefore take place within the organ. Very conflicting statements have been made as to the possibility of preparation of
a soluble enzyme from the liver cells, but there seems to be
a gradually accumulating body of evidence in favour of its
existence.

Leaving that question for the moment and turning to other
body fluids in which the existence of diastasic enzymes would
seem less probable than in saliva and the pancreatic and
intestinal juices, we find a certain amount of evidence of their
existence in blood and lymph. Bial in 1893 stated that diastase
could be extracted from the serum of both these fluids but not
from their corpuscles. Röhmann about the same time made the

observation that if glycogen is injected into the lymphatics of
an animal and the lymph from the thoracic duct then allowed to
run into alcohol, the percentage of sugar in the lymph is found
to have risen. The glycogen in his experiments was injected
in suspension in a small quantity of solution of common salt,
about ·6 % in strength (normal saline solution). The salt
solution produced no effect when injected alone. Röhmann
also found diastase in blood, existing side by side with other
enzymes, but being present only in relatively small amount.
Dastre also found diastase in lymph, which he says contains
·097 parts of glycogen per 1000. If the lymph is examined after
standing 24 hours, the glycogen has disappeared. In the paper
already quoted Hamburger also states that an enzyme capable
of producing sugar exists in the blood. He speaks of it as giving
rise very slowly to a sugar having a reducing power greater
than that formed by saliva. It is probable, as he points out,
that this sugar is not the product of the action of diastase
alone, but that the body formed by the latter is further changed
by another enzyme also present in the blood. He shows clearly
however that blood has the power of transforming starch into
sugar, which can only be due to diastase.

Foster has found that pericardial, pleural, and peritoneal
fluids have the power of converting starch into sugar, and
Grohe says that chyle behaves similarly. According to Cohnheim and Béchamp fresh filtered urine has the same power,
containing enough diastase to allow the enzyme to be obtained
from it by precipitation. Its presence in the urine probably
indicates that it is in process of being excreted from the blood,
as the amount is increased after meals. Foster says it can be
prepared from the natural deposits of urates, with or without
previous washing with alcohol.

The diastase which is present in saliva, pancreatic juice, and
succus entericus, is like that of the germinating barley grain,
a secretion by a definite tissue. Saliva in mammals is formed
in certain structures, known as salivary glands. There are three
pairs of these in the body, the parotids situated on the sides of
the face, in front of the ears, the submaxillary, just below the
base of each inferior maxilla, and the sublingual, placed along

the floor of the mouth between the tongue and the gums of
the lower jaw. Each is composed of a number of tubes opening
into ducts which ultimately coalesce to form a single large
outlet through which the whole secretion of the gland flows.
The tubes are much convoluted and bound together into lobes
by connective tissue, so that the whole gland forms a solid mass
of small size. The act of secretion is carried out in the
ultimate tubes or alveoli of the gland. Each of these consists
of a transparent basement membrane, on the inner face of which
the secreting cells are placed. There is a very small lumen or
canal in the centre, formed by the free margins of the cells.
Each alveolus is bathed by the lymph which exudes from a
close plexus of blood capillaries, situated in the connective
tissue which is outside the basement membranes and which
helps to bind the whole gland together. The cells thus receive
supplies of lymph at their bases, while the secreted saliva is
poured out from them into the delicate canal which is the
centre of the alveolus. Each gland is supplied with a double
set of nerve fibres, one from the cerebro-spinal, the other from
the sympathetic system.

The structure of the secreting cells differs in one important
respect from that of the cells of the scutellar epithelium of the
barley. There is no cellulose membrane investing them and
their protoplasm is therefore constantly bathed by the lymph
which reaches them freely. They are composed of protoplasm
which can be seen to be arranged in a network; this is fairly
regular and close in some cells, while in others the size of
its meshes varies in different parts of the cell. The outer
part or limiting layer of the cell has much closer meshes than
any other, and forms an almost continuous layer. Between the
meshes of the protoplasm there is a more transparent, possibly
fluid material, often known as paraplasm. This may contain
various substances and may be crowded with granular matter.
There is always a conspicuous nucleus present but its position
varies in cells of different glands.

The process of formation of the diastase, or *ptyalin*, as it is
frequently called, can be traced by a microscopic examination
of the cells under different conditions of the animal. If

examined after a period of fasting the meshes of the protoplasm are found to be filled with granules to such an extent
that the outlines of the cells are hardly if at all visible and the
nucleus is altogether obscured. If the gland is taken from the
body immediately after a copious outflow of saliva has taken
place, the granules are smaller and sparser, and are accumulated
on the sides of the cells next to the lumen, the outer border
appearing clear. The whole cell is somewhat smaller, as if
shrunken, and the nucleus is plainly visible. The examination
can be made upon the fresh tissue, or more easily after treatment of it with osmic acid, which stains the granules black.

The process of secretion appears to consist of three stages,
absorption of nutritive matter from the lymph by the cell, the
manufacture and deposition of the granules in its interior, and
the subsequent solution of these and their coincident extrusion
from the cell into the duct. All these processes are under the
control of the nerves which are supplied to the glands, each of
which has a definite part to play in the sequence of events.
In the case of the sub-maxillary gland, the formation and
solution of the granules appear to be regulated by a branch of
the cervical sympathetic nerve, and their extrusion from the
cells by the chorda tympani, a branch of the 7th cranial nerve.
The whole nervous action is controlled by the brain.

The exact nature of the granules is not completely known.
They are probably not diastase itself, but an immediate antecedent which becomes diastase in the process of dissolving and
leaving the cell. This view is however based rather upon
analogy with what happens in certain other glands than upon
direct observation of the salivary ones.

In some of the salivary cells the granules are certainly not
all diastase, as the secretion of the gland contains other bodies
as well, notably mucin. Still the granularity and the amount
of diastase obtainable from a gland appear to vary together.

The granular appearance thus noticed in the animal cell
recalls the fact that a similar granularity may be observed in
the scutellar epithelium of the barley grain while diastase is
being formed there.

Similar appearances to those described for the salivary

gland may be observed in the cells of the pancreas, which has a very similar structure. The cells of this organ however secrete, in addition to diastase, other enzymes which will be discussed later. The granules of these cells have been ascertained not to be composed of the enzymes themselves, but of antecedents of them known as *zymogens*. Possibly the granules differ from each other, some being those of one zymogen, others those of another.

The amount of diastase in saliva and pancreatic juice varies according to the animal from which those fluids are taken. It is least in the juices of the herbivora, the saliva of the horse being practically devoid of it. There is very little again in the secretions of young animals so long as they are suckled.

The diastase of blood and lymph may be compared with the translocation diastase of plants. It acts upon substances in those fluids and is not brought into contact with carbohydrates in the alimentary canal. Its mode of formation is unknown, the microscope affording no information on this point.

Turning now again to the question of the storage of reserve carbohydrates in the body and the possibility of their translocation by means of diastase, we find our information is not so complete as that concerning the secretions of the glands described. The most prominent organ in the transformation of such reserve-carbohydrates is the liver, in which it is easy to detect large quantities of glycogen which vary considerably from time to time. Glycogen in the animal organism appears to take the place of starch in the vegetable one, being the form in which carbohydrate material, largely derived from the alimentary canal, is stored till needed in the processes of nutrition.

It has been known since the time of Claud Bernard that the blood coming from the liver contains a variable amount of sugar, and that during a period of fasting it contains more than does the blood of the portal vein which carries to that organ the products of digestion. When the liver of an animal is removed from the body, the post-mortem changes which take place in it are always accompanied by the formation of sugar, and as the glycogen gradually disappears it is an indication that the latter is the source of the former.

The question of the existence of a diastasic enzyme in the liver cells is one which has excited a great deal of controversy. Most of the older experimenters failed to extract such a body, and the change was supposed therefore to be effected by the protoplasm of the cells, both during life and during the post-mortem changes.

In considering whether the evidence that has been accumulated points to the existence of an enzyme it should be borne in mind that in many cases it is by no means easy to extract such a body from the protoplasm of a cell by the mere process of maceration with water or other extracting fluid. In many cases it has been ascertained that the enzyme is held by the substance of the protoplasm with very great tenacity. This appears very evident on comparing the results of Wortmann on the diastase of foliage leaves with those of Brown and Morris which have already been alluded to. Another instance of this difficulty is found in the recent experiments of Buchner, in which he has extracted an alcohol-producing enzyme from the body of the yeast cell by the employment of strong pressures. These will be discussed in a subsequent chapter, but the difficulty of extraction is apparently not sufficient proof that an enzyme may not be present, though simple maceration fails to extract it.

Of the earlier writers on the subject of the metabolism of the liver, Von Wittich and Claud Bernard both claim to have extracted an enzyme from the hepatic cells. Tiegel on the other hand failed to obtain one, and Seegen and Kratchmer, after many experiments, say that it is impossible to obtain satisfactory evidence of its existence.

An extended series of observations was made by Miss Eves in 1882. This author dried the livers of various animals at a moderate temperature and then reduced them to a powder in a mortar. The dried powder was subsequently extracted with salt solution and the liquid filtered. Extracts so prepared were found to be capable of effecting starch transformation, but the diastasic powers they possessed were very small in comparison with those of saliva; so small indeed that her conclusion was that certainly the post-mortem formation of sugar and disappearance of glycogen could not be attributed to an enzyme in the cells. She attributed the small amount obtained in her extracts to the blood which had remained in the organ.

The subject has been investigated by many experimenters during the present decade, and there is a growing opinion that a diastase is present in the liver, though there is not yet an agreement as to the part which it plays in the metabolism of the organ. About the year 1890 Kaufmann published the results of some experiments made with the bile secreted by various animals. He says that the bile of the pig, the sheep, and the ox are all strongly diastasic, that no enzyme can be detected in that of the dog, while that of the cat contains a relatively small amount. Kaufmann infers from these results that the liver cells do normally form diastase.

Very exhaustive experiments upon this subject have been carried out by Pavy. He removed the liver from rabbits immediately after death and rapidly reduced them to pulp by passing them through a mincing machine and subsequently pounding them in a mortar. The pulp was stirred up with a large volume of absolute alcohol and allowed to stand under this liquid for six months, by which treatment the proteids of the tissue were completely coagulated. At the expiration of this time, the alcohol was strained off and the liver substance washed successively with fresh alcohol and with ether. It was then dried, powdered, and passed through a fine sieve. Two grams of this liver powder were thrown into boiling water, while another portion of the same weight was incubated for 4 hours at 46° C. with 20 c.c. of 1 % sodium chloride solution. Both were then titrated with Fehling's fluid. The boiled digestion contained ·46 per cent. of reducing-sugar, the incubated one 4·27 per cent. As the original liver contained both glycogen and a small amount of sugar, the increased quantity found after incubation clearly shows the transformation of the glycogen under the influence of liver diastase.

Pavy has shown further that the diastase so detected will survive prolonged heating to 80° C. if perfectly dry. He heated some of the liver powder suspended in absolute alcohol for 5 minutes over boiling water, then dried it again. Experiments with this material carried out on the lines of the one just quoted gave the same result, except that the portion not incubated but boiled at once in water caused no reduction of Fehling's fluid. This shows that the reduction found in this portion in the first experiment was due to sugar present at death in the liver tissue; in the second experiment this had been extracted by the alcohol in which the powder was heated. Hot alcohol is of course much more efficacious than cold in removing sugar from the tissue, which holds it with some tenacity.

Miss Tebb has also been successful in preparing a diastasic extract from liver. She dried the liver of the pig at 35—40° C., shredded it finely and dialysed away the sugar which it contained. The liver substance was then dried and powdered. She found it capable of hydrolysing both starch and glycogen. An extract of the dried powder was prepared with a 5 per cent. solution of sodium sulphate and dialysed till free from sugar. It was then digested with a 4 per cent. solution of glycogen for 21 hours at 37° C., at the end of which time phenyl-hydrazine showed the presence of sugar in some quantity.

Bial holds that both during life and during post-mortem changes the conversion of the glycogen into sugar is due to an enzyme. He found that the process was not prevented by a 1 per cent. solution of sodium fluoride, and he considers this fact to prove the presence of diastase, as this salt is fatal to the vital action of the cell. He does not decide whether the liver cells secrete this enzyme, or whether it is derived by them from the blood and the lymph which bathe them. He agrees that the sugar which is formed is glucose, and on this ground says that the diastase which is formed is the same as that existing in the blood and lymph. There are reasons however for thinking that this is not due to the diastase being of a different kind from that of the saliva, but that there is another enzyme also present which converts maltose into glucose.

Salkowski, from the results of an independent investigation, coincides with the views of Bial.

Schwiening also has made some experiments upon the post-mortem changes in the liver, especially upon the influence of chloroform upon them. As already mentioned, this reagent kills protoplasm but does not interfere with the action of diastase. He made an extract of liver with chloroform water and boiled half of it as a control. In both his preparations liver cells were present; in one they had been killed by chloroform, leaving any diastase unaffected; in the other not only were the cells killed but the diastase was destroyed by the action of the high temperature. Both contained the glycogen originally present in the tissue. After a certain time of incubation the unboiled extract was found to contain no glycogen, but a considerable quantity of sugar, while the boiled one contained glycogen with only traces of sugar. He concluded in consequence that the action was due to an enzyme, the direct action of the protoplasm being impossible under the conditions of the experiment. The traces of sugar found in the boiled extract may very well have been due to a small quantity existing in the liver at the time of its excision.

Schwiening says further that in the boiled extract there was a conversion of glycogen into sugar after a prolonged incubation, a fact which he holds to point to a continuous post-mortem formation of diastase. This however seems a little difficult to understand, as the boiling not only killed the liver cells, but presumably destroyed any antecedent of diastase as it did the enzyme itself.

The liver however is not the only seat of stored or reserve carbohydrate, for the muscles also contain a considerable amount of glycogen. Nasse has shown that these organs are also the seat of an amylolytic or diastasic enzyme which can be detected in the juice extracted from them by pressure. Halliburton confirms Nasse's observation. He has found that a watery extract of the dried precipitate produced by treating muscle juice with alcohol can change both starch and glycogen into a reducing-sugar with the concomitant production of an intermediate body possessing the characters of a dextrin. He finds the action on starch very slow, no sugar being noticeable till the digestion has proceeded for 5 or 6 hours.

Boudourg has observed the presence of diastase in the pyloric tubes of certain Teleostean fishes.

Among the Invertebrata diastase appears to be of very general occurrence; it can be detected even in unicellular organisms, although from their structure it exists side by side with other enzymes in the general protoplasm.

Fredericq has extracted a juice from the cells of several sponges, which acts upon starch, fats and proteids. He obtained a similar extract from several of the *Echinodermata*, which had the same properties but was slower in its action. In *Uraster* this secretion was found in the pyloric cæca, situated in each ray of the animal. An extract of that part of the alimentary canal of *Lumbricus* contained in the anterior portion of the body, up to the 6th segment, will convert starch into sugar. Its intestine is almost surrounded by a glandular tissue which has the same properties, but the conversion of starch by this tissue will take place only in neutral solutions. Other worms also form diastase in similar positions. In Rotifers two large glandular tubes open into the anterior portion of the stomach, and yield a secretion which acts on starch and on proteids.

It is very common in all the higher Invertebrata to find both diastasic and proteolytic enzymes yielded by the same gland, just as they are in the pancreas of vertebrates. The so-called liver of the Crustacea and Mollusca is apparently a hepato-pancreas and both enzymes can be prepared from it.

The glands which are situated towards the anterior region of the alimentary canal in many of the lower forms are only diastasic. Among the Insecta *Blatta* is known to be furnished with well-marked salivary glands, secreting an alkaline fluid which converts starch into sugar with great facility. The enzyme can be extracted by precipitating an infusion of the glands with dilute phosphoric acid and lime water, when it is carried down by the calcic phosphate. The mixed precipitate when washed with distilled water gives up the enzyme to the solvent and it can be thrown down in a fairly isolated condition by the addition of a large excess of alcohol.

In the Lepidoptera salivary glands are present in the imago of many species; they possess also glandular follicles

in the stomach, the juice of which behaves like that of a
vertebrate pancreas. Biedermann has found in the intestine
of the larva of *Tenebrio Molitor*, the meal worm, a secretion
which contains diastase.

The spiders are similarly supplied with salivary and pancreatic glands.

As has been mentioned the "liver" of Mollusca has pancreatic functions. In *Helix aspera* there are true salivary
glands as well, which open into the mouth. In *Helix pomatia*
the diastasic function disappears from the pancreatic organ
during the winter sleep. In the Cephalopoda the secretion of
the salivary glands is diastasic only, while the "liver" besides
hydrolysing starch, emulsifies fats, and saponifies them, makes
milk transparent and acts proteolytically on albumin. The
Lamellibranchiata are stated to possess no special salivary
glands. Diastase has been said to exist in the mantle of the
oyster so long as it is uninjured and to occupy different cells
from those which contain glycogen.

Abelous and Heim demonstrated the presence of diastase
in the eggs of certain Crustaceans, among which may be
mentioned *Maia squinado*, *Platycarcinus pagurus*, *Portunus
puber* and *Galathœa strigosa*. They were able to prepare it
from either fresh eggs or such as had been preserved for some
time in 95 per cent. alcohol, using as a solvent either water
or glycerine. They found it acted most energetically at a
temperature of 35° C. and in a medium containing ·1 per cent.
of hydrochloric acid.

Müller and Masayama state that the yolk of the ordinary
hen's egg contains an enzyme capable of transforming starch
into dextrin and sugar.

There is no doubt that the diastase which occurs in the
animal organism is identical with that which is of vegetable
origin. Brown and Heron have shown that the course of the
action of pancreatic and malt diastase is the same, that they
work under identical conditions and are influenced by various
external circumstances in precisely the same manner.

CHAPTER IV.

PREPARATION OF DIASTASE AND ITS COURSE OF ACTION.

THE isolation of diastase in a state of purity has not yet been accomplished. Indeed so little is known about its composition that no test of its purity has been satisfactorily established. Beyond saying that it is undoubtedly a nitrogenous body, and that it has much in common with the proteids, it is at present difficult to go. This indeed is the case with all the enzymes at present discovered.

The usual method of preparing a diastasic extract is to mince or triturate in a mortar the tissue in which it exists and to macerate the débris for some time with an appropriate solvent, after which the solid matter is removed by filtration. Such an extract of course can only be regarded as a preparation containing diastase, together with such constituents of the tissue as are soluble in the extracting fluid.

Many attempts have been made to prepare the enzyme free from such admixture, and though several methods purify it from many of the more general contaminations none can be regarded as wholly satisfactory.

One of the earliest of these methods was employed by Cohnheim, who took advantage of one of the properties of the enzymes as a class, viz. their tendency to be carried out of solution by any inert precipitate formed in the liquid in which they exist. His method was the following :—he added to saliva a certain quantity of phosphoric acid and then neutralised the mixture with lime water, which caused a copious precipitate of phosphate of calcium. This precipitate carried down with it a large proportion of the proteids of the saliva together with the diastase or ptyalin. The precipitate was then collected by filtration and extracted with a volume of water equal to that of the saliva originally used. The diastase is slightly more soluble than the proteids and accordingly passed into solution first. He repeated this process several times and finally precipitated the last extract by the addition of alcohol. This precipitate was collected, washed with alcohol and dried over sulphuric acid, appearing then as an amorphous powder, white in colour and freely soluble in water. Cohnheim held it to be free from proteids as its solution did not yield the usual reactions characteristic of these bodies.

Another method of preparation of salivary diastase has been described by Krawkow. Saliva is diluted with an equal volume of water, and saturated with neutral ammonium sulphate. The precipitate which is caused by the saturation is collected on a filter and washed for a short time with strong alcohol. It is then allowed to stand under absolute alcohol for one or two days, and finally dried at a temperature of 30° C. On extraction with water it yields a solution which is strongly diastasic and which gives no proteid reactions.

Von Wittich adopted the plan of treating a diastase-secreting tissue with glycerine and subsequent precipitation with alcohol, purifying the product by repetition of the process.

Mialhe obtained it from saliva by merely precipitating it by means of alcohol. Probably neither he nor v. Wittich ever obtained it in anything like a pure condition.

From several animal tissues a larger yield has been obtained by using for the first extraction a solution of common salt of about 10 per cent. concentration instead of water.

Lintner has prepared a relatively pure diastase from malt by the following method. One part of either green or air-dried malt was extracted for 24 hours with 2 to 4 parts of alcohol of 20 per cent. strength. The spirit was then filtered off and mixed with two and a half times its volume of absolute alcohol. A precipitate settled out, which was collected on a filter and washed with absolute alcohol. It was then transferred to a mortar and well washed, with stirring, with a mixture of absolute alcohol and ether. The liquid having been decanted off, the residue was dried in vacuo over strong sulphuric acid, and remained as a very active yellowish-white powder. To purify it, it was redissolved in water and again precipitated with alcohol, after which it was submitted to dialysis.

Analysis of it in this condition showed it to contain nitrogen and a certain amount of ash, which was chiefly composed of calcium phosphate. The process of purification of the first product increased the percentage of nitrogen, while it diminished the amount of ash. At the same time it increased the diastasic activity. A sample which on first preparation contained 8·3 per cent. of nitrogen, after purification by two reprecipitations contained 9·06 per cent., and after subsequent dialysis, 9·9 per cent., while the percentage of ash went down from 10·6 to 4·79.

The diastasic activity was determined by allowing 3 c.c. of a solution of 1 gramme of the diastase in 250 c.c. of water to act on 10 c.c. of a 2 per cent. solution of starch for one hour at the ordinary temperature of the laboratory, comparing it with a control made of a standard diastase and used in the same proportions and under identical conditions. The same sample as that mentioned above had a diastasic activity of 96 before purification and of 100 afterwards.

Another method of preparing diastase has been used by Loew, which is not however so trustworthy as the last-mentioned one. The tissue which contained the enzyme was extracted by water, and soluble calcium salts and salts of lead were added to it. Then it was made alkaline with caustic soda, which produced a precipitate, and the latter carried down the diastase with it. The precipitate was filtered off and dissolved in dilute acetic acid, and the lead removed by a current of sulphuretted hydrogen. Alcohol of 95 per cent. strength was next used to precipitate the enzyme from the filtrate from the sulphide of lead. This precipitate was filtered off, dried over sulphuric acid and extracted with glycerine.

Musculus employed a method which was a modification of that of Payen and Persoz. A quantity of powdered germinated barley was macerated for an hour with twice its volume of water

and strained. To the turbid liquid its own volume of alcohol
was added, and the bulky precipitate filtered off. To the filtrate
its own volume of alcohol was again added, when a much less
bulky precipitate fell. This was collected on a filter and dried
at a gentle heat. The paper retained the precipitate, which was
found to be strongly diastasic. So prepared it could be preserved
for a considerable period.

Loew prepared the enzyme in a condition which he considered relatively pure by the following process:—A measured
quantity of germinating barley, softened by a little water, was
digested for two days with 40 per cent. alcohol, with frequent
stirring. After filtration, the liquid was precipitated by a
mixture of two volumes of alcohol and one volume of ether,
and the precipitate collected and washed with absolute alcohol.
It was then dried over sulphuric acid and extracted with water.
Sub-acetate of lead was added, which yielded a precipitate ;
this was filtered off and suspended in water. A current of
sulphuretted hydrogen was then passed through it to remove
the lead, and to the filtrate was added a mixture of absolute
alcohol and ether, which precipitated the diastase. This was
collected on a filter, washed with absolute alcohol and then with
ether and dried. This process was attended with a loss of a good
deal of the diastase and it did not yield a pure product.

Wroblewski has recently published an account of some
researches which he carried out with a view to ascertaining the
nature of diastase, in the course of which he has adopted a
somewhat different method of preparation. He ground some
malt to a very fine powder and washed it repeatedly with alcohol
of 68 per cent. concentration, in which diastase is not soluble.
The enzyme will dissolve in a mixture of alcohol and water
which only contains about 45 per cent. of spirit. He therefore
extracted the powder twice with an alcoholic solution of that
concentration, mixing the two extracts after filtering them
separately; a further quantity of alcohol was then added to the
mixed filtrates till the concentration again reached 68 per
cent. when the diastase which had been dissolved was precipitated. It was collected on a filter, redissolved in 45 per
cent. spirit, and again reprecipitated. The precipitate was then

dissolved in water, and the solution saturated with magnesium sulphate, which again threw down the diastase. The latter was again dissolved in water and the solution freed from the salt by dialysis. A mixture of alcohol and ether was then added to precipitate the enzyme. Prepared in this way it was found to be associated with a carbohydrate body, *araban*, yielding arabinose on hydrolysis. This substance has been found associated also with emulsin by Hérissey. Bertrand observed the occurrence of arabinose as one of the products of the hydrolysis of laccase, an enzyme which will be described later.

Wroblewski has described another process for the preparation of diastase which avoids the use of alcohol till the last stages. To a malt extract or other aqueous solution of the impure enzyme, ammonium sulphate was added till a turbidity leading to a precipitate was reached. This consisted mainly of the araban, and the diastase was almost entirely left in solution. After filtration, more of the salt was added to the filtrate till another precipitate fell. This was composed mainly of pentosan, but some diastase was mixed with it. It was filtered off and the filtrate saturated with the salt, when a third precipitate fell, which was almost pure diastase. This was dissolved in water and freed from the ammonium sulphate by dialysis, after which the enzyme was thrown out of solution by addition of a mixture of alcohol and ether.

Action of diastase on starch.

When a thin starch-paste, preferably containing 1—2 per cent. of starch and prepared with boiling water, is mixed with a small quantity of a diastasic extract a definite sequence of changes can be noticed. In a few minutes the liquid loses the opalescence due to the suspended starch, which goes into solution and the whole becomes limpid. A few drops removed at this stage become blue on the addition of iodine, showing that the starch is chemically unaltered. After a little longer interval the iodine reaction changes, a sample removed from the digestion becoming purple with that reagent. Later the colour given is a deep red-brown. Finally iodine ceases to produce any colour. Periodical examination with Fehling's solution during this process shows that a power of reducing cupric oxide which is not possessed by either starch or diastase is rapidly developed and increases continuously as the action proceeds.

The series of changes so set up in the starch-paste by the enzyme are thus evidently of a complicated character. The action is one of *hydrolysis* or the decomposition of starch together with the incorporation of water into its molecule, leading ultimately to the formation of sugar. But the process is a long and gradual one, and starch and sugar are not the only two bodies which are met with during the transformation. Certain other products known as *dextrins* are found to occur, and one of these is the cause of the brownish-red colour which the liquid gives on the addition of iodine at a particular stage.

The first attempt to ascertain the stages of the action was made by Payen and Persoz in 1833. These authors recognised one of the dextrins and described it as a body soluble in water and in weak alcohol and not coloured by iodine. In 1860 Musculus recognised that dextrin and sugar are produced simultaneously by diastasic action. He described the dextrin a few years later as not colourable by iodine and having no power of reducing cupric oxide. He thought it was not further acted on by the diastase, but remained as a final product. Payen says on the contrary that diastase saccharifies it easily.

About the years 1871–2 Griessmeyer, O'Sullivan and Brücke, working independently, ascertained that there are at least two dextrins formed, one of which is uncoloured by iodine while the other becomes brownish-red with that reagent. Brücke gave them the names of *achroodextrin* and *erythrodextrin*. O'Sullivan at that time thought they were really identical, and describes them as having no power of reducing cupric oxide and possessing a specific rotatory power of $[a]_j = +213°$. He said that malt extract converted them both into maltose.

Some years later Musculus and Grüber showed that achroodextrin was not a single body, but that there were at least three such dextrins, none of which reacted with iodine. They give for the specific rotatory powers of the three $[a]_j = +159°$, $+190°$, and $+150°$ respectively. Based upon their experiments these observers advanced a theory of the starch transformation

IV]   AND ITS COURSE OF ACTION.   51

which has only been slightly modified by subsequent workers. This was that the starch molecule breaks down by a series of hydrations and subsequent decompositions, maltose being formed at each splitting, together with a dextrin of less molecular weight.

In 1879 the subject was investigated by Brown and Heron. From their experiments they came to the conclusion that the molecule of starch has the formula 10 $C_{12}H_{20}O_{10}$, and that under the action of diastase groups of $C_{12}H_{20}O_{10}$ are successively removed, the residue in each case being a dextrin whose molecule becomes less complex at each stage. They suggested that there are eight of these possible dextrins. Each successive group removed then becomes hydrolysed to maltose.

O'Sullivan in the same year came to the conclusion that there are probably four dextrins formed, erythrodextrin and three achroodextrins. He says that these when pure have no power of reducing cupric oxide and have a specific rotatory power of $[\alpha]_j = +222°$.

Another view of the process was suggested about the same time by Herzfeld. According to this the transformation of starch by diastase is not a splitting up of the molecule, but there is consecutively a conversion of it into soluble starch, erythrodextrin and achroodextrin, and the last is subsequently converted into a body which he calls *maltodextrin* and maltose. When starch is transformed at a temperature lower than 65° C. maltose and achroodextrin are formed; above 65° C. erythrodextrin and maltodextrin in addition. Herzfeld suggested for the latter body a composition corresponding to two groups of dextrin united to one sugar group of the formula $C_6H_{12}O_6$. Herzfeld's views have not been generally accepted, the theory of Musculus and Grüber meeting with more favour. Probably his maltodextrin was by no means pure.

In 1885 a modification of the views of Musculus was put forward by Brown and Morris. They found that in a starch transformation the usual result was the formation of about 80 per cent. of maltose and 20 per cent. of dextrin, the latter being further hydrolysed with great difficulty. Among their products they found constantly occurring a certain amount of a body intermediate in properties between maltose and dextrin, and they gave to this the name *maltodextrin*, a term originally used by Herzfeld. The substance however differed considerably in its behaviour from that described by Herzfeld under the same name. Brown and Morris conclude however that the differences are due to the impurity of Herzfeld's preparation.

Starting with the observation that apparently only four-fifths of the starch is readily convertible into maltose they put forward the following theory. Starch has at least a formula of $5\,[(C_{12}H_{20}O_{10})_3]$ and consists of five amylin or dextrin-like groups, four of which are arranged symmetrically round the fifth. By the action of diastase the decomposition takes place in successive stages, one group of $(C_{12}H_{20}O_{10})_3$ being split off at each step, leaving a dextrin residue. Each group so split off is then hydrolysed by the addition of water, forming maltodextrin $\begin{Bmatrix} C_{12}H_{22}O_{11} \\ (C_{12}H_{20}O_{10})_2 \end{Bmatrix}$, and this is further hydrolysed by diastase below $65°$ C. into maltose, two molecules of water being taken up. The last group, around which the others are arranged, undergoes hydrolysis with great difficulty and consequently remains as a dextrin at the conclusion of the transformation. The maltodextrin being composed of maltose and dextrin they term an *amyloïn*.

The reactions of the amyloïn are given by them as follows:—

(1) It gives numbers on analysis which allow its composition to be expressed in terms of a mixture of maltose and dextrin.

(2) It cannot be separated into maltose and dextrin and is therefore not a mixture but a compound body.

(3) It is completely converted into maltose by the action of diastase, while a mixture of the two in the same proportion always leaves a residue of dextrin.

(4) It is unfermentable during the primary fermentation by yeast.

In a later paper in 1889, while adhering to their theory of the grouping of the starch molecule and the general course of the transformation, they come to the conclusion that instead of one maltodextrin there are several of them formed. They think that the groups in the starch molecule are considerably larger than their first experiments led them to suggest. The first decomposition may be expressed by the equation

$$5\left[(C_{12}H_{20}O_{10})_{20}\right] = (C_{12}H_{20}O_{10})_{20} + 4\left[(C_{12}H_{20}O_{10})_{20}\right].$$

starch   stable dextrin   amylin groups

The 4 amylin groups are capable of gradual and complete hydrolysis to maltose, a series of amyloïns of varying degrees of complexity being formed, there being a general tendency for the molecular aggregations to become smaller as the hydrolysis proceeds, until finally maltose is reached. The extreme stages of this hydrolysis can be represented by the following equations:—

$$(C_{12}H_{20}O_{10})_{20} + H_2O = \begin{cases} C_{12}H_{22}O_{11} \\ (C_{12}H_{20}O_{10})_{19} \end{cases}$$

and

$$(C_{12}H_{20}O_{10})_{20} + 19\,H_2O = \begin{cases} (C_{12}H_{22}O_{11})_{19} \\ C_{12}H_{20}O_{10} \end{cases}.$$

As the hydrolysis proceeds these complex amyloïns are broken up into smaller ones, two of which have been prepared by them; these they call maltodextrin $\begin{cases} C_{12}H_{22}O_{11} \\ (C_{12}H_{20}O_{10})_2 \end{cases}$ and amylodextrin $\begin{cases} C_{12}H_{22}O_{11} \\ (C_{12}H_{20}O_{10})_6 \end{cases}$. The latter body was first observed by W. Nägeli in 1874. He considered it erroneously to be identical with soluble starch.

A curious difference in the action is noticed when diastase is allowed to act on starch-paste in the cold. The optical activity of the final result of the transformation is lower than it should be on the assumption that only maltose and dextrin are present. After standing for some time, or after boiling, this discrepancy disappears, and the optical and cupric-oxide-reducing powers show the same relationship that they do when the starch transformation is carried out at higher temperatures. The explanation appears to be that at the lower temperature the maltose formed does not possess its full power of rotating the polarised ray, being liberated in a state of "half-rotation." Freshly prepared solutions of maltose always exhibit this peculiarity. When freshly dissolved the optical power of maltose bears the relation to that of an old or a boiled solution of about 133° to 150°. Brown and Morris have shown that this relation holds good in the case of freshly prepared products from cold starch-paste.

From these researches there seems little doubt that the hydrolysis of starch is a gradual and continuous process. Another view has been recently put forward by Mittelmeier which differs in many of its details from that of Brown and Morris. According to Mittelmeier the action of diastase on starch occurs in two stages; during the first a small portion of the starch is rapidly converted into amylodextrin, erythrodextrin and achroodextrin, and sugar, but these dextrins are not identical with those formed during the second stage. The author speaks of them respectively as primary and secondary dextrins. He claims to have isolated primary erythrodextrin and primary achroodextrin. The former is insoluble in water, has a specific rotatory power $[\alpha]_D$ = about $+170°$, does not react with phenyl-hydrazine, reduces cupric oxide, and yields dextrose and maltose when treated with diastase. He gives no reactions for the other primary dextrin. The sugar from the primary dextrins yields maltosazone, while that from the secondary ones gives an osazone which is oily in character, and melts at 145—148° C. Mittelmeier thinks it is derived from a new sugar for which he suggests the name *metamaltose*.

His results have not at present been confirmed.

The nature of the sugar formed by the action of diastase is according to most observers *maltose* when the transformation is allowed to take place under the ordinary conditions of the laboratory. Many writers have pointed out however the occurrence of *glucose* among the products under particular conditions, and doubts as to the identity of the enzyme in different secretions have been based upon these discrepancies in the final results of their activity.

These two sugars differ considerably in their composition, maltose being the most complex, and indeed in certain circumstances one molecule of it splits up into two molecules of glucose. Glucose, or as it is often called *dextrose*, is colourless and crystalline. The crystals when deposited from water occur in six-sided tables or prisms or are agglomerated into warty lumps. They are soluble slowly in cold, rapidly in hot water; slightly soluble in cold alcohol, more readily in warm. A solution of this sugar possesses a specific rotatory power of $(\alpha)_D = +52\cdot5°$. When heated with a solution of cupric sulphate in the presence of an excess of caustic alkali, it reduces the cupric oxide formed, to the cuprous condition. This power is made use of in estimating the quantity of glucose present in a solution, the latter being in a definite proportion to the amount of cupric oxide reduced.

Glucose when warmed with phenylhydrazine acetate forms an osazone which crystallises in yellow needles. It is almost insoluble in water and melts at about 205° C. Glucose is capable of undergoing alcoholic fermentation by yeast without previous decomposition. It has the composition $C_6H_{12}O_6$ and is represented by the formula $[COH - (CHOH)_4 - CH_2OH]$.

Maltose crystallises with difficulty from its watery solution, the crystals being in the form of fine needles. It is very soluble in water, but not so readily dissolved by alcohol as is glucose. Its specific rotatory power is much greater than that of the latter, being $(\alpha)_D = +140°$. Its power of reducing cupric oxide is on the other hand considerably less, being only about two-thirds that of glucose.

Maltose forms with phenylhydrazine acetate an osazone which crystallises in fine yellow needles resembling those of glucosazone, but they are soluble in about 75 parts of water at 100° C. Its melting point is 205° C.

There is a general agreement among chemists that the action of vegetable diastase on starch results in the production of maltose and not glucose. Somewhat conflicting statements have been made about the action of the animal enzyme. Musculus says that the vegetable diastase and the animal ptyalin produce the same sugar. Nasse on the other hand maintains that saliva forms a third sugar, differing from both glucose and maltose; he calls it *ptyalose*, and says its reducing power is only half that of glucose. The changes that take place in the liver both during life and after death lead to the production of glucose in its cells, yet the action of diastase on glycogen forms maltose. Nasse and some other observers maintain that it is ptyalose. Bial says that the serum of blood produces glucose and not maltose.

The discrepancy can be explained however in the light of more recent researches. The liver has been shown by Miss Tebb to contain in addition to the hypothetical diastase another enzyme, *glucase* or *maltase*, which converts maltose into glucose. The other animal tissues, which have been alluded to, also contain this enzyme, as do blood and lymph. The action of the diastase in all these cases gives rise to maltose, which is subsequently but rapidly converted into glucose by the maltase. This enzyme which has only comparatively recently been discovered will be discussed in a subsequent chapter.

There has been considerable discussion as to whether there are not at least two kinds of maltose formed during starch transformation by diastase. Fischer has prepared a sugar by the action of hydrochloric acid on glucose which possesses special characteristics, and he has given to it the name *isomaltose*.

In 1891 Lintner claimed to have identified this sugar in certain beer-worts. He said that on warming them with phenylhydrazine acetate an osazone was formed which in melting point and crystalline appearance corresponded to that of Fischer's new sugar. Later, in conjunction with Düll, Lintner stated that he had isolated this body, and the two authors published an account of its preparation from the products of the action of diastase on starch. According to them isomaltose is a sugar which ferments with yeast very much more slowly than maltose; it is very hygroscopic, and possesses a very sweet taste. It has a specific rotatory power like that of maltose $(\alpha)_D = +140°$, but its reducing power is only 83 per cent. of that of the latter. It is convertible into maltose by the action of diastase. With phenylhydrazine acetate it forms an osazone which melts at 150°—153°.

In a later paper in 1895, Lintner says that the conversion into maltose is incomplete, sometimes only 30 per cent. being so changed. He puts forward the further hypothesis that there may be two stereoisomeric isomaltoses. Lintner's views have been challenged by several observers. Ulrich repeated the experiments on which they were founded, but failed to prepare an osazone melting below 159°, which is several degrees higher than that of Lintner's body.

Ling and Baker prepared the latter, and from a consideration of its behaviour under different conditions came to the conclusion that it is not homogeneous.

Brown and Morris made a careful study of it and agreed with Ling and Baker that it is not a chemical entity, and does not correspond to Fischer's isomaltose at all. In their paper they say that it can be further split up by careful fractionation with alcohol and by fermentation with yeast, in such a way as to show that it is a mixture of maltose and dextrinous compounds of the maltodextrin or amyloïn class. The crystallisable osazone which Lintner describes as isomaltosazone, and on which he bases his views to a very large extent, is only maltosazone modified in its crystalline habit and melting point by the presence of small but varying quantities of another substance. Brown and Morris have shown this by recrystallisation of pure maltosazone in the presence of the non-crystallisable products of the action of phenylhydrazine acetate on the maltodextrin; they thus obtained an osazone having the properties of Lintner's body. Ling and Baker endorse this view, but they think that the substance mixed with the maltose is a simple dextrin having the formula $C_{12}H_{20}O_{10} + H_2O$.

Jalowitz confirms Brown and Morris, showing that pure maltose, when mixed with varying quantities of dextrin, yields osazones which not only differ in melting point, but also in their general appearance and crystalline form. With equal quantities of maltose and dextrin an osazone is obtained which melts at 150°—155° C., but the melting point alters after several recrystallisations.

Ost also controverts Lintner's views.

The balance of evidence is therefore very much in favour of the view that the only sugar which results from the action of diastase on starch is maltose.

The preparation of isomaltose by the action of animal enzymes has also been suggested, though not very clearly proved. Indeed it is probable that further investigation will show that isomaltose has not been isolated in this case either. Külz and Vogel have published the results of investigations into the action of the chief animal diastasic secretions, with special reference to the sugar formed. They used a five per cent. solution of rice starch, and identified the sugar by the phenyl-hydrazine reaction, analysing the osazones. They give as the results of their several experiments:—

Human parotid saliva— Isomaltose.
Mixed human saliva—   Isomaltose and later maltose,
                          with traces of glucose.
Dogs' saliva—   Isomaltose.
Ox pancreas—   Isomaltose.

Using glycogen instead of starch they found :—

With liver glycogen, human parotid saliva gave a mixture of isomaltose and maltose in the proportion of 1 : 2.

With muscle glycogen, the same saliva gave maltose only when the saliva was in excess; with less saliva they obtained isomaltose, with a little maltose and a little glucose.

With liver glycogen, ox pancreas gave isomaltose, with a trace of maltose.

With muscle glycogen, the same enzyme gave isomaltose, with a little glucose.

The products of the starch transformation by diastase in the animal body probably differ somewhat from those which are obtained in the laboratory. We have seen that in the latter case the final bodies are generally maltose and dextrin in the proportion of four parts of the former to one of the latter; the dextrin being the intractable group around which the four more easily hydrolysable groups are placed. In the body the final product seems to be maltose only. Lea, who definitely ascertained this fact, attributes it to the removal of the maltose from the region of diastasic activity as fast as it is formed. He says that when in the laboratory starch transformations are carried out in dialysing tubes, made of parchment paper and placed in water which is frequently renewed, the final product is maltose only. This is in accordance with the statement which many observers have made that any enzyme is much impeded if not actually inhibited in its action by the accumulation of the products of its activity.

The products which are formed in a plant during the action of diastase have not been at all fully investigated. There is no doubt that sugar is the most prominent of them if not the only final one, as the dextrins are as indiffusible as starch itself. The researches of Brown and Morris on foliage leaves leave no doubt that in *Tropaeolum* at any rate this sugar is maltose.

The writer in the course of some researches on the germination of the pollen grain was able to trace certain changes in the starchy reserves contained therein as the pollen tube developed. The plant whose pollen gave the most satisfactory results was *Lilium pardalinum*. The ripe pollen grains when treated with a solution of iodine in chloral hydrate, which renders them transparent, were found to be charged with starch-grains which stained the usual blue colour. Mixed with them here and there were a few grains staining like erythrodextrin. As the tube was put out from the grain these granules were gradually carried over into the protruding portion, and they flowed slowly down the tube as it extended. When the tube was as long as twice the diameter of the grain, if the chloral-hydrate-iodine solution was added they were found to be somewhat different in colour, becoming slightly purple with the iodine. With longer tubes, the grains in which were still travelling forward, this change was more and more marked, particularly near the tip of the tube. When a tube which had attained a length of 20—30 times the diameter of the grain was treated in the same way, the general effect of the iodine was markedly different. There were but few blue granules and those in the part nearest the pollen grain. The greater part of the length of the tube was studded thickly with purple grains and towards the tip they became nearly red. The starch was evidently in process of digestion under the action of diastase, which other experiments had shown the same pollen to contain. The granules did not change their

shape and showed no sign of corrosion even when magnified
very highly.   The alteration of the starch was apparently
uniform throughout the grain.

Du Sablon has examined the transformations accompanying
the digestion of starch in several bulbs and other underground
reservoirs, especially those of the Lily, Tulip, Hyacinth, Arum,
and Colchicum.   He finds it transformed into a sugar, dextrin
being an intermediate product.

These experiments, though far from exhaustive, appear to
show that the course of starch conversion inside the plant is
the same as that which has been demonstrated in the laboratory.

Before leaving the subject of the transformation of starch
some work published in 1890 by Wijsman calls for a passing
notice.   In all the foregoing experiments it has been taken for
granted that diastase is a single enzyme.   Wijsman holds that
this is incorrect and that under the one name two bodies have
been included; one of them he calls *maltase*, which is not
however the enzyme several times alluded to above, but more
nearly corresponds to diastase itself in that it converts starch
into a mixture of maltose and erythrodextrin; the second one
he terms *dextrinase*, and says that the formation of the
maltodextrin of Herzfeld and of Brown and Morris is due to
its action.   He agrees with the other authors that maltodextrin
is converted into maltose by his maltase, but says that erythrodextrin is converted by dextrinase into a dextrin which he
calls *leucodextrin*, which does not reduce cupric oxide and is
not coloured by iodine.   To show the presence of two enzymes
in the diastasic solution prepared from malt extract he adopted
the method of allowing it to diffuse into a gelatinous mass
made by adding gelatin to a solution of Lintner's soluble
starch.   When a little diastasic solution was placed upon a
layer of the gelatin mixture the progress of the hydrolysis
could be traced by subsequently moistening it with iodine
solution.   After 1—2 days' action, the area so treated formed
a colourless zone bordered by a violet ring, while the gelatin
with unaltered starch was coloured blue.   From this Wijsman
concludes that the two enzymes diffuse into the gelatin at

different rates according to their concentration. The violet-coloured ring showed where the maltase had penetrated beyond the dextrinase, while in the uncoloured area both were present. When a portion of the violet-coloured zone was removed and placed on a fresh film of the starch-gelatin, and the enzyme allowed to diffuse, no non-coloured zone was observed, but all was coloured violet by iodine.

Wijsman's results appear however to be quite in keeping with the view that only true diastase is concerned in the transformation; its progressive action would explain why the centre of the zone of influence was uncoloured, while the violet-coloured margin would indicate the earlier stages of the hydrolysis. The portion of such a zone removed would contain relatively little diastase and hence the violet reaction would be given for a considerable time when such a piece was allowed to act on the starch gelatin.

There appears therefore very little evidence to support his hypothesis.

Duclaux also holds the view that the ordinary diastase of malt is a mixture of two enzymes. Of these the first, which he terms *amylase*, liquefies starch paste, converting the starch into dextrin. The second, *dextrinase*, transforms this by hydrolysis into sugar. There is not much evidence again in favour of this view, except that, as will appear later, the power of liquefying starch paste and that of hydrolysing soluble starch do not always go together. Duclaux' dextrinase does not appear to be identical with the hypothetical enzyme of Wijsman bearing the same name.

Action of diastase on glycogen.

The hydrolysis of glycogen by dilute acids appears to resemble somewhat closely the corresponding process in the case of starch. According to the recent work of Miss Tebb when glycogen is heated, not necessarily to boiling point, with 1 per cent. of hydrochloric acid, the opalescent liquid rapidly becomes clear and limpid. If the heating is stopped at that stage, and the liquid neutralised, the latter is found to give a red colour with iodine and to reduce Fehling's solution. The sugar can then be removed by dialysis, and subsequent saturation with ammonium sulphate throws out a gummy precipitate which consists of soluble glycogen. After complete removal of this body, dialysis for several days is necessary to remove the ammonium salt. Concentration and subsequent addition of excess of alcohol leads to the precipitation of an erythrodextrin.

More prolonged action of the acid, extending over 1—2 hours, transforms both the soluble glycogen and the dextrin so far that iodine ceases to give a colouration. Addition of alcohol then causes the precipitation of another dextrin, belonging to the achroo-group, which like the erythrodextrin first formed is not precipitated by ammonium sulphate.

Miss Tebb gives the following table, showing the percentages of alcohol necessary to precipitate starch and glycogen and the intermediate products of their hydrolysis.

	Percentage of alcohol necessary to	
	commence precipitation	complete precipitation
Starch and its products		
Starch-paste	5	27
Soluble starch	12	60
Erythrodextrin	45	90
Glycogen and its products		
Glycogen	35·5	55
Soluble glycogen	44	50
Erythrodextrin	44	90
Achroodextrin	65	90

The action of the diastase of saliva or malt-extract differs in some particulars from that of dilute acids. There is no appearance of soluble glycogen or of erythrodextrin, but achroodextrin is always formed. This may only mean that the action is so rapid in the early stages that the first two are directly hydrolysed further. In addition, a new dextrin belonging to the achroo-group appears, which is peculiar in resisting conversion into sugar. This body Miss Tebb has called *dystropodextrin,* identifying it with a substance previously obtained by Seegen from glycogen and called by him by the same name. It appears to differ from the other achroodextrin by requiring greater concentration of alcohol for its precipitation.

The enzyme prepared from the liver acts similarly.

The sugar formed from glycogen by diastase alone appears to be maltose, according to Miss Eves and other observers. The liver enzyme yields dextrose as it does with starch, but as already mentioned this is probably due to glucase being present as well as diastase in the extracts of this organ.

CHAPTER V.

CONDITIONS OF THE ACTION OF DIASTASE.

THROUGHOUT the foregoing discussion of diastasic hydrolysis we have considered the action of diastase on starch in a general sense, but there are many points which need a further examination. Comparing the action of the two varieties of vegetable origin we have seen that diastase of translocation dissolves the starch grain without corrosion, while the so-called secretion-diastase disintegrates the grain before hydrolysing it. When these two kinds of diastase are compared with regard to their action on starch-paste and on a solution of *soluble* starch we find this difference of behaviour is emphasised. The enzyme of the scutellar epithelium is capable of liquefying starch-paste with great rapidity, while that of the general vegetative body of the plant has very little power of doing so, though it can hydrolyse soluble starch with the greatest ease. The power of liquefying starch-paste and that of eroding the starch grain seem to go hand in hand and to be a property pre-eminently of secretion-diastase. This power can be estimated separately therefore from the hydrolytic action which is common to both varieties. We may indeed speak of the liquefying power and the power of hydrolysing starch as in a way separate from each other, though we have no instance of a digestive solution which will carry on the former process and fail to secure the latter. Both these properties of diastase are capable of being modified by variation of the conditions under which the action is carried out.

The progress of the hydrolysis is further dependent to a considerable extent upon the kind of starch undergoing decomposition. Baranetzky has carried out many investigations upon this point, with the result of establishing the fact that ungelatinised potato-starch is highly resistant to diastase, while most of the starches of seeds are readily acted upon. Buckwheat starch appears to be perhaps the most easily hydrolysed.

Diastases from different sources appear to possess very different powers of inducing hydrolysis. Roberts has compared equivalent quantities of many preparations, with the results given in the following table, taking the pancreatic diastase of the pig as a standard.

Pancreatic secretion of pig	100
Human saliva	10–17
Pancreatic juice of sheep	12
„ „ ox	11
Malt diastase at 60° C.	10
„ „ 40° C.	4–5
Human urine	·03–·13

The susceptibility of diastase to differences of temperature is a feature which it shares, as already pointed out, with most other enzymes. In each case the action of the ferment is not perceptible at 0° C.; gradually rising with elevation of the temperature it reaches an optimum point, or point at which it works with greatest freedom; beyond this its activity falls with further warming until at another point, generally rather high, it ceases to work and is destroyed. We note a *minimum*, an *optimum* and a *maximum* temperature therefore in each case. These can be expressed in the form of a curve, the abscissæ of which represent degrees of temperature and the ordinates the hydrolytic power possessed at each temperature, estimated by the amount of cupric-oxide-reducing power possessed by the products of digestions carried on at such temperatures for the same amount of time.

If we compare the curves of the two varieties of vegetable diastase we find that of translocation diastase rises more sharply from its minimum point than that of the other form, being as high at 4° C. as is the latter at 14·5° C. It reaches its optimum point at 45°—50° C., and thence declines. The curve of secretion-diastase, though rising slowly, shows that the optimum point of action is not reached till 50°—55° C. Complete curves of the action of the two varieties have been published by Lintner and Eckhardt.

Müller has quoted figures representing the relative starch-transforming powers of diastase at different temperatures. He carried out digestions at 0° C., 10° C., 20° C., 30° C. and 40° C., and found the relative rates of sugar formation were respectively 7, 20, 38, 60, and 98.

The optimum point of the action of the enzyme of the saliva and pancreatic juice differs again from those of both the vegetable diastases, being apparently about 38°—40° C., the temperature of the body.

Some investigations made in 1892 and subsequent years by Effront make it evident that the action of diastase is capable of being materially modified by changes in the medium in which it is at work. He has ascertained that the enzyme of malt is favoured in its action by weak traces of mineral acid, a fact which was noted earlier by Baranetzky. Slightly larger quantities of neutral salt, such as sodium chloride, have the same effect. Effront has found further that there are three classes of bodies which have the power of markedly increasing this diastasic activity. These are salts of aluminum, compounds of phosphoric acid, and various amide bodies, among which asparagin is conspicuous. In his first paper Effront details some experiments with these bodies which are extremely striking.

The method he adopted was that of preparing a malt extract by macerating a given weight of ground malt with 40 times its weight of water. The filtered solution was then allowed to hydrolyse a solution of soluble starch, the various bodies influencing the digestion being added at the same time. The following table gives the comparative results of several of them :—

Malt extract	Starch solution	Percentage of foreign body used	Maltose formed per 100 c.c. starch solution
1 c.c.	200 c.c.	0	8·63 gms.
1 „	200 „	·7 Hydric ammonic phosphate	51·63 „
1 „	200 „	·5 Calcic phosphate	46·12 „
1 „	200 „	·25 Ammonia alum	56·3 „
1 „	200 „	·25 Potash alum	54·32 „
1 „	200 „	·25 Acetate of aluminium	62·4 „
1 „	200 „	·02 Asparagin	37·0 „
1 „	200 „	·05 Asparagin	61·2 „

The same relative results were obtained whatever was the temperature of the digestion.

It made no difference whether the diastase was submitted to the action of the reagents before being added to the starch, or whether they were added to the digesting mixture.

In a later paper Effront says that an infusion of raw grain, made in the cold, filtered, and boiled to destroy any diastase derived from the grain, has a slighter but still similar effect, often tripling the diastasic power of the malt.

In these experiments the difference which has been already alluded to between the two effects of diastase comes out very conspicuously. The reagents employed affect the hydrolytic action in the manner described, but they have little or no effect upon the power of liquefying starch-paste.

The influence of these reagents can only be detected in the earlier stages of the transformation, the maximum being reached when about 25 per cent. of the starch has been converted into maltose. Beyond this point it is much less and in the presence of such a quantity of diastase as will rapidly cause the saccharification of 60—70 per cent. of the starch it becomes very slight.

Their utility has consequently by no means such a practical application as at first seemed probable.

Effront points out that the liquefying and saccharifying powers of malt extracts are not in fixed proportion, but vary very much in different cases, depending upon the methods of manufacture of the malt.

It has been noticed by other observers that addition of proteids to malt extracts increases their hydrolytic powers. Yeast cells, after their vitality has been destroyed by boiling, have a considerable intensifying action, which is presumably due to their proteid constituents.

A similar influence has been observed in the case of the diastase or ptyalin of human saliva. Chittenden and Ely ascertained that peptone has a remarkable effect upon starch transformation. In a series of experiments carried out in 1881 these observers worked with quantities of 100 c.c. of liquid containing 1 per cent. of starch, 1 per cent. of peptone and 25 c.c. of filtered saliva. Controls were digested simultaneously in the preparation of which the peptone was omitted. The mixtures were maintained at 40° C. for 45 minutes and then boiled to stop the action of the enzyme. After being diluted with water to 500 c.c. they were filtered and an aliquot part of each was titrated with cupric-potassium-tartrate. The percentage of starch converted into sugar was 4 per cent. greater in the presence of the peptone. Similar experiments with glycogen instead of starch yielded the same results. The quantity of peptone most favourable to the digestive process was found to be 1 per cent.; increase of the amount had little or no additional effect.

Chittenden and Ely found further that the addition of ·025 per cent. of hydrochloric acid to the digestion containing 1 per cent. of peptone was still more advantageous; the quantity of starch transformed amounting to 48 per cent. as against 41 per cent. when saliva alone was used.

Langley and Eves have endorsed with a certain reserve the statement of the influence of peptone, but they came to an opposite conclusion to Chittenden's when they added acid in addition. In their experiments they found that the maximum effect of peptone was reached with a rather low percentage of the proteid; when the saliva was diluted 10 times the most advantageous amount of peptone was ·1 per cent. They state in opposition to Chittenden and Ely that peptone combined with hydrochloric acid retards the action of the diastase, and the retardation is greater the nearer is the point of saturation of the peptone by the acid. Langley and Eves found also that myosin, alkali-albumin, and acid-albumin act like peptone and that albumin probably has the same effect.

It was mentioned above that according to Baranetzky small traces of neutral salts accelerate the action of malt diastase. Nasse states that this influence of certain neutral salts is maintained till the proportion of the salt reaches 4 per cent. Chittenden and Ely found that by using ·024 per cent. of sodium chloride the diastasic activity of saliva was increased about 6 per cent., and with ·03 per cent. of calcium phosphate it rose 3 per cent. With hydric-disodic phosphate there was on the other hand a slight diminution of the activity.

Osborne and Campbell also state that sodium chloride is beneficial, sometimes increasing the diastasic activity of a solution seven-fold.

Müller points out an interesting circumstance connected with the action of diastase in the vegetable cell. Under ordinary conditions the cell-sap in the vacuole of such a cell contains a considerable quantity of CO_2, corresponding in some cases to a pressure of several atmospheres. Even at ordinary pressures saturation of the liquid of a digestion with CO_2 may triple the energy of the diastase. In presence of CO_2 diastase can act on unboiled starch.

The evidence respecting the action of traces of free acid upon diastase is conflicting. Baranetzky pronounced the presence of such traces to be favourable. He has been confirmed by Effront and by Brown and Morris. Chittenden and Griswold have stated that the ptyalin of saliva acts most energetically in the presence of ·005 per cent. of hydrochloric acid. Astachewsky has pointed out that human parotid saliva is sometimes acid and says that then it is more active than the alkaline parotid saliva collected at other times.

On the other hand Langley and Eves found that the action of neutralised saliva was retarded by the presence of free hydrochloric acid, as little as ·0015 per cent. having a distinct effect in this direction, while ·005 per cent. was very strongly inhibitory. The retarding effect increased rapidly as the percentage of the acid was increased.

When the proteids of saliva were saturated with acid, they found a diminution of diastasic action, though no free acid was present. This diminution was made more marked by the addition of the smallest quantity of free acid.

Chittenden and Ely came to the same conclusion with regard to the influence of hydrochloric acid. With ·025 per cent. of the latter present in a digestion, they found that the amount of sugar produced was only about 8 per cent. of that formed in a similar digestion without the acid. Osborne and Campbell state that the same thing is true of citric acid, as well as hydrochloric.

Langley and Eves suggest that the beneficial effect of the small traces of acid (·005 per cent.) noticed by Chittenden and Griswold may be explained by the fact that saliva is normally faintly alkaline, and the acid would probably merely neutralise this alkalinity. Saliva, according to them, works most favourably in a completely neutral solution. Chittenden says however that its reaction is due to alkaline phosphates and not to free alkali.

The influence of the free acid appears not to be merely inhibitory, but the enzyme is destroyed by it. If acid is allowed to remain in contact with the diastase for some time and is then removed by dialysis, the enzyme has afterwards no more power to hydrolyse starch than it has when the acid is added to the starch solution in which digestion is going on.

According to Kjeldahl, malt diastase is made more active by the addition of 20 milligrammes of sulphuric acid to each litre of solution. This quantity must not be exceeded, for 30 milligrammes and upwards affect it prejudicially. Fernbach makes a similar observation, but he attributes the improvement caused by the smaller quantity to the action of the acid on the phosphates present in the malt extract. Under normal conditions the latter contains bibasic phosphates which inhibit the action of diastase. The acid converts these into monobasic phosphates, which accelerate it. After sufficient acid has been added to effect this change in the phosphates, any further quantity remains free and inhibits the diastase, as the other observers quoted have pointed out.

Weak alkalis also have a deleterious influence on the action of diastase. Chittenden and Ely found that when ·6 per cent. of sodium carbonate was added to a starch digestion, the diastasic activity of the saliva was only about one fourth what it was in the absence of the salt. With 3 per cent. the power was about half and with ·1 per cent. about two thirds of that possessed by the unaffected saliva.

Langley and Eves found a distinctly inhibitory action to be manifested by the presence of as little as ·0015 per cent. of sodium carbonate. Sodium hydrate was still more deleterious than the carbonate.

Comparing the action of acids and alkalis, the rate of decrease of the action of ptyalin under the influence of the latter is slow compared with that of the free acid.

Another difference between the two is that while both are strongly inhibitory, the weak alkali is not destructive of the enzyme. If it is removed by dialysis the diastase is found to possess nearly its original starch-transforming power. Sodium carbonate therefore works by impeding the action of the enzyme, while hydrochloric acid destroys it.

Chittenden and Ely have ascertained that the inhibitory effect of both acids and alkalis is diminished by the coincident presence of peptone. Langley and Eves, while disputing this in the case of acid, agree with the other observers in the case of alkali. They say that the favourable influence of peptone makes itself felt with very high percentages of the alkaline salt.

Brown and Morris state that the presence of tannin is very prejudicial to the action of diastase; this fact has no doubt considerable importance when an attempt is made to ascertain the amount of diastase in such vegetable tissues as contain tannin. It is not only inhibitory of the diastasic action, but its presence is a formidable obstacle to its extraction from the tissue.

Loew has indicated hydrocyanic acid in 25 per cent. solution, and atropine in small doses, as both deleterious to diastase.

It has been mentioned in a preceding chapter that it is only with considerable difficulty and delay that the whole of the starch in a digestion is converted into maltose. Different observers agree in noticing this, though they differ as to the amount which resists the change. Musculus only succeeded in transforming 33 per cent. of the starch he used; Payen was able to convert 52·7 per cent., while Brown and Morris give as the usual result of a starch transformation 80 per cent. of maltose, and 20 per cent. of dextrin.

As the digestion proceeds more and more difficulty is found to attend its progress. This is possibly due to an inhibitory action exerted by the products of the hydrolysis. Sheridan Lea has thrown some light upon this question by carrying out starch transformations simultaneously in flasks and in dialysing tubes suspended in a stream of running water. In the former apparatus the difficulty noted by other observers was met with, but in the latter nearly all the starch was converted into sugar. Lea states that under such conditions the rate at which the digestion takes place is increased, and the total amount of starch converted into sugar is much greater, while the residue of dextrin is much less than under conditions, otherwise similar, when the products are not removed. Brown and Morris deny that the inhibition is due to the presence of maltose. They found that when maltose was added to a mixture of starch and diastase, the course of the action was not affected but that 80 per cent. of the starch still underwent complete hydrolysis.

When a digestion is prolonged, this inhibition is seen even in glass vessels to be only a retarding of the action. Brown and Morris quote one experiment in which digestion was continued for 20 days, chloroform being used to prevent putrefactive changes. At the end of that time 92·4 per cent. of the starch dissolved had been converted into maltose.

Another fact of great interest is that diastase is not exhausted by continued activity. This we have seen to be a property of enzymes in general. It was first established by Foster in the course of an examination of the properties of saliva. Moritz and Glendinning have noticed the same peculiarity in the case of malt diastase, which they say is as potent at the end of a fermentation as at the beginning, so long as the temperature is at or below the optimum point. If the digestion is carried out at a point higher than this, the enzyme

loses power, but they consider this to be due to the temperature
and not to anything of the nature of exhaustion.

Conditions of the secretion of diastase.

In the simplest organisms in which the secretion of diastase
occurs, such as bacteria or moulds, or among the higher plants in
such simple structures as pollen grains, the formation of the
enzyme is found to have a direct bearing upon the question of
the nutrition of the organism.   Lauder Brunton and Mac
Fadyen found that a microbe upon which they were experimenting secreted diastase when it was cultivated upon starch
paste, but did not do so when the culture medium was meat-
broth.   In this case it gave origin to a peptonising enzyme.
Pfeffer has found that in the case of several of the mould fungi
the secretion of diastase depends upon similar conditions.   In
the case of the pollen grains of several species of plants the
appearance of diastase only becomes noticeable after the
absorption of a trace of sugar by the grain.   During the
growth of the pollen tube the writer has noticed that the
quantity of diastase is continuously increased, presumably in
consequence of the growing amount of nutritive material acting
as a stimulus to the protoplasm of the pollen.   This is very
noticeable in the case of special species of *Lilium*.   In *Zamia*
no diastase can be detected in the ungerminated pollen grains,
but on the absorption of either cane sugar or glucose, even
before visible extrusion of the pollen tube has taken place, a
small amount of the enzyme can be found to have been
secreted.   The secretion of diastase by pollen is not altogether
dependent upon such absorption, for in the case of Lilium there
is a considerable formation when the grains are germinated in
pure water.   Here the necessary stimulus for the increased
secretion may have been afforded by the transformation of a
small quantity of the reserve starch existing in the resting
grain by the diastase also present in its ungerminated condition.

The secretion of diastase by the scutellar epithelium of the
grasses does not appear to be caused by similar conditions.
Barley embryos cultivated by Brown and Morris upon gelatin
containing starch grains did not form more diastase than similar
embryos cultivated upon gelatin alone.   The influence of the

culture medium in this case does not appear to be the same as in the case of the micro-organisms already mentioned.

In some other experiments the same authors found that while certain barley embryos attacked and dissolved starch grains suspended in gelatin, others, precisely similar, left the starch untouched when a small amount of an assimilable sugar had been added to the gelatin-starch medium. In this case, so long as any sugar remained unabsorbed by the young plantlet, the starch grains escaped disintegration. They were attacked however as soon as the supply of sugar was exhausted.

The authors ascertained that in these cases there was no question of inhibition of the *action* of diastase by the sugar, but that in the presence of the latter, the epithelial cells of the scutellum actually did not *secrete* the enzyme. This inhibition of secretion was only caused by such carbohydrates as were directly assimilable by the embryos, other carbohydrates, especially mannitol and lactose, which are of no use in nutrition, being incapable of exerting any inhibitory power.

Not only the complete embryo but sections of the scutellum covered by the epithelium showed this inhibition of secretion by assimilable carbohydrates.

The phenomenon of secretion in the case of the scutellar epithelium appears to be one associated with starvation. As long as readily assimilable substances are supplied the secretion of diastase does not take place, but when no such substances are available, diastase is formed at once.

When barley is examined during the process of germination, the epithelium is found not to begin immediately to secrete the enzyme. Petit, who has studied the question, has ascertained that diastase does not appear in germinating barley till the fourth day. It reaches a maximum almost at once and then gradually diminishes till the ninth day, when the grain contains about one twentieth part of what it possesses at the maximum. This fact supports Brown and Morris's view of the stimulus to secretion being one of starvation. During the early part of the period of germination the embryo is feeding upon the carbohydrates it contains; its needs at that period are small, and it is not till growth has become vigorous that it needs to draw upon the reserves deposited in the endosperm. After the ninth day these reserves are considerably depleted, and the young embryo having by this time developed its rootlets is able to draw in nutriment from without.

Brown and Morris also found that the secretion gradually increases during germination. They differ from Petit however in saying that the maximum secretion is not reached so soon. They found a decided increase from the fourth to the seventh day.

Evans obtained slightly different results during the process of malting a Californian barley. The diastasic power at the end of 24 hours from steeping was 27·8; it rose gradually for three days, when it was 95·4; it then sank for three days, falling to 50; from this time it rose till the ninth day, when it was 86·5 and then gradually fell till germination ceased, being 62 on the eleventh day and 28·3 on the fourteenth, when the malting was completed.

The stimulus of starvation has been observed also by Wortmann, who has found that a certain microbe has the power of excreting a starch-dissolving enzyme when starch grains are the only available food, but that no secretion takes place if sugar or tartaric acid is offered to the organism along with the starch.

The same thing has been observed in connection with other enzymes.

In the animal body the formation of diastase is largely dependent on nervous impulses. The salivary glands are those most easily examined. Each in the higher mammals is supplied by nerves derived from the cerebro-spinal and from the sympathetic systems. Stimulation of either nerve is followed by an increased flow of saliva and consequently by an increased formation of the characteristic enzyme. The stimulation may be either direct or reflex, the latter method being that which is called into play during normal secretion. Sapid substances placed on the tongue, or even the smell of food, will cause an increased flow of saliva. A detailed examination of these phenomena would however be beyond the scope of the present work. It will suffice to say that if salivary glands are examined before, during, and after excitation of the nerves with which the glands are supplied, the appearances lead to the conclusion that such stimulation produces the changes in the gland cells which indicate a formation of the solid constituents of the secretion from the protoplasm of the cells, and a subsequent extrusion of them together with a certain flow of fluid derived from the lymph with which the secretory tubules are bathed.

The physiological object of the secretion of diastase must in nearly all cases be the digestion of solid starch or glycogen with a view to the absorption and utilisation of the soluble and diffusible products which the enzyme forms. This digestion may be extracellular, as in the cases of the secretions of saliva, pancreatic juice, secretion-diastase as supplied by the scutellar epithelium, and certain bacterial diastases. Again it may be intracellular as in the cases of the liver, and the ordinary vegetable reservoirs, such as potato tubers. The translocation of starch in the cells of foliage leaves and other green parts of plants belongs to the same class of phenomena. The immediate result is in both cases one of nutrition, either of the whole organism, or of the cell in which the changes may be seen.

In the guard-cells of stomata the transformation of starch appears to have a different meaning. The function of these cells is primarily if not entirely the regulation of transpiration, or the escape of watery vapour from the interior of the plant. When the guard-cells are turgid from the absorption of water into their interior, they become convex and separate slightly, allowing the aperture between them to be enlarged. The working of the stomata consequently depends upon their power of absorbing water from the cells on which their guard-cells abut. This water is attracted into them by osmosis, and this physical process depends upon their containing substances which have an attraction for water, or an *osmotic affinity*. The affinity possessed by starch is practically nil, but sugars and the acids derived from the latter possess a considerable power of attracting water. Digestion of the starch by the diastase in these cells consequently leads to the development in them of osmotic powers which must be of great value in causing their change of shape in the ordinary course of the transpiration process.

CHAPTER VI.

INULASE.

BESIDES starch and glycogen carbohydrate reserve-material occurs in the vegetable kingdom in the shape of inulin. It usually exists in a condition of solution in the sap of the cells which constitute the reservoir, and may be precipitated therefrom by the addition of alcohol.

Inulin as a reserve-material has a wider distribution than was till recently known to be the case. It has long been recognised in a group of plants which are prominent among the Compositae, the chief of which are the genera *Dahlia, Helianthus, Inula* and *Atractylis*. These plants possess either tubers, or tuberous or fleshy roots, and in these the stores of inulin are located. More recently Parkin has ascertained that inulin is a very common reserve material in many Monocotyledons, occurring in great abundance in many of the bulbous Liliaceae and Amaryllidaceae as well as in other Natural Orders related to them. Here it may be found in the leaves and other parts concerned in active vegetative metabolism as well as in the seats of storage of reserve materials, the bulbs or corms which are so characteristic of these Natural Orders.

Inulin can be detected in such structures as the tuber of the artichoke (*Helianthus tuberosus*) by steeping pieces of the tissue in alcohol for some time and then immersing sections of the hardened material in water, when it is found to crystallise out in the form of sphaerocrystals, sometimes of large size. These are seen to consist of needles symmetrically arranged so as to radiate from a common centre. They may be found in the interior of the cell, or may extend through several cells, forming a comparatively large aggregation, visible in such a case to the naked eye. The fleshy roots of the Dandelion (*Taraxacum*), the Helicampane (*Inula Helenium*), also yield these crystals on similar treatment.

In the cells of the tissue of the bulb-scales of the snowdrop (*Galanthus*) and other Amaryllidaceae the crystalline appearance is not so readily presented, but the inulin may be detected in a somewhat amorphous form.

Inulin is insoluble in cold water, but may be prepared from these plants by boiling the tissue containing it with large quantities of water, concentrating the decoction so obtained, and allowing it to stand till it deposits a sediment. This must then be redissolved in a small quantity of hot water, decolourised by boiling with animal charcoal and again concentrated, when fairly pure inulin is gradually deposited. It can be further purified by redissolving it and again evaporating, when the inulin is once more precipitated; finally it should be well washed with cold water containing a little alcohol.

Inulin so prepared is a white powder which is readily soluble in warm water and which remains in solution when the liquid cools. Very little can be dissolved by cold water, and it is insoluble in alcohol. A solution of it can be precipitated by the addition of the latter reagent, the precipitate occurring when the percentage of alcohol in the liquid reaches 65.

The composition of inulin, like that of starch, cannot yet be said to be definitely ascertained. The formula usually given for it is $C_{12}H_{20}O_{10}$ $2H_2O$. Under the action of hydrating agents such as mineral acids it is converted into fructose, a sugar which has a laevo-rotatory power $[a]_D = -106°$. Inulin itself has a specific rotatory power of $[a]_D = -39 \cdot 9°$.

Inulin can be converted into fructose by boiling its watery solution under pressure, or by allowing it to remain in contact with cold water for a lengthened period, care being taken to prevent the access of micro-organisms.

In the process of germination of the tuber, or in the resumption of growth of the plant from the fleshy roots, the transformation of inulin into sugar is brought about by the action of an enzyme to which the name *inulase* has been given. The existence of this enzyme was first demonstrated by the writer in 1887, when the germination of the tuber of the artichoke was made a subject of examination.

Examination of the young shoots and roots arising from the germinating tuber showed that while inulin was present in them, they contained also a far larger proportion of sugar than was present in the resting tuber of the plant. The sap of the tuber when expressed was almost neutral; very sensitive litmus paper showed a faint trace of acid, but too little to account for the conversion of inulin into sugar by its instrumentality.

Tubers of the artichoke were allowed to germinate till the young plants arising from them had attained a height of about six inches above ground. The tissue of the tuber by this time had been considerably altered, the interior having become spongy and the cells almost empty, while on the outside and for some distance inwards it was firm and succulent. Tubers in this condition were minced finely and extracted with glycerine, in which inulin is insoluble. After twenty-four hours the glycerine was strained off and dialysed in several changes of water, till the dialysate failed to reduce Fehling's solution.

A small quantity, about 10 c.c., of such an extract was then mixed with about 4 times its volume of a one per cent. solution of inulin and put into parchment dialysing tubes, which were suspended in about 200 c.c. of distilled water, thymol being used to prevent the action of micro-organisms. A control experiment was set up at the same time, in which the extract of the tubers employed had been boiled to destroy any enzyme that might be present.

The dialysates were changed at intervals of 24 hours and each as it was removed was tested for sugar.

Hydrolysis of the inulin in the tubes containing the unboiled extract, began gradually and proceeded regularly for several days, the difference between the dialysates of the boiled and unboiled tubes becoming more and more emphasised as the digestion proceeded. That in which the boiled extract was suspended contained no sugar after one or two changes, while that surrounding the unboiled extract continued to give a copious reduction of Fehling's solution.

Similar results were obtained when the digestions were carried on in glass vessels, a continuous formation of sugar being indicated.

Coincidently with the appearance of sugar in the dialysates, the amount of inulin in the parchment tubes underwent a regular diminution. The amount of precipitate thrown down by alcohol from the fluid containing the unboiled extract of the tubers became less and less as time went on, measured quantities being withdrawn for each observation. Similar quantities taken from the controls showed no such diminution.

From these results it is evident that the germinating artichoke contains an enzyme by whose instrumentality inulin is transformed into some form of reducing-sugar. The time taken up in the experiments is possibly to be accounted for by the very small quantity of the enzyme present and its dilution in the process of extraction. In the artichoke it is probable that it only exists at any particular time in the cells whose contents are actually being hydrolysed, and as it takes several weeks for the conversion into sugar of the inulin present in any tuber, there must be very little of the enzyme to be found at any one moment.

This enzyme, *inulase*, cannot be found in the tuber in the resting condition, nor is it easy to demonstrate its presence until the young stems begin to emerge. Once found, however, its presence can be demonstrated so long as any inulin remains in the germinating tuber.

During the progress of the growth of the young stems, they may be found to contain inulin, which occupies the centre of the shoot and leaves the circumference free. Inulin possesses a feeble power of dialysis, and it is no doubt by means of this that it passes from the tuber into the shoot. The inulin cannot be traced up to the growing point, but stops abruptly just behind the actively growing zone. It is accompanied in its progress by sugar, which extends further forwards towards the growing point. The transformation under the action of inulase is not therefore confined to the tuber, though it originates there. The enzyme is present in the growing shoot, and continues to hydrolyse the inulin as growth proceeds.

The occurrence of inulin in this region may be due to another cause than dialysis or transport of the store in the tuber. It may be brought about by the supply of sugar being in excess of the needs of the growing cells, the surplus being temporarily converted again into inulin. Such a reconversion is a matter of constant occurrence in the case of supplies of starch in other plants where the latter is the form in which the reserve store of carbohydrate material occurs.

The occurrence of inulase has been observed by Bourquelot in the fungus *Aspergillus niger*. When cultivating its spores in different media during the year 1885 he found that they were capable of growing in a solution containing inulin as freely as in others which contained glucose or cane-sugar. At that time the enzyme had not been described, but on returning to the subject in 1893 Bourquelot was able to prepare it from the mycelium of the fungus. Aspergillus contains several enzymes, but he was able to convince himself that inulase was present and was a distinct entity, quite unlike diastase.

Chevastelon has shown that the inulin which can be prepared from the Natural Orders of Monocotyledons already alluded to is capable of hydrolysis by the inulase derived from Aspergillus. He worked especially with the inulin contained in garlic.

The presence of this carbohydrate in such abundance in Monocotyledons suggests that inulase is present in them also. Up to the present year however it had not been identified there. Recently the writer has ascertained that it can be obtained from the bulbs of *Leucojum* and *Scilla*. Its existence is most easily demonstrated in the germinating bulb, at the time when the flowers are just formed and before they have expanded. The method used by the author was exactly the same as that described above, and the enzyme behaves in the same way as that of the artichoke. It is more difficult to extract on account of the mucilage present in the cells.

The enzyme appears in all cases to be mainly if not entirely confined to the cells in which the actual conversion of the inulin is taking place.

The action of inulase can best be studied by carrying out the digestion in dialysing tubes. In the author's experiments such digestions were carried on under antiseptic precautions for several days, the dialysates being changed every 24 hours. The first three days' dialysates were rejected, to ensure that whatever might be found in the later ones besides inulin should be only the products formed by the action of the enzyme. The fluid was of course bulky and contained only a relatively small quantity of carbohydrates. The later dialysates were mixed and concentrated over water baths till of small volume, when they had a syrupy appearance and consistency. The syrupy liquid was found to contain three bodies which could be separated from each other by treatment with alcohol. The first of them was a sugar which was separated by extracting the residue with absolute alcohol. About half of the syrup dissolved, and on concentrating the alcoholic liquid over a water bath, it again became syrupy and remained so, refusing to crystallise, even when exposed over strong sulphuric acid.

After separation from the sugar the residue was treated with cold and subsequently with hot water. A good deal of it dissolved in the cold water, which indicated that it was not composed of inulin, as this is insoluble until the water is warmed. The residue insoluble in cold, but soluble in hot water, was found to consist of inulin, which having a feeble power of dialysis had passed through the membrane during the later stages of the digestion.

The two solutions were mixed and alcohol added gradually. When the liquid contained 65 per cent. of alcohol the inulin separated out as a crystalline precipitate. This was filtered off and more alcohol added. When 82 per cent. of spirit was present the liquid became opalescent and gradually a very finely-granular precipitate settled out. In 100 parts of the residue there were about 62·5 parts of unchanged inulin and 37·5 parts of this second body, which appeared to be an intermediate body resulting from the action of the inulase.

On concentration of the watery solution of this substance it deposited a quantity of crystalline matter. The crystals differed remarkably from the sphaero-crystals of inulin, appearing generally as plates, sometimes pentagonal, sometimes rhomboidal or oblong, with here and there needle-like prisms forming part of a rosette.

On examining several commercial samples of so-called inulin, it was found that variable quantities of this intermediate body were present. It can be separated from inulin by fractional precipitation by alcohol or less easily by dialysis. It has a greater power of dialysis than inulin; one sample of the commercial inulin examined contained 12—14 per cent. of it as ascertained by the alcohol method; after dialysis for five days the dialysate when concentrated contained both, but the intermediate body formed sixty-six per cent. of the total precipitate thrown down from it by alcohol.

This product then differs from inulin in the following particulars :—

(1) It is more soluble in cold water.

(2) It has a greater power of dialysis.

(3) It has a different crystalline form.

(4) It is soluble in alcohol of sixty-five per cent. strength, not being precipitated by less than eighty-two per cent.

It is clear that this body occurred in the dialysates of the digestions in consequence of its formation by the inulase and not from having been present in the inulin used, for the dialysates of the first three days were rejected. If any had been mixed with the inulin taken for experiment this would have escaped during that time, as its dialysing power is so great compared with that of inulin. Its occurrence as a consequence of the activity of the enzyme recalls the occurrence of dextrins during the action of diastase on starch. It somewhat resembles dextrin in its behaviour with alcohol, being soluble in stronger percentages of this reagent than the original carbohydrate, but not soluble in a greater percentage than eighty-two, at which point dextrin also is precipitated.

The sugar formed has been investigated by Bourquelot. He quotes a series of observations made with a solution of inulin

containing 1·32 per cent., which was mixed with inulase and left at the temperature of the laboratory till digestion was complete. As it proceeded, the rotation of a ray of polarised light by the digesting liquid was observed at definite intervals of time, the temperature being noted at each observation.
The results are expressed in the following table.

Duration of digestion.	Rotation observed.	Temp. of liquid.	Percentage of reducing-sugar calculated from the observed rotation.
	− 1·06	17° C.	0
12 hours	− 2·03	17° C.	0·871
36 „	− 2·43	17·5° C.	1·283
64 „	− 2·5	19° C.	1·371
84 „	− 2·53	19·5° C.	1·403

The inulin used had a specific rotatory power of $(\alpha)_D = -39\cdot9°$

From this calculation Bourquelot concludes (1) that the proportion of reducing-sugar actually formed attained nearly that which could theoretically be yielded by the inulin originally present; (2) that, reckoning the influence of the temperatures noted on the rotatory power of levulose (fructose), the observed rotation in the last observation corresponded very closely to that which would be given by a solution containing a proportion of levulose equal to that of the reducing-sugar found.

It thus appears that under the action of inulase, the inulin used (prepared from *Atractylis*) was converted almost entirely into levulose (fructose).

Bourquelot calls attention to a further fact of some interest. When inulase is added to a solution of inulin, alcoholic fermentation by yeast can be induced in the mixture. Now inulin does not undergo the alcoholic fermentation, while levulose does so readily.

Inulase is not present in the resting tuber of the artichoke, as we have seen above. It can however be prepared from such tubers before germination begins. If they are minced and kept for 24 hours at a temperature of 35° C. the extract prepared from them possesses the power of hydrolysing inulin. Without the preliminary warming such an extract is quite inert. The inulase appears to exist in such tubers in the form of an antecedent body or *zymogen*.

Inulase has no action on starch; it exists however in some plants in the same region as diastase, with which there is some danger of confusing it. In *Aspergillus* it is found associated with several other enzymes, with one of which, *trehalase*, it may be confounded. It can be distinguished from the latter by the temperature at which it is destroyed.

Inulase works most advantageously in a neutral or very faintly acid medium, the best proportion of acid being about ·001 per cent. of hydrochloric acid. Stronger acids than this are prejudicial, and exposure for an hour at 40° C. to an acidity equal to ·2 per cent. of hydrochloric acid destroys it altogether. Alkalies are also deleterious, no hydrolysing power surviving an hour's exposure to a strength equal to 1·5 per cent. of sodium carbonate. The rapidity with which the destruction of the enzyme by acid takes place is dependent on the temperature at which it is kept during the time the two are in contact. At a low temperature it is much less affected than at 40° C., but after an hour's exposure at only 10—15° C. its working power is much impaired.

The energy of inulase shows the same variation with the temperature, being much greater at 40° C. than at the ordinary temperature of the laboratory. It survives exposure to a temperature of 64° C., at which point trehalase is destroyed, but it loses its activity if heated to about 70—75° C.

CHAPTER VII.

In the seeds of many plants, especially among the Mono-cotyledons, the carbohydrate reserve food-material for the nutrition of the embryo during germination takes the form of extremely thickened cell-walls, usually those of the endosperm in which the young plantlet is embedded. In other cases cell-walls of comparatively little thickness serve the same purpose, though to a much smaller extent, as may be seen in the barley and other grasses. During the germinative processes these cell-walls are dissolved and disappear, undergoing transformations comparable with those of starch so far as to yield ultimately some form of sugar. It has been usual to regard these cell-walls as composed of *cellulose*, but recent researches show that they are of a more complex nature than has been supposed.

The composition of the cell-wall varies very greatly in different plants, and indeed in different parts of the same plant. In addition to cellulose it contains bodies known as *pectoses*, which include several very complex substances. The term *cellulose* itself is now taken to cover a very large class of plant constituents, the members of which show considerable differences in properties and in facility of decomposition. The gradual utilisation of the cell-walls of different endosperms involves very complex changes, in which both celluloses and pectoses are concerned.

The cellulose bodies according to Cross and Bevan fall into three distinct categories as under:—

(1) Those which offer a maximum resistance to hydrolytic action, and which contain in their molecule no directly active

CO groups. These are represented by the cellulose of cotton fibres.

(2) Those of less resistance to hydrolysis, which contain active CO groups. These are perhaps best regarded as *oxycelluloses*. They appear to constitute the main mass of the fundamental tissue of flowering plants, and they exist in conjunction with lignin in the walls of wood-cells.

(3) Those which hydrolyse with some facility, being more or less soluble in alkalies and easily decomposed by acids, with formation of carbohydrates of low molecular weight. Included among these is the cellulose of the walls of the cells of seeds.

The members of these three groups behave very differently with hydrolysing agents, and the ultimate products of their decompositions vary considerably.

The pectose group of the constituents of the cell-wall has been investigated by many writers, among whom may be mentioned Payen, Vauquelin, Mulder, Fremy, Kutsch, Vogl, and Wiesner.

The most recent observations have been made by Mangin, who has given the fullest account of them and their chief reactions. According to him they fall into two series, one comprising bodies of a neutral reaction, while the others are feeble acids. In each series there are probably several members, which show among them every stage of physical condition between absolute insolubility and complete solubility in water, the intermediate bodies exhibiting gelatinous stages, characterised by the power of absorbing water in a greater or less degree.

Of the neutral series the two extremes are presented by *Pectose* and *Pectine*. The former is insoluble in water and is closely associated with cellulose in the substance of the cell-membranes; the latter is soluble in water and forms a jelly with more or less facility. In the other series the two most noteworthy members are *Pectic* and *Metapectic acids*. The former generally exists in the cell-membranes in combination with the metals of the alkaline earths, especially calcium; when in the free state, it is insoluble in water. Metapectic acid is soluble in water, but does not form a jelly.

The two series are closely related to each other, for by the action of heat, acids, and alkalies, the various members of both can be prepared from pectose. The final product of the action of the reagents is the freely soluble metapectic acid.

Mangin gives their characteristic peculiarities as under:—

Pectose. The actual properties of this substance are not at all easy to ascertain, nor can they be said to be well known. The material is so closely associated with cellulose that it has not yet been obtained pure. The reagents which separate it from cellulose convert it either into pectine or into pectic acid, the former being soluble in water, the latter in alkalies. The cell-wall can be shown to contain the two constituents by the action of Schweizer's reagent (ammonio-cupric sulphate) which, when used with proper precautions, dissolves out the cellulose and leaves the framework of the cell apparently unaltered; it then consists, however, not of pure pectose, but of a compound of pectic acid with the copper of the reagent.

Pectine. This body swells up and dissolves in water, forming a viscid liquid which soon becomes a jelly. It exists in considerable quantity in many ripe fruits and in some mucilages. It gives no precipitate with neutral acetate of lead, but is thrown down by the basic acetate in the form of white flocculi. If boiled for several hours in water, it is converted into an isomer, *parapectine*, which is precipitated by neutral lead acetate. Further boiling with dilute acids converts it into *metapectine*, which is precipitated by barium chloride.

Pectic acid. This body is insoluble in water, alcohol, and acids; it forms soluble pectates with alkalies and insoluble ones with the metals of the alkaline earths, of which calcic pectate is most widely distributed. It dissolves in solutions of alkaline salts, such as the carbonates of sodium and potassium, stannates, alkaline phosphates, and most organic ammoniacal salts, forming with them double salts, which gelatinise more or less freely with water. Its solution in alkaline carbonates is mucilaginous, but when ammonic oxalate is the solvent it is perfectly limpid.

Metapectic acid. This is a body with an acid reaction, freely soluble in water, and forming soluble salts with all bases, especially those of calcium and barium, which precipitate pectic acid. Metapectates warmed with an excess of alkali are coloured yellow. This body and its compounds approach the gums in their composition. When acted upon by dilute sulphuric acid it splits up into a dextro-rotatory crystallisable sugar, apparently identical with *arabinose*, and into a little-known organic acid, indicating by this behaviour some relationship to the group of the glucosides.

The cellulose and pectose constituents of the cell-wall show considerable differences of behaviour. The former are soluble, the latter insoluble in Schweizer's reagent; when oxidised with nitric acid the former yield oxalic, the latter mucic acid. The celluloses when partially hydrated stain blue with iodine; the pectoses give no coloration with this reagent. They behave differently also to staining reagents, and to dilute acids and alkalies. According to Tollens and Ganz the sugar produced by hydrolysing cellulose with dilute sulphuric acid is a dextro-rotatory *mannose*.

Under the action of strong acids and other hydrating reagents, a thickened cell-wall that has not been lignified or cuticularised swells up strongly and shows traces of lamination. The successive laminae correspond to the layers or shells of thickening matter which have been deposited upon the original membrane. Each layer contains a mixture of the substances described, and they differ somewhat from each other in the proportions of the constituents.

Between contiguous thickened cell-walls a very conspicuous layer frequently appears, known as the *middle lamella*. This can with difficulty be ascertained to exist in unthickened cell-membranes. It gives characteristic reagents, which show that it differs markedly from the rest of the cell-wall. Its composition will be discussed in a succeeding chapter.

Cytase.

The composition of the cell-walls that ultimately yield nutritive material for the young embryo during the germination of the seed is thus extremely complex, and the products of their digestion cannot be ascertained as easily as those which are derived from the transformation of starch. During germination, such transformations are brought about by the action of an enzyme, which has been called *cytase*, though how far this is a single enzyme, and if so which of the constituents it attacks, cannot at present be said to be fully ascertained. A similar if not identical enzyme has been discovered to exist in certain fungi. It may be that we have several such enzymes, or there may be only one.

The existence of cytase has not been known for many years, our acquaintance with it dating back only to 1886, when it was discovered by De Bary. In some investigations into the life-history of a peculiar fungus, *Peziza sclerotiorium*, he found that while normally a saprophyte, it could under certain conditions become a parasite, and that then it frequently attacked reservoirs of reserve-materials, such as the roots of carrots and turnips. In its behaviour it appeared to live saprophytically, though infesting the living tissue. Its hyphae or some of them excreted something which poisoned the cells, the latter then yielding a nidus in which the hyphae spread and multiplied. In its attack upon the living tissues this Peziza formed dense masses of mycelia in the interior of the roots, softening them as if they had been boiled. Besides assailing reservoirs like the roots, it was capable of infesting the stems of living plants, and produced there the same curious softening of the substance. Microscopical investigations showed that the hypha grew especially between the cells, destroying the middle lamella and converting the cell-walls into a semi-mucilaginous material. On studying cultures of the ascospores of this fungus De Bary observed that each developed a mycelium of branched septate hyphae, and that these formed curious organs of attachment composed of branched hyphae in the shape of a kind of tassel, which eventually gained the power of penetrating through cellulose. If the culture was made upon a young seedling, *Petunia violacea* being peculiarly susceptible to its attacks, these organs of attachment ultimately killed certain cells of the epidermis and effected an entrance into the sub-jacent tissues, in which the hyphae subsequently ramified. The cells in the neighbourhood of the hyphae were curiously affected, losing their water and collapsing. The destruction appeared to be caused by something excreted from the hyphae and to prepare the way for their subsequent growth.

De Bary then examined the tissues attacked by this excreted material; he expressed the juice from them, and discovered that when pieces of the internodes of the stem of the bean, or fragments of fresh carrot, or of seedling cabbages, were steeped in it, they underwent rapid destruction, the cells becoming plasmolysed, and their walls swollen, while later the middle lamellae were dissolved.

In the sap of the carrot expressed after the fungus had attacked it De Bary thus discovered the presence of something having the property of causing cell-walls to swell and of dissolving a certain portion of them. When the expressed sap was boiled it was found to lose this power. The fluid which possessed the peculiarity of dissolving the cell-walls could also be expressed from the sclerotia of the Peziza, so that the substance in question was a secretion of the fungus and not a product of the disintegration of the tissue of its host plant. This substance was the enzyme now under discussion, to which the name *cytase* has been given by subsequent observers.

In 1888 this enzyme was discovered by Marshall Ward in a fungus belonging to the genus *Botrytis*, which infests certain lilies, especially *Lilium candidum*, and which presents many features of resemblance to that investigated by De Bary. The germinating spores of this Botrytis may be cultivated in a hanging drop of culture-fluid suspended on a cover-slip over a moist chamber in such a way as to be easily observed under a microscope. The spore puts out hyphae which develop into a mycelium ramifying through the culture-fluid; from this short branches grow out at right angles to the mass of the mycelium, and run up to the lower surface of the cover-slip. As each branch comes in contact with the glass it swells and softens, and its outer contour becomes surrounded by a glairy film, which is derived from some deliquescent constituent of the cell-wall. In many cases these erect hyphae branch, either at or near the extreme tip, or further behind. In some cases branches from the mycelium grow from the surface of the hanging drop towards the cavity of the moist chamber, and coming into contact with the sides of the latter attach themselves to the glass just as do those first described. These attaching organs in many respects resemble those described by De Bary.

It may frequently be noticed that when attachment is not made, the tips of the hyphae exude small drops of a translucent, more or less viscous fluid, which contains a number of minute brilliant granules, the protoplasm of the hypha becoming extremely vacuolated. The exudation may go on for a few hours, the drop becoming more and more granular, and its colour changing to a pale brownish-yellow. The drops when tested by microchemical reagents appear to contain quantities of proteid matter, suggesting, in connection with the vacuolation, that they are composed either of the protoplasm itself, or some substance derived immediately therefrom.

These drops contain the enzyme cytase, which thus appears to be a secretion of the protoplasm of the hyphae. Such granules as have been described are often found to be associated with the formation of enzymes in other cases. The secretion is not the pure enzyme, but is complex in composition, containing various proteid bodies, as already demonstrated. The evidence which points to the existence of the enzyme in the drops is twofold. The tips of the hyphae which extrude the drops have been observed to penetrate into and through the cell-walls of thin sections of parenchymatous tissue placed with them on a glass slide on the stage of a microscope. Further, a watery extract of such a mass of hyphae will cause unchanged cell-walls to swell up when thin sections of parenchyma are placed in drops of it.

The localisation of the enzyme in the tips of the hyphae suggests that the softening of those tips when they come into contact with the glass of the cover-slip, or the side of the moist chamber, may itself be due to the action of the ferment, the deliquescent matter which has been described as surrounding the tips of the attached branches being transformed cellulose, made mucilaginous in order to adhere to the surface.

The action of the cytase may be seen also in the process of the growth and branching of the hyphae. The ferment is

secreted mainly at or near the ends of the filaments; accumulating there, it causes a softening of the cell-wall either at
the apex or just behind it, enabling the hydrostatic pressure
within the hypha to originate a new branch, or to continue the
apical growth of the original filament.

When the mycelium is allowed to grow upon the surface of
a piece of tissue such as a leaf or bud of a lily, contact of the
erect branches with the epidermis of the latter creates a stimulus
to which the fungus responds by secreting the enzyme. It is
then discharged from the tips as in the cases described above,
but it does not attach the hypha to the epidermis as it does to
the glass slip. On the contrary it softens the walls of the tissue
and, slowly dissolving the cellulose, allows the hypha to make
its way into the interior; subsequent continuous formation and
excretion of the enzyme enables the mycelium to develop at
the expense of the cell-walls of what has now become the host
of the parasite, together with whatever nutrient matter the
cells themselves contain.

It is interesting to notice in the behaviour of this fungus,
that the stimulus of contact with a living leaf seems to lead to
an increased formation of the enzyme. The secretion appears
to be a starvation phenomenon in the main, the fungus
appreciating the presence of food-material in the substance
touching it, and at once endeavouring to appropriate it. If
in a culture carried on under microscopic examination, a little
fresh soluble nutritive material is added to the culture fluid
when the mycelium has begun to attack a piece of vegetable
tissue by means of the exudation of these drops, the formation
and extrusion of the latter is at once suspended, and the
mycelium goes on growing and branching at the expense of the
new food supplied. When this store is exhausted the drops
are again formed and the tissue present in the culture fluid is
again attacked.

The cytase does not attack all parts of the cell-wall with
equal facility; though it softens the outer layers of the epi-
dermal cells and penetrates transversely through the walls of
the subjacent parenchyma, it runs more frequently in the plane
of the middle lamella between the cells. The composition of

these three modifications of cell-wall we have seen to be very
varied, and it does not appear very evident what is the particular chemical change which is brought about. To this
point reference will be made later.

It was stated above that a watery extract of an active
mycelium possesses the power of causing the cell-walls to swell
up when a piece of tissue is placed in it. The enzyme can be
thus extracted from the fungus by artificial means; it is
moreover actually excreted from the mycelium into the
nutrient fluid in which the latter is cultivated. If either the
culture-medium or the expressed extract is boiled the power of
dissolving cellulose is lost, the cytase being destroyed. This
destruction of activity on heating may indeed be taken as
evidence that the action is really due to an enzyme.

Microscopic examination of a thick cell-wall exposed to the
action of cytase shows that the action of the latter resembles
that of dilute mineral acids, the successive laminae of the wall
becoming evident and each lamina in turn swelling up, and
ultimately dissolving.

Cytase may be prepared from such cultures as have been
described by grinding up the mycelium with sand, and pressing
out the fluid from the resulting pulp. It should then be
filtered and the filtrate allowed to fall into a large excess of
alcohol, when a precipitate will be thrown down, flocculent in
character, partly amorphous, and partly crystalline. In this
condition it is very impure, being mixed with several other
bodies. On dissolving the precipitate in water the solution will
be found to possess the property of the original liquid, and to
dissolve cell-walls with some facility. The middle lamella
between the cells of a piece of tissue placed in either is the
first part to disappear; the cells become separated and their
walls show evident lamination and swelling.

Cytase is seen from these observations of De Bary and
Marshall Ward to be secreted by fungi, and to play an important part in their development and in their nutrition.
Confirmation of their results has been obtained more recently
by Kean, who has separated a variety of cytase from *Rhizopus
nigricans*. What part of the fungus it resides in has not been

determined, nor is more said of its action than that it is capable
of softening cellulose cell-walls. In studying this fungus as
cultivated in gelatin media, Arthur has noticed that here and
there upon the sides of the hyphae there is an extrusion of
drops something like those of Ward's *Botrytis*, and further that
branches of the hyphae originate at the spots where the drops
had been observed. This is to a certain extent a corroboration
of Kean's statement.

Grüss has detected cytase in the mycelium of *Penicilium*.

Cytase has been discovered in the higher plants, but so far
as is known it has not the wide distribution of diastase. Its
detection is due chiefly to the investigations of Dr Horace T.
Brown, undertaken in some cases jointly with Dr Morris and
with Mr Escombe.

The first experiments which may be discussed were undertaken by Brown and Morris in connection with a study of the
process of germination of the barley and other cereals. This
work has already been referred to as leading to the discovery of
the secretion of diastase by the epithelium of the scutellum.
It has already been said that cytase as well as diastase is
present in the material formed and excreted by these cells,
though not so much as by those of the aleurone layer.

Brown and Morris have described in detail the changes
which can be noticed in the endosperm immediately underneath
the scutellum. Between the starch-containing cells of the
former and the scutellar epithelium is a comparatively thick
layer of emptied and compressed cells forming a somewhat
transparent band of endosperm tissue, the starch these cells
originally contained having been utilised by the embryo
during its early growth and before it entered upon its resting
period. After 24 to 36 hours' germination the cell-walls of
this band of tissue are softened and partially dissociated. As
the germination proceeds, a dissolution of the walls of the
starch-containing cells also can be observed, which starts
under the aleurone layer and proceeds progressively from the
proximal to the distal regions of the endosperm. No action of
the diastase upon the starch is perceptible so long as these
walls are intact, indeed the diastasic digestion of the starch
follows continuously upon the disappearance of the cell-walls.

Brown and Escombe have observed similar changes taking place in the endosperm of barley grains from which the embryo has been carefully removed. The changes begin immediately underneath the aleurone layer and spread thence into the interior of the grain. The first effect of the germinative process is the separation of the aleurone layer from the cells beneath it by the solution of the cell-walls of that region.

If the progress of the dissolution of the cell-walls is carefully followed, the course of events is seen to be the following: at first the cell-wall swells up slightly and its stratification becomes very apparent; the separate lamellae are then gradually disintegrated, the middle lamella being the most resistant. Ultimately the whole of the cell-wall is broken down into very minute spindle-shaped fragments, with their longer dimensions arranged tangentially to the original cell-wall. Finally these fragments also disappear. The whole contents of the region of the endosperm affected become mealy, even before the starch begins to be dissolved.

The same course of action can be traced in other grasses, notably *Bromus mollis*, and *B. asper*, where the walls of the endosperm cells are very considerably thickened.

The extent to which the dissolution of the cell-wall precedes that of the cell-contents was measured by Brown and Morris in some barley grains that had been germinated for 5 days. At that period all traces of cell-walls had disappeared from most of the tissue in the proximal region of the endosperm; indications of action on the starch-grains did not then extend further than 0·15 mm. from the scutellum.

The extraction of cytase from the germinated barley can be effected with considerable ease. If an extract of air-dried malt is prepared and sections of barley immersed in it, the parenchymatous tissue becomes disintegrated in about 24 hours, the cell-walls either disappearing or remaining in a much swollen and altered form. Previous heating of the extract to a temperature of from 60°—70° C. destroys its activity, the tissue then remaining in it unaltered for an indefinite time. Nor is the action confined to sections of barley, for pieces of potato immersed in it lose their coherence in a few hours, and fall to

pieces with great readiness. The cell-walls may be seen to
swell up and become differentiated into numerous very thin
laminae, which later break up into spindle-shaped fragments
and ultimately disappear, with the exception of a thin layer
which probably represents the middle lamella. Brown and
Morris found that the parenchyma of different plants did not
all behave in the same way. The artichoke, carrot and turnip
were like the potato in this respect, but the cell-walls of the
beet were affected only slightly, and those of the apple not at
all. Nor does cytase attack the thickened cell-walls of the
endosperm of Palms. These facts however need not surprise us
when we remember the probable differences in such cell-
membranes as stated at the commencement of this chapter.

Brown and Morris prepared the cytolytic enzyme by the
same method as they adopted in the case of diastase. They
precipitated an extract of malt by alcohol, dehydrated the
precipitate by absolute alcohol and dried it in a vacuum. This
precipitate contained however both cytase and diastase.

Cytase is destroyed by heating its solution in water to a
temperature of 60° C., while diastase is unaffected at this
point.

The action of cytase is accelerated if the medium in which
it works is made slightly acid with formic or acetic acid.

Cytase only attacks unchanged cell-walls, having no action
on those which have undergone partial lignification or cuti-
cularisation.

Brown and Morris have shown that the secretion of cytase,
like that of diastase, takes place in the cells which constitute
the aleurone layer and to a less extent in those of the scutellar
epithelium. Their method was the same as in the case of the
diastase and has already been described in a previous chapter.
The secretion of both enzymes seems to be dependent on the
necessity for obtaining nutriment for the embryo; that is, it
is a phenomenon attending starvation.

In the paper by Brown and Escombe already quoted it has
been shown that the aleurone layer takes a much larger share than
the scutellum in the formation of cytase. In the earlier stages
of the germination the mealiness of the grain is due almost

entirely to the enzyme secreted there, the secretion of the scutellar epithelium being at that time of subsidiary importance.

Cytase exists not only in germinating barley, but in certain varieties in the resting grain also. There appears to be some relation between its occurrence and the climatic conditions under which the cereal is cultivated. The more perfect the latter, the less of the ferment is present. Brown states that the presence of the enzyme can be demonstrated in a cold-water extract of the grain, and its influence in the self-digestion of the cell-walls is often distinctly noticeable when the coarsely ground meal is macerated with water for a short time at a temperature of 35°—40° C.

Brown has found evidence of the presence of cytase in rye and in oats, the latter being especially rich in it. An aqueous infusion of raw oats is in fact much more active in swelling and dissolving cellulose than an extract made from an equal weight of air-dried barley malt.

Cytase is not capable of resisting so high a temperature as diastase, being destroyed at 60°—65° C., while diastase needs to be heated to 70° C.

Though the grasses have been shown to be especially rich in cytase, this enzyme can be shown to be present in other plants. It has been examined by Gardiner in the endosperm of *Tamus communis*, where it can be observed to attack the thickened cell-walls during the germination of the seeds. The cell-membranes of this endosperm exhibit the continuity of the protoplasm of the contiguous cells, threads of it being traceable through the walls even before they become thickened. The cytase is secreted in the interior of the cells and passes along the connecting threads into the substance of the wall soon after germination has begun. The penetration begins at several spots in the wall, and when the cytase has once entered them its action can be traced through the substance of the membrane, quickly causing its complete disorganisation. The walls on being attacked do not swell so markedly as in the cases already mentioned, but their stratification or lamination becomes very evident. The middle lamella in this case is the least resistant to the action of the enzyme.

Newcombe has confirmed the results of Brown and Morris completely, as to the occurrence and mode of action of cytase in the barley grain. He has discovered its presence also in germinating seedlings of *Lupinus albus*, and in the cotyledons of *Pisum sativum* and *Fagopyrum esculentum* (Buckwheat).

In his experiments he made a watery extract of the material under investigation, and precipitated the cytase by the addition of alcohol. He purified it to some extent by redissolving it in water and reprecipitating it. Delicate sections of vegetable tissue, generally starch-free sections of barley grains whose protoplasm had been killed by heating them, were placed in a faintly acidulated solution of the precipitate, and after a few hours the cell-walls were found to have become hyaline, and to be gradually dissolving. He found that the first part of the wall to disappear was a strip on each side, intermediate in position between the middle lamella and the free edge. The middle lamella was the most resistant.

Newcombe also detected cytase in the so-called *Taka-diastase*, which is a preparation of the fungus *Aspergillus* (*Eurotium*) *oryzæ*. The commercial product is a powder which dissolves in thirty times its weight of water. Such a solution, slightly acidulated with hydrochloric acid, was able to dissolve the cell-walls of the barley endosperm more rapidly than the preparations made from the seedlings already alluded to. The course of solution was however very different, the middle lamella being the first part to be attacked.

All Newcombe's experiments were carried out in the presence of chloroform.

There are several other plants in which the presence of cytase is probable, especially the Palms, in which the carbohydrate reserve-material of the seeds takes the form of very much thickened cell-walls. The thickening matter is composed of a variety of cellulose mixed probably with pectic bodies. Careful search has been made for an enzyme during the germination of the seeds of several species, but it is only recently that satisfactory evidence of its existence has been obtained. The process of germination is however so slow that if the enzyme

7—2

exists, it is in very small quantity at any time, and hence may well escape the ordinary process of extraction.

The first of the Palm seeds investigated was the Date (*Phœnix dactylifera*), upon the germination of which Sachs published some observations in 1862. The embryo lies embedded in about the middle of the hard seed, and on germination it remains attached to it by a portion of the cotyledon, which becomes transformed into an absorbing organ. This continually encroaches upon the endosperm as the young plant grows, until the store of cellulose is exhausted. The surface of this absorbing organ is covered by an epithelium which strongly resembles the scutellar epithelium of the grasses. Sachs suggested that the cells of this layer excrete a cytase, but he did not definitely prove it.

Grüss states that there is an enzyme present in these seeds, which acts on the reserve cellulose, but very slowly. It probably yields mannose as the final product.

The writer made a series of observations in 1887 on the germination of the seed of *Livistonia humilis*, which has a similar structure to that of Phœnix. The progress of events was the same as that observed by Sachs. The epithelium cells were very granular and stained deeply with iodine and with Hoffman's blue. By the use of these and other staining reagents some additional facts were obtained bearing on the behaviour of the cellulose. After two months' germination the absorbing organ had penetrated about half-way through the endosperm. The cells of the latter in contact with the cotyledon were broken away at their edges, and when chlor-zinc-iodine was applied to the affected area, the cells nearest the cotyledon stained a deep violet, while those farther back were coloured only a pale blue. Two well-marked zones thus became evident, which differed in the chemical or physical constitution of the cell-walls. An aqueous solution of iodine stained the inner zone a pale blue, but did not affect the outer one. When Hoffman's blue was used as a reagent the inner zone was coloured very deeply towards its outer part, while the coloration was gradually paler and paler as it was traced towards the unaltered endosperm. The outer zone, slightly affected by the chlor-zinc-iodine, was not coloured by the blue dye. The inner deeply-staining part was much swollen and the outlines of the cells were indistinct. The appearances led to the view that the cellulose was being gradually hydrated, the change starting from the absorbing organ and proceeding outwards.

When sections of the disintegrating endosperm were boiled with Fehling's solution there was a copious deposition of cuprous oxide on the surface of the inner zone and particles of the same substance were seen to be embedded in the almost deliquescent region abutting on the cotyledon. These gradually became less numerous towards the interior of the inner zone and did not extend beyond it. The cavity of the endosperm after removing the absorbing organ showed a deposition of cuprous oxide over its whole surface. There was an evident gradual conversion of the cellulose into something that reduced Fehling's fluid, presumably some form of sugar.

The appearances were consistent with the view that the epithelium cells of the cotyledonary absorbing organ excreted a cytasic enzyme, which gradually corroded and dissolved the endosperm, forming a sugar capable of absorption.

More definite evidence has recently been secured by New-combe. He cultivated date-seedlings in earth till the endosperm was nearly half-consumed, and then collected the cotyledons and the endosperm in which they were plunged. They were separately either dried and ground in a mill and subsequently extracted with water, or ground up with water and infusorial earth while still moist. After a short extraction with water, five times its volume of alcohol was added to each extract, when a precipitate was obtained, which was filtered off and dried. A very strong extract of the precipitate was made with a ·05 per cent. solution of hydrochloric acid and a little chloroform was added as an antiseptic. These extracts were found to be capable of dissolving the cell-walls of barley endosperm after the sections of the latter had been killed. The middle lamella was found to be the most resistant layer to this preparation of cytase.

The enzyme was found to be capable of dissolving the cell-walls of the cotyledon of the lupin and those of the endosperm of the date. In the latter case the process was one of corrosion followed by solution, as in the case of Livistonia.

In some investigations upon various gums and the trees yielding them Lutz has found that in *Acacia* gum-reservoirs occur in the bark and in the pericycle. These consist of lacunæ caused by an enormous swelling and ultimate deliquescence of the cell-walls. His experiments have led him to suggest that this formation is due to a kind of cytase.

Elfving's observations on the germination of the pollen of the grasses point to the existence of cytase in the pollen tube, which was found to penetrate the tissue of the style by boring its way through the middle lamellæ of the cells, and not entering their cavities.

De Bary describes the behaviour of a micro-organism, *Bacillus amylobacter*, which is an agent in the destruction of decaying parts of plants and which acts by destroying the cell-membranes. He says it decomposes the cellulose, forming dextrin and glucose, and that it does so by disengaging an enzyme. It does not attack suberised membranes, nor those of bast-fibres, of submerged water-plants, of Mosses and many Fungi, but it rapidly decomposes the cell-walls of fleshy and juicy tissues, such as those of leaves, herbaceous stems, and the softer kinds of wood.

The digestion of cellulose in the intestines of animals is probably brought about by similar organisms. No definite evidence of the presence of cytase in these microbes is at present forthcoming. De Bary does not quote experiments proving the formation of the enzyme in the cases he mentions, and for the present it can hardly be considered established. The course of the decomposition too as suggested by him does not rest on an experimental basis.

On the other hand certain experiments point to the existence of cytase in some animal secretions. Griffiths has found that the digestive fluid which is present in the intestine of the earth-worm dissolves cellulose. More recently Biedermann and Moritz have shown that the secretion of the so-called liver of the snail (*Helix pomatia*) contains a very active enzyme which attacks not only the hemicelluloses of the cell-wall, but the more insoluble constituents which have not been actually lignified. The cytase is present in the liver as an antecedent of the enzyme and is developed from the latter as the secretion is poured out into the intestine.

Vignal has shown that *Bacillus mesentericus vulgatus* secretes a cytase which dissolves the middle lamella of vegetable cells.

The products of the action of cytase have not been satisfactorily examined up to the present time. The decompositions are no doubt very complex, as we should infer from the varying character of cell-walls. The cellulose constituents of the latter ultimately yield some form of sugar, while the pectose is hydrolysed to metapectic acid, which also possibly forms a similar body. By the action of acids, especially sulphuric acid, cellulose itself can be hydrolysed through a series of dextrins to glucose. Some of its modifications yield on hydrolysis pentose sugars as well as hexoses. The pectoses yield a series of gelatinous hydrates, resembling in this behaviour the mucilage-yielding constituents of many seeds. Ultimately metapectic acid is formed, which is apparently identical with the arabic acid found in great abundance in gum arabic. From the latter by continued hydrolysis, galactose (a hexose), and arabinose (a pentose), can be prepared.

According to Tollens and Ganz cellulose can undergo transformation into a sugar which they call *seminose*, which is apparently identical with mannose. They prepared it by warming mucilage with dilute acids. Reiss prepared the same sugar from reserve-cellulose by the action of dilute sulphuric acid.

Cross and Bevan say that when the oxycelluloses, such as occur in lignified cell-walls, are boiled with sulphuric acid, at first concentrated and subsequently diluted, they are resolved into carbohydrates of low molecular weight, glucose and sometimes also mannose being formed. The pecto-cellulose group, of which the cell-walls of ordinary parenchyma consist, yield in addition galactose and the two pentoses xylose and arabinose. Galactose has been obtained from the cell-walls of the seeds of *Lupinus luteus, Soja hispida, Coffea arabica, Pisum sativum, Cocos nucifera, Phœnix dactylifera,* etc. Mannose is obtained in relatively large quantity from many seeds, especially those of *Phytelephas*, and pentoses from the seeds of the cereal grasses and of several leguminous plants.

Reiss has stated that in the hydrolysis of cellulose obtained from the seeds of *Phœnix* and *Phytelephas*, a body corresponding to dextrin is first formed. It is lævo-rotatory. By further hydrolysis it yields a dextro-rotatory sugar, identical with the seminose of Tollens and Ganz. Schulze obtained both galactose and a pentose, probably arabinose, from the thickened walls of the cells of the cotyledons of *Lupinus luteus*.

Bourquelot and Hérissey have recently described a variety of cytase to which they have given the name *pectinase*, and which they have found capable of acting on pectine, destroying its power of gelatinising, and giving rise to a reducing-sugar. They prepared it from germinated barley which had been slowly dried at a temperature of 30°—35° C., by extracting it with chloroform water and adding alcohol till a precipitate fell. The precipitate was collected and dried *in vacuo*. In three vessels they prepared the following mixtures:

(1)   15 c.c. of a 1 per cent. solution of the precipitate, and 15 c.c. of a 2 per cent. solution of pectine.

(2)   A similar mixture, but the enzyme solution was boiled.

(3)   A mixture like the first, with the addition of ·05 grm. of chalk.

These were then digested at the laboratory temperature for 42 hours. At the end of this time mixtures (1) and (3) had lost the power of gelatinising, which was not affected in (2). Alcohol gave an abundant gelatinous precipitate with (2) but not with either of the others. Titrated with Fehling's solution, (2) gave no reduction of the cupric oxide, while (1) showed the presence of 11 mgrms. of sugar computed as glucose; (3) contained 36 mgrms. of the same sugar. They attribute the difference between (1) and (3) to the feeble acidity of the pectine, which retards the action of the enzyme. This acidity was neutralized in (3) by the chalk.

It is not clear that this enzyme is different from that discovered by Brown and Morris.

From a consideration of all these experiments there appear

to be two varieties of cytase, one attacking most readily the middle lamella and the other the layers deposited upon it. The first includes the enzymes discovered by De Bary, Marshall Ward, Gardiner and Vignal; the second is the cytase of the grasses and palms. Newcombe has prepared representatives of both.

Caroubinase.

An enzyme which has much in common with cytase, if it is not identical with it, has been described by Effront. It occurs in the seeds of the Carob-bean (*Ceratonia siliqua*) which are sometimes used for fodder. These seeds contain a peculiar carbohydrate, resembling a variety of galactan, which is different from either starch or cellulose, though it seems to have a composition similar to the latter. When the seeds, freed from their coats, are digested with hot water over a water-bath, they yield a transparent jelly which when diluted is somewhat syrupy in consistence and can be filtered through silk. Addition of twice its volume of alcohol or baryta water precipitates from this jelly a carbohydrate, which can be purified by re-dissolving it and again precipitating it with alcohol. When this product is dried at 100° C. it is a white spongy, friable substance, giving no coloration with iodine, and having the formula $C_6H_{10}O_5$. If soaked in water or weak solution of soda it again forms a jelly or a viscid transparent mass: 3—4 grams dissolved in a litre of water yield a viscous fluid of a syrupy consistency. It is soluble in hydrochloric acid in the cold, and the solution does not reduce Fehling's solution, and has no action on polarised light. If oxidised by nitric acid, it gives only small traces of mucic acid, showing that it is not of a pectic nature. This new carbohydrate has been observed by Effront not only in the Carob-bean, but also in the grains of rye and barley. It appears to be a member of the group of celluloses, and to resemble mucilage in its reactions. During the germination of the Carob-bean this body, to which Effront gives the name *caroubin*, is readily hydrolysed by the action of an enzyme, which he calls *caroubinase*. The hydrolysis takes place feebly at first, but becomes vigorous as soon as chlorophyll is developed in the young seedling.

In his experiments Effront proceeded as follows:—Seeds of the Carob were bruised in a mortar, and extracted with water to which a few drops of chloroform had been added, and they were allowed to macerate for 24 hours at 30° C., after which they were pressed and the extract filtered. The latter was then examined for the enzyme. Glass vessels were taken and in each were placed 50 c.c. of water, ·1 c.c. of normal formic acid and 1 gram of powdered caroubin. After mixing, a measured quantity of the extract was added to each, and a little chloroform used as an antiseptic. Controls were prepared in each case, in which the extract used had been exposed for half-an-hour to a temperature of 90° C. The activity of the extract was estimated by the degree of gelatinisation remaining in each case after a definite interval. The controls were found to remain completely gelatinised, while the mixtures prepared with unheated extract were liquefied more or less completely, the amount of viscosity remaining being inversely proportional to the quantity of extract used. Tested in this way, the latter was found to liquefy the caroubin and subsequently to convert it into a reducing-sugar. At the moment of liquefaction only traces of the latter were present, but from that point onwards the cupric-reducing power of the digestion continued to increase as long as the experiment was continued. When an excess of alcohol was added to the liquid at the moment of liquefaction, it threw down a precipitate consisting of an intermediate product, which was strongly dextro-rotatory and freely soluble in water. This body was easily hydrolysed into sugar by the action of dilute mineral acids.

The sugar resulting from the action was found to be un-crystallisable, and to be a hexose. It was fermentable by yeast, and with phenyl-hydrazine acetate it yielded an osazone which melted at 183° C. The sugar had the same reducing power as glucose, but its specific rotatory power was $(a)_D = +24°$.

Effront states that the enzyme can be precipitated by addition of five times its volume of alcohol to the extract of the seeds. In his experiments he was able by the method described to compare the amount of caroubinase present in seeds at different stages of germination.

Caroubinase acts energetically at 40° C., but its optimum point is between 45° and 50° C. At 70° C. it acts with difficulty, and is destroyed at 80° C. It acts feebly in neutral solution, and more energetically in the presence of ·01—·03 per cent. of formic acid.

More recently Bourquelot and Hérissey have examined the reserve carbohydrate of the Carob-bean, and state that it is a manno-galactan which on hydrolysis by sulphuric acid yields a mixture of mannose and galactose. They have found similar manno-galactans in the seeds of *Trigonella fœnum-grœcum*, and of *Medicago sativa*, which yield similar products of hydrolysis. There are probably several bodies of this sort in the different seeds, as the relative amounts of mannose and galactose vary in different cases. The lucerne (*Medicago*) yields 1·223 grm. of mannose to 1·178 grm. of galactose ; the fenugrec gives a proportion of 1·249 grm. of the former to ·978 grm. of the latter. Bourquelot and Hérissey have extracted the enzyme from these germinating seeds. They suggest for it the name *seminase*.

The sugar, called by Effront caroubinose, has been examined also by Marlière, who believes it to be a mixture of glucose, fructose, and galactose.

Hadromase.

The varieties of cytase so far described act upon either the pectose or the cellulose constituents of the unaltered cell-wall. The more permanent tissues of the plant, especially the woody ones, undergo profound changes of composition as they take on their adult structure. Wood is marked by the presence of lignin in the cell-wall. This new substance appears to arise in consequence of a chemical change in the substance of the membrane, which becomes marked as soon as the wall begins to increase in thickness. The exact nature of the change is not very definitely known, but it appears to be due to a chemical alteration of some of the constituents of the wall, which are converted into lignin. The change takes place gradually, and by appropriate solvents the newly-formed matter can be separated from the rest of the membrane, leaving the latter apparently intact. Lignification is thus not a mere infiltration of lignin into the body of the wall. The substance of the woody wall seems to be a compound of cellulose and lignin having the composition of an ether. The lignin can be recognised by turning red when treated with phloroglucin in the presence of a weak mineral acid.

Many fungi have been found to live upon lignified cells, their mycelium penetrating the woody walls and deriving a certain amount of their nourishment therefrom. Among these fungi may be mentioned *Merulius lacrymans, Pleurotus pulmonarius,* and *Stereum hirsutum.*

Czapek has found that an enzyme is present in the mycelia of these plants, and that it can be extracted from them by the ordinary methods. The extract so prepared can effect the decomposition of woody substance if macerated with it for some days, and can bring the lignin constituent into solution, leaving the cellulose framework intact. He made a watery extract of the tissue of the fungus, and after adding to it a little chloroform and some shavings of wood, set it aside at 28° C. to digest. After about a week's exposure he found the wood was undergoing decomposition. When a sample of the digestive liquid was evaporated to dryness, the residue gave up to alcohol the material which becomes red in colour when treated with phloroglucin and hydrochloric acid. The action went on for several days, and at the end of the experiment the residue of the woody matter only gave the cellulose reaction with iodine and sulphuric acid.

A boiled control showed the wood unaltered at the end of the same time, and an alcoholic extract of it gave no reaction with phloroglucin.

The enzyme, which Czapek has named *hadromase,* can be precipitated by alcohol from a watery extract of the fungus as a grey powder. A solution of this in water possesses the same power as the original extract.

Hadromase thus separates the woody constituent of a

lignified cell-wall from the cellulose basis in which the lignin is formed, dissolving the former and leaving the latter unattacked. Czapek found that it was associated in the fungi with cytase, which subsequently dissolved the cellulose

Gum ferment of Wiesner.

In 1885 Wiesner described an enzyme existing in gum-arabic, which he says transforms cellulose into gum or mucilage and which has the power of converting starch into dextrin but not into a reducing-sugar. In his experiments he used 5 c.c. of a 2 per cent. solution of gum-arabic and 1 c.c. of thin starch-paste. After 6 hours' exposure to a temperature of 23° C. the solution of starch became limpid; after 24 hours it gave the erythrodextrin reaction with iodine, and subsequently this disappeared and achroodextrin only remained. No reducing sugar was formed in 24 hours.

Reinitzer subsequently examined the action of gum-arabic, and came to conclusions opposed to Wiesner's. He found that the solution of gum had no action on cellulose, and that it converted starch-paste into a reducing-sugar. He held it therefore to be only diastase.

CHAPTER VIII.

SUGAR-SPLITTING ENZYMES.

THE enzymes which we have discussed so far have been shown to have one peculiarity in common; they set up different series of transformations of various carbohydrates, and the final product of each action is a sugar, although the same sugar is not formed in every case.

Before Fischer started his classical researches on the sugars, our acquaintance with them hardly extended beyond such as contained within their molecule 6 carbon-atoms or a multiple of them. Of those with 6 carbon-atoms we knew glucose, fructose, galactose and sorbose. Kiliani showed that another sugar, arabinose, has only 5 carbon-atoms. Fischer has shown that the simpler sugars are to be regarded as aldehyde-alcohols, or as ketone-alcohols, and that the simplest of them is glycollic-aldehyde $C_2H_4O_2$. We can now distinguish a series of these sugars, the successive members of which differ from each other by CH_2O. Fischer has shown that the synthesis of these can be as easily accomplished as that of the organic acids and other compounds. On the hypothesis of Le Bel and Van t'Hoff on the asymmetric carbon-atom, there should be various isomerides of each of these sugars except the first, the number varying according to the configuration of the molecule. Sixteen different isomeric sugars should theoretically exist having the formula $C_6H_{12}O_6$, and the synthesis of twelve of these has already been accomplished.

Besides these sugars another series exists, the *polysaccharides*, which may be regarded as derived from the first series by the elimination of one molecule of water from two or more molecules of the simpler sugar. When they undergo hydrolysis through the dilute action of mineral acids the change which takes place is represented by the equation $C_{12}H_{22}O_{11} + H_2O = C_6H_{12}O_6 + C_6H_{12}O_6$. The resulting hexoses may be the same, or may be different ones; cane-sugar yields glucose and fructose; milk-sugar glucose and galactose; maltose gives two molecules of glucose only.

In the living organism, whether animal or vegetable, we find enzymes which are capable of effecting the hydrolysis of these polysaccharides with considerable facility. Six of them have been discovered up to the present time, and have been named *invertase* or *invertin, glucase* or *maltase, lactase, trehalase, raffinase* and *melizitase.*

Invertase (Invertin, Sucrase).

As long ago as 1847 Dubrunfaut discovered that cane-sugar could be decomposed by the action of dilute mineral or organic acids, which cause it to take up water, and split into equal quantities of glucose and fructose, the former rotating a ray of polarised light to the right, the latter turning to the left. The process of the decomposition has been termed *inversion* and has given its name to the enzyme under discussion. Cane-sugar itself has a dextro-rotatory power, but the mixed sugars resulting from its transformation turn the ray of polarised light to the left. This inversion of the direction of the displacement of the polarised ray is due to the fact that the levo-rotatory power of fructose is considerably greater than the dextro-rotatory power of glucose, and as the two hexoses are formed in equal quantities, the resulting mixture of the two must deflect the ray to the left.

In his experiments upon this process of inversion Dubrunfaut succeeded in separating the two hexoses from each other by treating a solution of the mixture, now commonly known as *invert-sugar,* with a salt of calcium. This forms a compound with both glucose and fructose, but that with the former is much the more soluble of the two. When a mixture of these sugars is treated in this way the fructose-lime compound slowly crystallises out and can be separated from the other by filtration. The crystals can be dissolved in an excess of water, and the two sugars can both be freed from calcium by passing a stream of CO_2 through the respective solutions. The precipitate of calcium carbonate can be removed by filtration and the separated sugars left in solution.

Dubrunfaut did not discover the enzyme which can set up this change. The latter remained unknown until 1860, when it was isolated from yeast by Berthelot. Its discovery is to be associated with the fact that cane-sugar is not fermentable by yeast, but that the alcohol produced by the latter arises from the sugars into which cane-sugar splits on hydrolysis. The earlier workers attributed this inversion to a certain acidity of the yeast, Pasteur suggesting that it arose from the succinic acid which he proved the latter to produce during fermentation. Berthelot was able however to show that it was quite independent of the reaction of the wort. He proved it to be due to a soluble principle which is present in the water in which yeast has been washed, and which is therefore presumably excreted by the cells of the organism. Berthelot gave this new body the name "ferment glucosique."

In his experiments he pressed the yeast till it was free from the wort in which it had been fermenting, and after washing the powder, allowed it to macerate for several hours with twice its weight of water. After filtering he found that the filtrate contained 1·5 per cent. of soluble matter. His first point being the investigation of Pasteur's hypothesis that the active agent in the hydrolysis of cane-sugar was the succinic acid of the fermentation, he added some of his extract to a solution of cane-sugar, containing 20 per cent. of the latter and 2 per cent. of sodium bicarbonate. Under such conditions he found hydrolysis took place and the alkalinity of the solution was not disturbed. The change was therefore clearly not brought about by any acid.

Berthelot further isolated the enzyme, though he did not prepare it in anything like a pure condition. His method consisted in precipitating it from an extract of yeast by the addition of an equal volume of alcohol. The precipitate took the form of white flocculent aggregations which sank to the bottom of the flask. They were separated from the liquid by decantation, and washed with alcohol at the ordinary temperature, when a yellowish horny substance was obtained, weighing about one-fiftieth of the original soluble matter. Berthelot found that a repetition of this treatment considerably impaired the hydrolytic power of the product. When prepared by a single precipitation only, between 50 and 100 parts of cane-sugar could be inverted by one part of the enzyme.

In 1864 Bechamp published the results of some investigations which considerably extended the state of knowledge of the distribution of this ferment. Besides existing in yeast, he says it is present in several moulds and in certain micro-organisms, though he does not specify them. If these moulds are bruised in solutions of cane-sugar and the resulting mass filtered, the filtrate can soon be found to contain glucose. When sufficient quantity of the mould can be obtained, an extract can be prepared of it in the same way as Berthelot's extract of yeast, or the active principle can be precipitated from an infusion of the mould by the method adopted by Payen and Persoz in the case of diastase. In either case it is found capable of hydrolysing solutions of cane-sugar of various degrees of concentration, but it loses the power if heated to a temperature of 60°—70° C.

Besides investigating its occurrence in fungi, Bechamp sought for it in the higher plants. Supposing that the presence of chlorophyll indicated a radical difference between the nature of the ultimate metabolism of the green plants and that of the colourless fungi, he formed the idea that those parts of the higher plants which had no chlorophyll would probably show the same phenomena as the fungi. Acting on this idea, he sought for the hydrolysing principle of the moulds in the coloured parts of various flowers. Though his fundamental position was unsound, he was nevertheless fortunate in his pursuit of resemblances of behaviour, for he discovered the enzyme in the petals of *Robinia viscosa, R. pseudo-acacia, Papaver Rheas,* and some varieties of *Rosa*; also in the coloured bracts of *Bougainvillea spectabilis.* He has described his experiments with the petals of *Robinia viscosa* in detail. Taking a quantity of flowers, he bruised the petals and

expressed the sap from them; dividing the liquid into two
parts, he added to one a certain weight of cane-sugar and a
definite quantity of water and observed the rotation of a ray of
polarised light sent through a length of 200 mm. of the
mixture. The other half was then boiled and made up to the
same strength of cane-sugar as the first and its rotatory power
observed. The two coincided, showing that the boiling of the
juice was not accompanied by any change that would interfere
with subsequent comparison of the two digestions. He protected himself against disturbance by micro-organisms by the
addition of two drops of creosote to each 100 c.c. of the
digesting liquids. The two preparations were then set aside at
the ordinary temperature and the rotatory power was examined
at intervals. The results were as under :—

	10 c.c. sap, 20 gms. cane-sugar, 2 drops creosote, made up to 100 c.c. with distilled water.	Boiled control.
Initial rotation	+ 29·52°	+ 29·52°
Rotation after 24 hours' digestion	+ 28·8°	+ 29·52°
„ „ 20 days „	− 4·8°	+ 29·28°

Bechamp failed to find the enzyme in the leaves of plants,
a fact which seemed to lend a certain plausibility to his
hypothesis. The optimum temperature for its working he
determined to be 40°—50° C.

Bechamp thought that the enzyme from fungi was not
quite the same as that existing in the flowers, the latter in his
opinion not being so active as the former. He gave to the
first the name *zymase*, which a little later was changed to
zythozymase, while he called the second *anthozymase* to indicate
its localisation.

He identified a similar body in the fruit of the Mulberry,
but finding that the juice of the latter was at the same
time capable of saccharifying starch solutions, he thought it
contained a new enzyme possessing both properties. He
named this new ferment *morozymase*. In the light of subsequent discoveries we can say with tolerable certainty that it
was a mixture of his zymase and diastase.

The name *zymase* was soon abandoned, as it began to be

applied to enzymes in general. The term *invertin* was substituted for it by Donath in 1875, and this name has been used in connection with the enzyme until quite recent years. It is indeed still employed by writers on animal physiology to denote an identical enzyme which has its origin in animal secretions. In vegetable physiology it is becoming replaced by *invertase*.

The recognition of the existence of this enzyme among the secretions of the alimentary canal of animals was first made by Claud Bernard. He ascertained that cane-sugar when present in the blood of an animal is of no value for purposes of nutrition. When injected into a vein it is almost immediately excreted from the body by the kidneys and appears unchanged in the urine. When taken into the alimentary canal by the mouth, however, it has a distinct nutritive value. Bernard was able to localise the region in which it undergoes digestion, and he discovered that it is hydrolysed in the course of its progress through the small intestine, and gives rise to the same reducing-sugars as are formed by the extract of yeast. If a solution of cane-sugar is placed in a loop of the intestine, or if it is mixed with an infusion of the mucous membrane of that region, after a very short interval the liquid is found to have acquired the power of reducing Fehling's fluid, which cane-sugar is unable to accomplish. The power of rotating a ray of polarised light possessed by the original solution at the same time becomes changed, and a comparison with a similar cane-sugar solution treated with a dilute mineral acid shows that the products of the action are the same in both cases. Bernard showed the transformation thus taking place in the intestine to be due to an enzyme identical with that extracted by Berthelot from yeast. He established the existence of invertin in the intestines of rabbits, dogs, birds, and frogs. Other subsequent investigators have shown that it exists also in the alimentary canal of some of the Invertebrata; Balbiani proving its presence in the digestive apparatus of the silkworm. Von Planta says the stomach of the bee contains a juice in which invert-sugar is present, while its food consists of cane-sugar.

Bernard's discovery that the animal organism is not able to assimilate cane-sugar as such, led him to the view that this

8—2

form of carbohydrate is probably not of immediate nutritive value to vegetable protoplasm either, a similar parallelism having been established in the case of starch. With the object of testing this hypothesis he instituted a series of researches upon the chemical processes involved in the resumption of development of the Beet after the formation and the resting period of the fleshy root. In the life-history of this plant two well-marked periods can be observed. During the first year of its life it produces only leaves; at the end of the summer the root becomes enormously enlarged and the sub-aerial parts die down. The succulent tissue of the root is found to contain a very large amount of cane-sugar which has been transferred to it from the leaves during their summer activity. After the winter, growth is resumed, and a new stem is put out which bears leaves and gives rise during the next summer to the flowers, fruit, and seeds. This renewed growth takes place at the expense of the reserve-materials stored in the root, the chief of which is the cane-sugar. Bernard discovered that at the onset of this second period of active life the cane-sugar in the root begins to diminish and gets continually less and less as the growth proceeds. Glucose at the same time makes its appearance there and can be traced upwards along the stem to the growing region. Glucose also appears in the leaves, while cane-sugar is altogether absent from them. Bernard further ascertained that a similar enzyme to that he had discovered in the animal intestine could be extracted from the beet-roots on which he was experimenting and he consequently proved that his hypothesis was well-founded, and that in the vegetable, as well as the animal organism, cane-sugar needs to undergo a process of digestion or hydrolysis before it is capable of affording nutritive material to the active protoplasm. The agent in both cases can be extracted from the seat of the digestion and is undoubtedly the same enzyme.

In recent years a very important modification of the generally received ideas of the primary formation of carbohydrates in leaves has been advanced by Brown and Morris, in the course of a discussion of a series of experiments they carried out on the physiology of these organs. These researches have already been referred to in connection with the occurrence of translocation diastase in the cells of the mesophyll. The authors suggest that cane-sugar and not glucose is the earliest-formed sugar and that the starch which rapidly appears in the chloroplasts is the expression of the production of an excess of this carbohydrate. As it is almost certain that cane-sugar is not immediately available for the nutrition of protoplasm, and as the cells in which the carbohydrate material is being elaborated are engaged in a very active metabolism, it is probable that some other form of sugar is produced to supply their immediate needs. There are two possible sources of such sugar,—the starch, which can be converted into maltose, and the cane-sugar, which may undergo hydrolysis as in other parts of the plant. It appears most probable, as Brown and Morris suggest, that the newly formed cane-sugar supplies this nutriment, the surplus only being stored away as starch. On this assumption, invertase should be present in the cells to hydrolyse the cane-sugar, so as to present a nutritive sugar to the protoplasm. It is far from unlikely, too, that such preliminary hydrolysis takes place in the case of that portion of the cane-sugar which is the antecedent of the starch, as the latter is more nearly related to the simpler hexoses than to the complex polysaccharide. The hypothesis of the first formation of cane-sugar, with the subsequent transformations suggested above appears to call therefore for the demonstration of the existence of invertase in foliage leaves.

Brown and Morris made some experiments to test this point. They took some air-dried leaves of a species of *Tropoeolum* and powdered them in a mortar. Two grammes of the powder were digested, with antiseptic precautions, with 100 c.c. of a solution containing 4·881 grms. of cane-sugar and were kept at a temperature of 30° C. for seven days. At the outset the solution had no action on Fehling's fluid but at the end of the week the 100 c.c. reduced 2·064 grms. of CuO. The opticity changed during the time from 9·4° to 7·2°. These numbers indicate an invertive action of the leaf equal to the production of 0·894 grm. of invert-sugar, an amount equal to nearly 45 per cent. of the weight of the dry leaf used.

Kosmann also has detected invertase in the buds and leaves of young trees, and Gonnermann has found it in the leaves of the sugar beet.

The enzyme has also been discovered in the rootlets of germinated barley, but its extraction presents considerable difficulty. The discovery was first made by Kjeldahl, whose work has been supplemented by the researches of J. O'Sullivan. The latter observer removed weighed quantities of moist rootlets from germinating barley and immersed them in a solution of cane-sugar, allowing them to digest for 24 hours at the laboratory temperature. Controls were at the same time prepared, in which water was substituted for the cane-sugar solution. The amount of invert-sugar present in the latter at the end of the experiment was a measure of the quantity present in the rootlets themselves. Deducting this amount he found in a series of three experiments :

(1) 1·7 grms of moist rootlets inverted 1·22 grms. of cane-sugar.

(2) 0·5 grm. of moist rootlets inverted 1·37 grms. of cane-sugar.

(3) 0·5 grm. of moist rootlets inverted 0·15 grm. of cane-sugar.

While inversion took place in all, the activity was not the same in each case, a fact possibly due to the material used containing different proportions of the enzyme.

The existence of the enzyme and the difficulty of extracting it are also established by the following experiment which O'Sullivan quotes—

(1) 0·75 grm. of rootlets were digested in the cold for two hours with 10 c.c. of water ; the extract was filtered and the residue washed on the filter till filtrate and washings amounted to 50 c.c. To 40 c.c. of this extract 25 c.c. of a 10 per cent. solution of cane-sugar were added. (2) The residual rootlets on the filter were added to another 25 c.c. of the same sugar solution. (3) 0·75 grm. of the same rootlets were added to a third 25 c.c. of the sugar solution. The three sets were then digested at 55°—57° C. for eight hours, and after standing at the laboratory temperature for a further eight hours, all were

made up to 100 c.c. The cane-sugar inverted was calculated from the cupric-oxide-reducing power of each 100 c.c. and was found to be as under:—

(1) Extract 0·75 grm. of rootlets inverted 0·19 grm. of cane-sugar.

(2) Residual rootlets of (1) inverted 1·9 grams of cane-sugar.

(3) Original 0·75 grm. of rootlets inverted 1·6 grm. of cane-sugar.

The cupric-oxide-reducing power of the sugar in the rootlets themselves was ascertained from the remaining 10 c.c. of the original extract in (1), which was reserved for that purpose.

The invertase of the embryo of the barley was not confined to the rootlets, but a certain amount was obtained from the plumule. The endosperm of the grains did not contain the enzyme.

These experiments taken in conjunction with those of Brown and Morris, point to cane-sugar, and consequently invertase, having an important part to play in the nutrition of actively growing vegetable protoplasm.

Invertase has been described by Atkinson and by Kellner, Mori and Nagaoko as existing in rice, and in *Koji*, a peculiar preparation of that cereal which is much used by the Chinese in the preparation of fermented liquids. The rice is treated under certain conditions with the fungus *Eurotium oryzæ*. It does not appear very evident whether the latter is not the source of the enzyme, rather than the grain.

Frankfurt has detected the enzyme in isolated wheat-embryos. Invertase is said by Mieran to be present in the ripe fruit of the Banana. An aqueous extract of the fruit, made at ordinary temperatures and allowed to stand for some time at 55° C., was laevo-rotatory, the rotation being given by Mieran as $(a)_D = -7\cdot65°$; a similar extract made at 100° C. was dextro-rotatory, the rotation being $(a)_D = +17\cdot49°$. When the ripe fruit was digested with a solution of cane-sugar the rotation of the polarised ray became reversed in direction. In one experiment it was observed to change from $+99\cdot6°$ to $-11\cdot22°$.

Mieran does not appear to have extracted the enzyme from the fruit.

Invertase is associated as we have seen with cane-sugar in reservoirs of reserve-materials in various parts of the plant. Cane-sugar is sometimes stored in other places than those mentioned, with a view to subsequent utilisation. Many pollen grains are known to contain it in small quantity and it is consumed in their germination. Van Tieghem first discovered it in these bodies in 1886; he germinated the pollen of various plants in a 10 per cent. solution of cane-sugar and detected a reducing-sugar in the liquid which he filtered off from his cultures. When the pollen-grains were placed in the same solution and germination prevented by the addition of a few drops of chloroform, the same result was arrived at, so that he was able to conclude that the enzyme was present in the pollen apart from germinative changes.

In 1893, the writer demonstrated its presence in the pollen of *Eucharis, Narcissus, Helleborus, Richardia, Lilium* and *Zamia*, and was able to extract it from the resting grains by bruising them in an agate mortar and extracting the resulting powder by various solvents. He was able to ascertain further that the quantity increased considerably during the process of germination, and that the increase was much greater when the pollen was allowed to germinate in cane-sugar solution than when it was sown in water only. A most striking experiment was made with the pollen of *Narcissus poeticus*. The pollen was collected from 906 anthers, and weighed ·3 grm. This was divided into three parcels of ·1 grm. each. One parcel (*A*) was steeped at once in 10 c.c. of chloroform-water; another (*B*) was made to germinate in water on a glass plate; and the third (*C*) similarly in 15 per cent. solution of cane-sugar. When germination was well advanced the cultures were carefully washed from the plates and all three made up to 15 c.c. with chloroform-water. They were then all filtered, each was mixed with 25 c.c. of 10 per cent. cane-sugar solution and all were allowed to digest for 93 hours at the ordinary laboratory temperature. The chloroform was subsequently removed from each digestion by boiling it for some time and finally all were titrated with Fehling's solution. The quantities of invert-sugar in the three cases were found to be as under:—

   A. (Ungerminated) ·19 grm.

   B. (Germinated in water) ·5 grm.

   C. (Germinated in cane-sugar solution) 1·37 grm.

The secretion or formation of the enzyme appears from this experiment to be materially increased by the absorption of the nutritive sugar, derived from the hydrolysis of the original carbohydrate.

Besides this very wide distribution in the higher plants, invertase has been found to exist in many fungi besides those already alluded to. Hansen has found it in several of the *Saccharomycetes*. Wasserzug discovered it in certain fungi belonging to the genus *Fusarium,* which are capable of growing in solutions of cane-sugar, and which lead to the formation of glucose in the culture liquid. He was able to detect an excretion of the enzyme by the mycelium by modifying the culture medium. When the latter was the bouillon prepared from veal, invertase was discharged into it in small quantity just at the time when the fungus was developing its conidia.

In 1878, Gayon found that when *Aspergillus niger* was cultivated in a medium containing cane-sugar a certain amount of invert-sugar was produced, a discovery which was confirmed by Duclaux in 1883. Fernbach in 1890 determined this to be due to the presence of invertase, and ascertained that the secretion of the enzyme was not confined to the period of fructification, but extended over the whole period of the life of the organism, the quantity yielded being approximately the same at all times. Bourquelot three years later found that an extract of the mycelium of Aspergillus effected complete inversion of a 2·5 per cent. solution of cane-sugar when digested with it for 24 hours at 22° C.

Bourquelot in 1886 detected the presence of invertase in *Penicilium glaucum* and was able to extract it from the mycelium, but only when the fungus had arrived at maturity. An allied species, *P. Duclauxi*, when cultivated in cane-sugar solution was found capable of inverting it, but the invertase was not discharged into the culture medium. The enzyme is

apparently present in the latter fungus throughout its whole
life, as its spores germinate readily in cane-sugar solutions.
Fischer and Lindner have noticed a similar occurrence of the
enzyme in *Monilia candida*, the fungus hydrolysing cane-sugar
but not yielding the ferment to an extracting fluid.

Invertase is not uniformly present in the fungi; *Polyporus
sulphureus* has been found by Bourquelot not to contain it.
Its distribution throughout the group however has not been at
all fully investigated.

Invertase appears to be very seldom met with in Bacteria.
Fermi and Montesano found that *Bacillus megatherium, B.
fluorescens liquefaciens*, the red Kiel *Bacillus* and *Proteus
vulgaris* were capable of producing it in bouillon to which cane-
sugar had been added. Van Tieghem has shown that it is
secreted by *Leuconostoc mesenterioides*. The writer has detected
it in cultures of a bacillus which is associated symbiotically with
a yeast in certain jelly-like masses infesting the sugar-cane. It
is excreted by the microbe into the surrounding liquid.

It is not certain how far the action of invertase in these
lower forms is intracellular. Different observers differ as to
whether in normal life the invertase is excreted into the
surrounding medium. The opinion is however gaining ground
that when the organisms are growing under favourable conditions
the enzyme passes out through the thin cell-walls and works
extra-cellularly. Its power of diffusion is not very great and
frequently so little of it is sent out of the cells that it may
easily escape observation. This is not difficult when the
method of examination used is that of precipitation of the
liquid by alcohol. This as we have seen above is deleterious to
invertase and probably destroys it entirely when it is present
in small amount. This may explain the failure of some
observers to obtain evidence of its excretion and extra-cellular
action.

The occurrence of invertase in the alimentary canal of
Mammals is chiefly confined to the small intestine, in which, as
already mentioned, Bernard first discovered it. Miura has stated
that it is present also in small quantity in the tissue of the
colon, stomach, and pancreas. He detected it not only in the

digestive tract of living animals, but also in that of animals
which were still-born.   Abelous and Heim found it with
diastase in the eggs of certain Crustaceans, and Biedermann
says it is present in the intestinal secretion of the larva of
Tenebrio molitor, the mealworm.

Various methods have been adopted for the preparation of
invertase from one or other of the sources already described.
Hoppe-Seyler killed yeast by ether and then extracted the
enzyme by water, subsequently precipitating it by alcohol.
Gunning extracted it from washed yeast by means of glycerine.
A more elaborate method was used by Zulkowsky and König;
these observers washed yeast with alcohol, dried and powdered
it and extracted the powder with water.   After filtration the
filtrate was shaken up with ether, which caused the invertase
to separate out in a mass resembling frog-spawn.   This precipitate was again extracted with water and the enzyme
thrown out of solution by alcohol.   So prepared it was when dry
a white substance, soluble in water, yielding a neutral solution.
It precipitated lead, copper, and mercurous salts, but had no
action on ferric perchloride or potassium ferrocyanide, and its
solution was not rendered turbid on boiling with dilute
acetic acid.   According to Bechamp it has an optical activity
$(a)_j = + 41°$.

O'Sullivan and Tompson have investigated the action of
invertase in considerable detail.   They prepared it in large
quantity from ordinary top fermentation pressed yeast.   This
was allowed to stand at a temperature of about 15° C. for
a month, when it had become partially liquefied.   The liquid
mass was then pressed in a screw filter-press and yielded
a clear yellow solution.   Alcohol was next added until 47 per
cent. of the spirit was present and the mixture was allowed
to stand two days.   There was a fairly bulky precipitate and
the supernatant liquid was quite inactive.   The precipitate
was allowed to settle in tall vessels and water was added to
redissolve the invertase.   When the strength of the alcohol
was reduced to 28 per cent. the enzyme had gone into solution
and a considerable amount of albuminous matter remained
undissolved.   The liquid was then filtered and again made up
to 47 per cent. of alcohol.   The precipitate reappeared, was

separated and washed with absolute alcohol, and dried in vacuo over sulphuric acid. So prepared the invertase was very active, but not quite pure; it contained about 5 per cent. of ash, which consisted of phosphates of potassium and magnesium. The authors regard the ash as an impurity and do not consider it as entering into the composition of the enzyme. Efforts to purify the latter beyond this point were futile, as it proved to be a very unstable body and decomposed during the further treatment. It possessed a specific rotatory power of $(a)_j = +80°$. Yeast appears to contain an amount of invertase equal to about 2—6 per cent. of its dry weight.

The action of the enzyme is the same as that of dilute mineral acids, a molecule of cane-sugar taking up a molecule of water and splitting into two molecules, one of glucose, the other of levulose (fructose). Evidence of the action is afforded by the change in the power the solution has of rotating a beam of polarised light and by the development of the property of reducing cupric to cuprous oxide when boiled in the presence of excess of alkali. Cane-sugar does not possess the latter property.

When invertase is digested with a solution of cane-sugar the action is most rapid at first and gradually diminishes as the cane-sugar disappears. The rate of the inversion may always be represented by a definite time-curve, which is practically that given by Harcourt as being the one expressing a chemical change of which no condition varies excepting the diminution of the changing substance. So long as the conditions under which the digestion is taking place remain unchanged this curve expresses the course of the transformation.

Among conditions which materially influence the progress of the action of invertase, we may mention the reaction of the liquid, the concentration of the cane-sugar solution, and the temperature at which the digestion is carried out.

Minute quantities of sulphuric acid accelerate the action in a very remarkable manner, but the actual amount required to produce the greatest effect varies with the amount of invertase present and with the temperature at which the hydrolysis is conducted. The more there is of the enzyme in the solution the greater is the amount of acid required to obtain the

maximum action.   When the temperature of the digestion was
50° C. O'Sullivan and Tompson found that with ·4 per cent.
of invertase present, the optimum amount of sulphuric acid was
12·5 parts per million of the solution; with 1·5 per cent. the
amount required for the greatest activity was 15 per million.
In another experiment, carried out at 15·5° C., when the
quantity of invertase amounted to 1·5 per cent., computed on
the weight of sugar used, the most advantageous quantity of
acid was 75 parts per million of the solution; with ten times
as much invertase the optimum amount of acid was 250
parts per million.   There appears to be no definite ratio between the quantities of invertase and acid which are necessary
to secure the maximum rate of hydrolysis.

If the amount of acidity was carried beyond the most
favourable point, even to a very slight extent, its effects were
very detrimental.   In one of O'Sullivan and Tompson's experiments, conducted at 60° C., ·34 per cent. of invertase being
used, an excess of only two parts of acid per million of the
solution lowered its activity elevenfold.

Müller found that so feeble an acid as carbonic acid
accelerated the hydrolysis.

The action of the caustic alkalis, even if they are present
in very small proportion, is very detrimental; to vegetable
invertase they are instantly and irretrievably destructive of
the enzyme.   This does not appear to be the case with the
invertin of the small intestine, which effects hydrolysis while
the reaction of the intestinal contents is still alkaline.

The most favourable concentration of the sugar solution has
been found by O'Sullivan and Tompson to be about 20 per cent.,
the digestion being conducted at 54° C.   Below this percentage
the speed of inversion rapidly declines; stronger solutions are
hydrolysed at only slightly lower rates, until the concentration
reaches 40 per cent.   In saturated solutions the hydrolysis is
very slow and feeble.

The activity of invertase can be detected at a temperature
a very little above 0° C. but it then works very slowly.   As the
temperature rises the action becomes more evident but rises
only gradually at first, and then more rapidly up to nearly

60° C. According to O'Sullivan and Tompson it doubles itself approximately for each rise of 10° C. The optimum point is somewhere between 55° C. and 60° C. Müller found the relative activities at 0°, 10°, 20°, 30°, and 40° C. to be respectively 9, 19, 36, 63 and 93, and the optimum temperature to be a little under 50° C. Above 65° C. the enzyme is slowly destroyed and at 10° higher the destruction is immediate.

The action of alcohol is deleterious, particularly if much is present. The enzyme is precipitated unaltered by 47 per cent. of the spirit but larger proportions decompose it. When alcohol is present in quantities too small to affect the composition of the enzyme it can diminish its hydrolysing power. If 5 per cent. only is present in a digestion, the activity of the invertase is reduced one half.

There is some difference of opinion as to the influence of the products of hydrolysis on the action of the enzyme. O'Sullivan and Tompson say that they have no influence on the rate of the change, while Müller states that they are distinctly inhibitory.

The power of the enzyme is practically inexhaustible; a sample which had induced inversion of 100,000 times its own weight of cane-sugar was found by O'Sullivan and Tompson to be still active; they showed moreover that invertase itself is not destroyed or materially injured by its action on cane-sugar.

Fernbach noted that his extract prepared from Aspergillus was less active in light than in darkness, and that the inhibitory effect of the illumination was greater as the extract was gradually made more acid.

A very remarkable feature of the invertase obtained by O'Sullivan and Tompson was that when it was heated in a solution of cane-sugar during a period of active hydrolysis it was able to withstand without decomposition a temperature of 25° C. higher than was sufficient to destroy it when in solution in water and consequently inactive. A discussion of this very interesting fact must be deferred however to a subsequent chapter.

The importance of invertase in both the animal and vegetable economy seems to be primarily what has already been indicated, viz. to convert cane-sugar into other sugars which can be assimilated by protoplasm. This species of digestion can take place either in the interior of the cells in which the enzyme is secreted, as in most vegetable sources, or outside them, as in the alimentary canal of the animal organism. In the case of yeast the excretion of the invertase appears to precede any digestion of sugar. Onimus has shown that when a solution of yeast in distilled water is allowed to diffuse through parchment paper into a solution of cane-sugar, inversion can be observed in the latter after 15 to 20 minutes. Yeast cells do not undergo rapid growth outside the dialyser until nearly 12 hours later. He suggests that the secretion of the invertase is necessary to prepare a nidus favourable to the development of the yeast organism. On this view the inversion of cane-sugar by yeast is extra-cellular.

Invertase has been shown to act chiefly if not entirely on cane-sugar. O'Sullivan however has stated that it is capable of hydrolysing raffinose, but with considerably less rapidity. The products appear to be galactose, glucose and fructose. Bourquelot has recently found it capable of hydrolysing gentianose, a peculiar sugar which is present in the roots of various species of Gentians. Like cane-sugar gentianose does not reduce Fehling's solution; it has a dextro-rotatory power on polarised light. On inversion a reducing-sugar is produced and the liquid becomes laevo-rotatory. Bourquelot thinks that gentianose is a very complex polysaccharide containing in its molecule cane-sugar and another polyglucose, and that invertase only attacks the former group.

CHAPTER IX.

Glucase. (*Maltase.*)

IN our discussion of the action of diastase on starch in an earlier chapter, attention was called to a curious discrepancy in the nature of the final product when diastase from different regions of the alimentary canal is taken as the hydrolysing agent. The enzyme of saliva converts starch into maltose; among the products of hydrolysis taking place in the intestine glucose undoubtedly appears; the sugar also which leaves the liver is the latter variety. A good deal of uncertainty has consequently been felt as to whether or no there might not exist more than one variety of diastase, characterised by the power of transforming starch into these different sugars respectively.

The study of the fate of maltose in the economy has shown that like cane-sugar, of which it is an isomer, it is not made use of by the tissues without preliminary hydrolysis. Philips showed in 1881 that when maltose is injected into the blood of an animal, a good deal of it is excreted unchanged in the urine, much as cane-sugar is, though not to so large an extent. Glucose on the other hand is used up by the organism if introduced into the body in a similar manner. Dastre and Bourquelot found in 1884 that if the quantity of maltose injected is not too large, a good deal of it is not voided by the kidney, and that therefore its fate differs somewhat from that of cane-sugar, part of it at any rate being utilised by the organism.

The similarity of its composition to that of cane-sugar, coupled with the ascertained fate of the latter and with the discovery that glucose appears among the products of the digestion of starch in the intestine, points to a process of inversion taking place before absorption. Maltose is capable of such hydrolytic conversion, for when boiled with dilute mineral acids it splits up according to the equation

$$C_{12}H_{22}O_{11} + H_2O = C_6H_{12}O_6 + C_6H_{12}O_6,$$

two molecules of glucose being formed, instead of one of glucose and one of fructose as in the case of the corresponding hydrolysis of cane-sugar. Maltose differs from cane-sugar and from glucose in its power of rotating a ray of polarised light, its specific rotatory power being $(a)_D = +140°$ in 10 per cent. solution at 20° C., while that of glucose is $(a)_D = +52.5°$. Maltose reduces alkaline solutions of cupric oxide, but only two-thirds as much as glucose. When hydrolysis takes place, therefore, the optical activity of the solution diminishes while the reducing power increases.

An investigation into the details of the digestive changes in the small intestine was undertaken by Brown and Heron, who published their results in 1880. They examined the behaviour of the pancreas of the pig and of the mucous membrane of the small intestine of the same animal. Taking a pancreas which was known to be in an active condition, they prepared an extract from it, which they found capable of hydrolysing maltose, with the formation of glucose, but which only acted very feebly. In their experiments with the small intestine, instead of using the fresh tissue, they dried the tube at 35° C. in a current of air, and shredded it, using the shreds with a solution of maltose kept at 40° C. They found that it was capable of converting the maltose into glucose with considerable ease, much more quickly indeed than the pancreas. They held that the most active region of the intestine was the part in which the so-called glands of Peyer occur, these structures being really lymphatic follicles. A piece of dried intestine from this region completely hydrolysed a solution of maltose in seven hours. They were thus led to the view that the digestion of starch in the small intestine consists of two stages, one, conducted by the pancreatic juice, being marked by the transformation of starch into maltose, the other, taking place subsequently, mainly under the action of the succus entericus, but with some assistance from the pancreatic juice, consisting of the further hydrolysis of the maltose into glucose. They further put forward the view already suggested that maltose is not directly assimilable by the living organism.

Brown and Heron further drew attention to the digestion of maltose as being similar to, but not identical with that of cane-sugar. They found the small intestine to be capable of inverting the latter, but not to be so active as in the case of maltose. The extract of the pancreas, while feebly hydrolysing maltose, had in their experiments no power of inverting cane-sugar, nor had the tissue of the gland itself.

In 1881, von Mering confirmed the observations of Brown and Heron on the part played by the pancreas. He found that an extract of that organ taken from a dog was capable of slowly hydrolysing maltose.

In 1883, Bourquelot carried out a series of researches on the rabbit, in which he confirmed and extended the conclusions of the previous observers. He used the fresh tissues and carried out his experiments under antiseptic conditions. His results may thus be summarised : (1) In the rabbit, during digestion, the pancreas and the small intestine both produce an enzyme which hydrolyses maltose. (2) This enzyme is much more abundant in the intestine than in the pancreas. (3) It is produced almost entirely in the median region of the small intestine, portions taken on the one hand near the pylorus and on the other near the large intestine being almost or entirely without action. (4) The enzyme is a different one from invertin, for the enzyme of the pancreas which hydrolyses maltose has no action on cane-sugar and the invertase from yeast (which is identical with invertin) has no action on maltose.

Bourquelot next turned his attention to the vegetable kingdom, and selected for experiment two fungi, *Aspergillus niger* and *Penicilium glaucum*, both of which thrive extremely well when cultivated in a solution of maltose. He triturated the mycelia with sand in a mortar, and extracted the pasty mass with a little water. After filtering his material, he added alcohol to the filtrate and obtained a precipitate which he separated by a further filtration, and dried at a low temperature. This precipitate was soluble in water, and the solution when mixed with maltose rapidly hydrolysed it.

On account of the fact that yeast can set up alcoholic fermentation in maltose solutions, as well as in those in cane-sugar, Bourquelot next directed his attention to this organism, thinking it probable that in both cases hydrolysis of the polysaccharide must precede the formation of alcohol. He failed however to obtain direct evidence of the presence of the new enzyme in the yeast-cells, but some experiments led him nevertheless to infer that it was present there. Adding yeast to a solution of maltose and inhibiting the growth of the yeast cells by the addition of chloroform, he found that there was very soon a diminution of the rotation of the ray of polarised light and a coincident increase of the power of reducing cupric oxide. Both these phenomena we have seen accompany the hydrolysis of maltose. Bourquelot concluded from these experiments that when yeast is present in a solution of this sugar, it secretes an enzyme which hydrolyses it and the resulting glucose is at once split up with the formation of alcohol, the alcoholic fermentation taking place as fast as the glucose is formed.

Fischer also has detected this enzyme in yeast.

The discrepancy between the results of Bourquelot and those of Philips, as to the fate of the maltose injected into the blood was explained by the discovery of Dubourg in 1889 that blood contains the enzyme under discussion. Dubourg took blood from a rabbit which had been fed on an amylaceous diet for 5 days, mixed 25 c.c. with 50 c.c. of a solution of maltose, and digested the mixture at 37·5° C. for 2 days. He found that after that period the optical activity was much diminished and that the cupric-oxide-reducing power was considerably increased. The action was not very energetic, but the experiment showed reason to believe that when, as in Bourquelot's experiments, a small quantity of maltose was injected into the blood there was sufficient enzyme present to hydrolyse it, so that no

maltose was excreted by the kidney; when, as in the work of
Philips, a good deal of the sugar was injected, part only was
hydrolysed and the rest appeared in the urine.

Bourquelot has given the name *maltase* to this enzyme.

In 1886, Cuisinier described an enzyme under the name of
glucase, which he claimed to have discovered in barley-malt and
in several of the cereal grains, which he said converted starch
into glucose. His conclusions were opposed a little later by
Lintner, but were reaffirmed in 1891 by Geduld, who extracted
from maize a soluble ferment capable of hydrolysing maltose.
Cuisinier's enzyme appears to have been a mixture of diastase
and glucase, which occur together as they do in the secretion of
the pancreas.

Geduld and subsequent writers apply the name *glucase* to
the enzyme which hydrolyses maltose, and which was previously
termed *maltase* by Bourquelot. Geduld claims to have extracted
it from maize in a fairly pure condition. It contains 8·12 per
cent. of nitrogen, is only slightly soluble in water, and gives a
blue coloration with tincture of guaiacum and hydrogen
peroxide. It is capable of hydrolysing only about one hundred
times its weight of maltose, so that it is feeble compared with
the invertase prepared by O'Sullivan and Tompson. It is most
active at a temperature of 57°—60° C.; above that point it is
weakened and at 70° it undergoes decomposition.

Since Dubourg's discovery of the presence of glucase in
blood, in 1889, many observers have conducted researches on its
occurrence there and in the tissues and juices of the animal body.

In 1893, Bial ascertained that the serum of both blood and
lymph contains enzymes which convert starch into glucose and
which act also upon dextrin and maltose. The ferments are
not present in the corpuscles. When alcohol is added to serum
and the resulting precipitate is allowed to stand for a time and
then extracted with water, the extract transforms starch into
maltose only, and has no further action on the latter. The
action noted seems therefore to be due to the presence of
diastase and glucase together in the serum, and to the latter
being destroyed by contact with alcohol.

In 1894, Röhmann made a comparative examination of the

digestive properties of serum, saliva, pancreatic juice and
succus entericus, and concluded that both diastase and glucase
exist in all four, but in very different proportions. Saliva
acting on starch forms mainly maltose and dextrin, but gives
also a trace of glucose. Pancreatic juice behaves similarly but
produces more glucose. Serum is more active in the latter
direction than either. On the other hand, while pancreatic
juice hydrolyses starch more rapidly than saliva, the latter is
much more efficient in this respect than serum. While all four
liquids contain both enzymes, diastase is most plentiful in
pancreatic juice and saliva, and in smallest quantity in serum,
but glucase is present in greatest amount in the latter and but
little of it occurs in the other fluids.

Bourquelot and Gley have also prepared glucase from serum.

Röhmann's results received confirmation in the succeeding
year by Hamburger, who compared the same four liquids. His
results show that when all four are allowed to act simultaneously on soluble starch, differences appear in two directions.
(1) The power of reducing cupric oxide attained in a digestion
of 24 hours' duration is different. (2) The time taken to attain a
maximum power varies. Taking the reducing power of glucose as
1, the reduction he obtained in 24 hours by the several fluids was
in the case of saliva ·31, of pancreatic juice ·36, of succus entericus
·26, of serum ·8. The maximum reducing-power was obtained
by saliva in one hour; by blood only after 24 hours' digestion.
Pancreatic juice acted more rapidly than saliva; intestinal juice
was slower than blood. While saliva acts quickly the reducing
power of the product is low; blood on the other hand acts
slowly, but the reducing power is high, a phenomenon which
Hamburger suggests is due to the different sugar formed, which
in the case of saliva is chiefly maltose, while in that of blood it
is glucose. He concludes that glucase is especially abundant
in blood; saliva contains more diastase than either blood or
succus entericus but scarcely a trace of glucase; pancreatic
juice contains more diastase than saliva, and glucase is present
in it in quite appreciable quantities. Succus entericus contains
less diastase than blood, and more glucase than saliva, but not
so much as either blood or pancreatic juice.

A search for glucase in the various tissues of the body, including several regions of the alimentary tract, was made in 1893 by Miss Tebb. Her mode of procedure was based on that of Brown and Heron, and like them she experimented with the tissues of the pig.

She sometimes used infusions made by steeping the dried and disintegrated organs in saline solutions, and sometimes the dried tissue itself. The temperature of drying was 37—40° C. In all cases she checked her results by control experiments in which the dried tissue or its extract had been boiled. The saline solutions were usually 5 per cent. solutions of sodium sulphate, and the extracts were always made with antiseptic precautions. She found the power of hydrolysing maltose to be possessed by dried pancreas, the mucous membrane of the small intestine, Peyer's patches, lymphatic glands, salivary glands, liver, kidney, stomach, spleen, and striated muscle. The relative activity possessed by equal weights of the dried tissue of these organs was found to be very different. Her results may be represented proportionately in the following table:

Mucous membrane of small intestine	3·21
Spleen	1·35
Lymphatic glands	0·93
Liver	0·80
Peyer's patches	0·64
Kidney	0·66
Stomach	0·45

The figures represent the proportion of glucose formed to one part of maltose left unchanged at the conclusion of the experiment.

Besides these various tissues Miss Tebb found glucase to be present in the serum from pig's blood, and in bile collected from the gall-bladder of the same animal.

Pregl found glucase in the succus entericus of a young lamb and Pautz and Vogel in the mucous membrane of the alimentary tract of dogs and new-born children. The most active region was the jejunum, but some inversion of maltose was effected by the stomach, ileum and colon.

A good deal of interest attaches to the discovery of glucase
in the liver. As we have already seen, the sugar leaving this
organ is not maltose, but glucose. In the earlier work on the
diastase of the liver most writers agree in saying that if an enzyme exists there at all it differs from the diastase of the saliva
and pancreatic juice in forming glucose and not maltose from
glycogen. At any rate the sugar in the blood of the hepatic
vein is glucose. It is probable from Miss Tebb's work that the
so-called liver-diastase is a mixture of diastase and glucase, the
first hydrolysing glycogen to maltose, and the second completing the action by further hydrolysing the maltose.

There is some controversy still as to whether glucase is
present in germinating barley and other cereals. Morris holds
that it is an enzyme peculiar to maize, as he was unable to
obtain any evidence of the inversion of maltose when the latter
was mixed with a cold-water extract of either barley, malt, oats,
rye or wheat. On the other hand Ling and Baker prepared a
ferment-extract from kiln-dried malt, which formed a certain
amount of glucose as well as maltose when allowed to digest
with soluble starch. They suggest that the diastase was altered
during the process of kiln-drying. When a little of this so-
called diastase was allowed to act on maltose alone, it converted
some of it into glucose. Kröber is of opinion that glucase
exists in normal malt.

Various species of yeast stand out conspicuously as sources
of glucase. Fischer and Lindner found it to be present in
Saccharomyces octosporus, the extract of which hydrolyses
maltose but has no action on cane-sugar. *S. Marxianus* on the
contrary hydrolyses the latter, but has no action on maltose.
C. J. Lintner found that in the case of other yeasts both
enzymes are present, glucase being less soluble than invertase.
Fischer has arrived at similar results in the case of Froberg
yeast.

Lintner and Kröber have studied the action of glucase
prepared from yeast by extracting the dried cells with water
at ordinary temperatures. They have found its optimum
point to be 40° C., which indicates a difference between it and
the enzyme prepared by Geduld from maize. The latter works

most advantageously at 57°—60° C. Yeast glucase differs also in this respect from invertase, which Lintner and Kröber found to have an optimum point of 52°—53° C. Their glucase was destroyed by heating it to 55° C. At temperatures up to 35° C. its hydrolytic activity was proportional to the temperature, the quantity present and the time of digestion remaining constant. An increase of the quantity of the enzyme did not proportionately accelerate the inversion.

Besides acting on maltose Fischer has found that glucase is capable of effecting the decomposition of several artificial glucosides which he prepared from various sugars, methyl and other alcohols, and hydrochloric acid. He has based upon these experiments a theory of the action of enzymes in general, which will be discussed in a subsequent chapter.

He has ascertained also that glucase is capable of acting upon certain natural glucosides which yield glucose on hydrolysis. Of these the most noteworthy is *amygdalin*, which occurs in certain plants belonging chiefly to the *Rosaceae*. As we shall see later, amygdalin undergoes hydrolysis under the action of an enzyme known as *emulsin*, the change being expressed by the equation

$$C_{20}H_{27}NO_{11} + 2H_2O = C_6H_5COH + HCN + 2(C_6H_{12}O_6)$$
Amygdalin   Benzoic   Prussic   Glucose
                    aldehyde   acid

Glucase attacks this body in a different way from emulsin, and instead of splitting it up as represented above, it only causes the separation of one molecule of grape-sugar, leaving another glucoside containing only one glucose group in its molecule. The sugar of amygdalin appears to be maltose, and to be capable of hydrolysis to glucose, either while in combination in the glucoside, or when the latter is completely decomposed.

Trehalase.

The sugars which exist in the members of the group of Fungi include representatives of the polysaccharides as well as of the simpler hexoses. Of the former the most characteristic is that to which the name *trehalose* has been given. This sugar was first described by Berthelot in 1857 as occurring in

Syrian manna. It was discovered almost at the same time by
Mitscherlich, who obtained it from a specimen of rye which was
infested by ergot (*Claviceps purpurea*). Müntz found it in
several species of *Hymenomycetes* in 1873; in recent years it
has been described by Bourquelot as present in many species
belonging to several different groups, though it is not of universal distribution throughout the fungi.

The formula assigned to it by Berthelot was $C_{24}H_{22}O_{22}$; he
described it as crystallizing easily, forming large crystals and
appearing much like sugar-candy, but not having such a sweet
taste as the latter. More recent investigations point to its
possessing the formula $C_{12}H_{22}O_{11} + 2H_2O$. It appears to resemble maltose by splitting up on hydrolysis into two molecules
of glucose.

It is not present throughout the life of the different fungi
in which it occurs, but is at a maximum just before the period
of fructification or spore-formation. If a fungus containing it
is plucked and allowed to remain in a fairly warm temperature
the sugar disappears. It is extruded in the sap of *Lactarius
piperatus* if this fungus is exposed in a closed vessel to the
vapour of chloroform.

Bourquelot found in the course of an extended research
into the sugars contained in fungi that the appearance of
glucose is always preceded by that of trehalose, and was thus
led to the hypothesis that the latter is hydrolysed by a definite
enzyme, just as cane-sugar is by invertase and maltose by
glucase. He tested the truth of the hypothesis by experiments which were at first confined to three species, *Aspergillus
niger, Penicilium glaucum*, and *Volvaria speciosa*. The organisms
were cultivated for about four days in a fluid containing cane-
sugar, when they had put forth abundant sporangiophores.
They were then ground up in a mortar with dry sand and
allowed to stand under alcohol for about six hours. After
filtering, the residue was pressed between folds of filter-paper
and dried in vacuo. The mass was next macerated for some
time with water, filtered, and the filtrate mixed with strong
alcohol. A precipitate fell, which was collected on another
filter, washed with alcohol and dried in vacuo.

The powder so prepared was known from other experiments to contain both invertase and glucase.

A simpler method of procuring the ferment was sometimes used, which consisted in replacing the culture-fluid by distilled water which was renewed after 12 hours. The fungus excreted its enzymes into the water, which on filtration after two or three days was found to possess a very decided power of hydrolysing cane- and malt-sugars. The first extract was never very active, owing probably to traces of acid developed during the growth of the organism in the original culture-liquid.

The extract prepared in either of these ways was then allowed to digest with trehalose obtained from *trehala*, a kind of waxy excretion, which a certain Coleopterous larva pours out to form its cocoon. In a typical experiment, 10 c.c. of the extract were added to 10 c.c. of a solution containing about 2 per cent. of the sugar. The rotatory power of the mixture was then ascertained to be 3° 36' of a polarimeter when examined in a tube 200 mm. long. Hydrolysis at once commenced and after 18 hours' digestion at 12°—15° C. the deviation was only 2° 20'. It continued to diminish steadily till the sixth day, after which no change took place. The final deviation was 1° of the polarimeter. The reducing-sugar was then titrated and found to be present in the proportion of ·98 grm. per 100 c.c. On the assumption that all the trehalose used was hydrolysed to glucose, the calculated results would be a deviation of the polarised ray equal to 1·013° of the polarimeter, and an amount of reducing sugar equal to ·962 grm. per 100 c.c., numbers which agree fairly closely with those obtained during the experiment.

The same results were yielded by the trehalose which was extracted from various fungi as already mentioned.

Bourquelot states that the enzyme which effects the hydrolysis of trehalose is different from either invertase or glucose, though it co-exists with them in the mycelium of the fungi which he describes. He has given it the name *trehalase*, and describes it as working most advantageously in a faintly acid medium, the best degree of acidity being about ·003 per cent. of sulphuric acid. Larger quantities than this are deleterious, the enzyme being almost without action in the presence of ·2 per cent. of the acid. This behaviour is very much like that of invertase under corresponding conditions.

As the extract of Aspergillus, and the precipitate obtained from it by the action of alcohol, act upon cane-sugar and upon maltose as well as upon trehalose it is necessary to carefully examine its behaviour in these three cases to determine whether the so-called trehalase is really not identical with either invertase or glucase. The former of these enzymes however can be prepared from several sources besides Aspergillus, and when perfectly isolated it has no power of hydrolysing trehalose. It is apparent therefore that the decomposition observed in the experiments described was not due to the invertase contained in the fungus.

There is more difficulty in deciding as to the identity of trehalase with glucase, as the latter enzyme cannot be readily obtained without admixture with others. Careful experiments upon the effect of heating the extract of Aspergillus to different temperatures and then digesting it with trehalose on the one hand and with maltose on the other, have led Bourquelot to pronounce unhesitatingly in favour of there being two separate enzymes hydrolysing these two sugars respectively.

He heated a quantity of the extract in a double water-bath, using a series of 18 test tubes each containing 24 c.c. The temperature of the bath was allowed to rise very gradually and the first of the series was withdrawn when it reached 44° C. No. 2 was taken out at 46° C. and the others in regular order at a constant difference of 2° C. The temperature was raised to 78° C. when the last was withdrawn. The tubes were then all cooled and 10 c.c. of the contents of each were mixed with 10 c.c. of a 2 per cent. solution of trehalose. Another 10 c.c. from each was mixed with 10 c.c. of a ·25 per cent. solution of maltose. The rotatory power of the trehalose mixtures was determined by a polarimeter and found to be 3° 40'; that of the maltose mixtures was 3° 16' of the same scale. They were all allowed to stand for 36 hours at the temperature of the laboratory and the rotation again observed. The results of 15 duplicate tubes are recorded in the following table:—

Temperature of heating.	Rotation of the Trehalose mixture.	Rotation of the Maltose mixture.
50°	1° 10'	1° 32'
52°	1° 10'	1° 32'
54°	1° 20'	1° 32'
56°	1° 24'	1° 32'
58°	1° 38'	1° 32'
60°	2° 18'	1° 32'
62°	2° 30'	1° 32'
64°	3° 40'	1° 32'
66°	3° 40'	1° 50'
68°	3° 40'	2° 05'
70°	3° 40'	2° 12'
72°	3° 40	2° 24'
74°	3° 40	2° 54'
76°	3° 40	3° 16'
78°	3° 40	3° 16

The influence of temperature consequently enabled Bourquelot to discriminate between the two enzymes. Trehalase begins to be affected at 54° C., while glucase is not weakened below 66° C. The activity of trehalase is destroyed at 64° C., while glucase survives up to 74° C.

In a subsequent paper Bourquelot has shown that trehalase exists also in *Polyporus sulphureus*, the extract of which is capable of converting the trehalose of trehala into a reducing-sugar during a digestion extending over 48 hours. Working with Hérissey he has detected it also in germinating barley, where it exists side by side with several other enzymes. In conjunction with Gley, Bourquelot has shown that intestinal juice is capable of hydrolysing trehalose, and that pancreatic extract and the serum of blood do not exercise such an action. Trehalase therefore appears to be present in animal as well as vegetable tissues. Its occurrence in the former is supported by the researches of Fischer and Niebel, who have ascertained that trehalose is hydrolysed slowly by an extract of the duodenum of some animals, while a similar extract prepared from others has no effect upon this sugar. The serum of certain fishes, especially that of the carp, has the same hydrolysing power.

Raffinase (Melibiase).

Another enzyme belonging to this group decomposes the sugar known as *raffinose*. This carbohydrate was discovered by Loiseau in 1876. It occurs in the root of the beet, the seed of the cotton-plant (*Gossypium*) and in barley and wheat during germination. Tollens has shown it to be identical with the so-called *melitose* of Eucalyptus manna.

Raffinose is a hexatriose having the formula $C_{18}H_{32}O_{16}$; on hydrolysis by dilute mineral acids it splits up into glucose, fructose, and galactose. According to Scheibler, by using very dilute acids at a low temperature, the hydrolysis can be shown to take place in two stages :—

$$C_{18}H_{32}O_{16} + H_2O = C_6H_{12}O_6 + C_{12}H_{22}O_{11}$$

Raffinose   Fructose   Melibiose

and

$$C_{12}H_{22}O_{11} + H_2O = C_6H_{12}O_6 + C_6H_{12}O_6$$

Melibiose   Glucose   Galactose.

Its hydrolysis has been attributed to invertase by the older writers, and recently by O'Sullivan, but Pautz and Vogel, and Fischer and Niebel, point out that the small intestines of the dog and of the horse, which contain invertase or invertin, cannot hydrolyse raffinose. The extract of yeast is able to effect this change as well as the inversion of cane-sugar. It is probable that the hydrolysis is due to the presence of a special enzyme which is contained in the yeast but not in the animal membrane.

Raffinose has a specific rotatory power of $(\alpha)_D = +103.12°$, and is not capable of reducing cupric oxide when boiled with it in alkaline solution.

Bourquelot found that its hydrolysis can be effected by an extract of *Aspergillus niger*. In his experiments he mixed 25 c.c. of a 2 per cent. solution of the sugar with 25 c.c. of the extract of Aspergillus, and digested it for 7 days at 12—15° C., taking precautions against the introduction of micro-organisms. The specific rotatory power of the solution was then only about $(\alpha)_D = +50°$, while 10 c.c. of Fehling's solution were completely reduced on boiling with 7·4 c.c. of the digestion.

The enzyme (which may be termed *raffinase*) has been shown by Bourquelot to be present also in baker's yeast and in low fermentation beer-yeasts. He digested 40 c.c. of the extract of each of these fungi with 40 c.c. of 2 per cent. solution of raffinose, keeping the vessels for an hour at 45° C. and then allowing them to stand at the ordinary laboratory temperature for 5 days, taking the same precautions as before against the introduction of micro-organisms. The rotation of the polarised ray in a 200 mm. tube was at first 2° 4′ of the scale of the polarimeter. After 20 hours it had declined to 1°; after 48 hours to 0° 58′. The baker's yeast was a little the more active. The reduction of Fehling's solution was about the same as in his experiments with the extract of Aspergillus.

The occurrence of this enzyme has also been observed by Fischer and Lindner, who were able to extract it from low fermentation yeasts of the Froberg and Saaz types, but not from the high fermentation forms. They dried the yeasts for 3 days in air at 20°—25° C. after draining them for some time on porous earthenware. When dry the yeasts were extracted by digesting them in water at 33° C. for 20 hours. The extract was found capable of hydrolysing melibiose.

Barr has extracted the enzyme from low fermentation Froberg yeast but he attributes to it only the final stage in the hydrolysis, the conversion of melibiose into glucose and galactose, and he has named it *melibiase* in consequence. Barr holds that the first hydrolysis of raffinose or melitriose into fructose and melibiose is effected by invertase. He differs from Fischer and Lindner in saying the *melibiase* is insoluble in water. This however appears very unlikely from a consideration of the properties of other enzymes.

Melizitase.

A sugar to which the name *melizitose* has been given was discovered in 1859 by Berthelot in Brançon manna, and was ascertained in 1877 by Villiers to exist in considerable quantity in the manna yielded by *Alhagi maurorum*, a leguminous plant of shrubby habit which grows in Persia and Bokhara. The manna is an exudation from the leaves and branches of the plant, and appears in hot weather in the form of drops which soon harden on exposure to the air and can be collected by merely shaking the branches. Like raffinose, melizitose is a hexatriose, having the formula $C_{18}H_{32}O_{16}$. It melts at 148° C., is without the power of reducing Fehling's solution, and has a specific rotatory power of $(\alpha)_D = +88\cdot15°$. When hydrolysed with dilute mineral acids it splits up into three molecules of glucose, showing in that way a resemblance to maltose.

Bourquelot found that a limited hydrolysis of this sugar can be effected by an extract of *Aspergillus niger*. He mixed 15 c.c. of this extract with 15 c.c. of a solution containing about 2·5 per cent. of melizitose and digested it for four days at the temperature of the laboratory, heating it every few hours to 50° C. and keeping it at that point for a few minutes. Bourquelot says that this treatment was effectual in preventing the access of micro-organisms. At the commencement of the experiment the observed rotation of the polarised ray in a 200 mm. tube was 2° 9′ of the scale of the polarimeter. After three days it had gone down to 1° 32′, and at the end of the experiment it was 1° 28′. The liquid then reduced Fehling's solution.

The hydrolysis of the sugar was not complete. The specific rotatory power fell from $(\alpha)_D = +88\cdot15°$ to $(\alpha)_D = +61\cdot2°$. The melizitose was converted into glucose and a hexabiose known as *touranose*, the decomposition being represented by the equation

$$C_{18}H_{32}O_{16} + H_2O = C_6H_{12}O_6 + C_{12}H_{22}O_{11}$$

Melizitose   Glucose   Touranose.

Touranose in turn can be further hydrolysed into two molecules of glucose, being isomeric with maltose. The enzyme of Aspergillus, which may be called *melizitase*, seems to be incapable of effecting this latter change, which is easily brought about by dilute mineral acids.

Lactase.

The nutritive value of a milk diet partly depends upon the sugar which it contains. This, which is known as *lactose*, or *milk-sugar*, is a member of the group of polysaccharides, and upon hydrolysis under the influence of mineral acids a molecule of it splits up into a molecule of glucose and another of galactose. It has only recently been ascertained that there exists in the animal body a soluble enzyme which is capable of carrying out a similar hydrolysis, and which may consequently be named *lactase*. Our knowledge of it is chiefly due to the researches of Röhmann and Lappe, who discovered it in the mucous membrane of the small intestine of calves and dogs. Prior to the appearance of these researches the transformations of lactose in the alimentary canal were ascribed to bacterial action and were thought to lead invariably to the formation of lactic acid. Röhmann and Lappe prepared extracts of the mucous membrane of the small intestine with antiseptic precautions and allowed them to act on solutions of lactose. After digestion lasting for several hours glucose was found to be present in the liquid by means of the phenyl-hydrazine-acetate reaction. The osazone produced gave all the reactions of glucosazone. Röhmann and Lappe precipitated the enzyme from the extracts of the intestine by the usual treatment with alcohol, and found that a solution of the precipitate possessed the power of hydrolysing milk-sugar.

The work of Röhmann and Lappe is supported by some researches of Pautz and Vogel, who ascertained that the mucous membrane of the jejunum of dogs and of new-born children is capable of slowly hydrolysing lactose. Pregl has found that lactase is not present in the intestinal juice of the lamb obtained by the Thiry-Vella method. Fischer and Niebel also have noticed that lactose is hydrolysed by extracts of portions of the small intestine, particularly of young animals. They have shown further that lactase is not present in blood serum.

Weinland also has found it in the intestine of sucking animals, and in that of the pig, dog, and horse. He states that after several months' feeding on a milk diet, the enzyme makes its appearance in the intestines of the rabbit and the hen. He has also found it in the pancreas of both young and adult dogs.

The extract of the *Kephir* organism has been said by Beyerinck to be capable of hydrolysing lactose. He claims to have separated a soluble enzyme from the organism, and to have shown that it possesses the same power. His work has been disputed by other observers, who say that his enzyme splits up cane-sugar and raffinose, but not milk-sugar. Fischer confirms Beyerinck as to its action on the latter, and says that it hydrolyses cane-sugar also. Besides being yielded by Kephir it can, according to Fischer, be extracted from certain yeasts when they are dried at the ordinary temperature, ground up with powdered glass and suspended in water. It differs by its greater stability from the glucase already described as occurring in some other yeasts.

Dienert also states that he has extracted lactase from certain yeasts that are capable of fermenting lactose.

Fischer says that the emulsin of the bitter almond is capable of hydrolysing lactose, a view which is supported by Hérissey. A careful examination of their experiments suggests however that their emulsin contained a small amount of lactase. Reference will be made to this more fully in a subsequent chapter. If this view is correct, lactase exists in the seeds of the bitter almond.

CHAPTER X.

GLUCOSIDE-SPLITTING ENZYMES.

MANY bodies of very complex character exist in plants, which have one property in common. When they undergo decomposition under the influence of either mineral acids or soluble enzymes, they yield a sugar, generally glucose, as one of the products of the action. The other bodies which are produced simultaneously are very varied, but one or more of them usually belong to the aromatic series of organic compounds. They are on the whole not very unlike the polysaccharides which have already been discussed, but while the latter are formed of molecules of sugars united to each other, in the former we find the sugar grouped with or united to other radicles. From the fact that the sugar is almost always glucose, these bodies have been termed *glucosides*. The best known of them are the *amygdalin* of the almond and other Rosaceous plants, the *sinigrin* of the Cruciferae, the *tannin* which is so widely distributed in the vegetable kingdom, the *salicin* of the willow, and the *coniferin* of the fir-trees. Many others however exist in other plants.

The decomposition which they undergo is generally of a hydrolytic character and frequently leads to the total disruption of their molecule. Sometimes it is less complete and consists only in the separation of part of their sugar, a less complex glucoside remaining. The latter decomposition is less frequent than the former and is only known in connection with some members of the group, particularly amygdalin. Mineral acids effect complete hydrolysis as they do in the case of the polysaccharides. The more limited decomposition is also known in the latter group, one member of which, melizitose, we have already seen, can be split up by one of the enzymes of Aspergillus, yielding glucose and touranose, the latter on final hydrolysis by acids yielding two more molecules of glucose.

While it would be too much to say that there is a special ferment for the decomposition of each glucoside it is yet certain that a great number of such enzymes exist. We are acquainted with *emulsin*, which hydrolyses amygdalin, *myrosin*, which decomposes sinigrin, *erythrozym*, which splits up the glucoside of the madder, *rhamnase*, which acts upon xantho-rhamnin, found in the seeds of the Persian berry, and *gaultherase*, which decomposes a glucoside of methyl-salicylic ether, occurring in *Monotropa hypopythis*, a parasite upon the roots of many trees. Others also are known to exist in various plants, though no definite examination of them has at present been made.

Emulsin (Synaptase).

The glucoside *amygdalin* was first obtained in 1830 by Robiquet and Boutron, who prepared it in crystalline form from the seeds of the bitter almond (*Amygdalus communis*). They showed this body to be the antecedent of the so-called essence of bitter almonds. In 1837 Liebig and Wœhler found that the transformation leading to the appearance of the latter was brought about under the influence of a certain albuminoid matter also existing in the kernel of the almond, and to this principle they gave the name *emulsin*. They ascertained that during the action of the latter on amygdalin, sugar and prussic acid were formed. Robiquet in the following year suggested that the action belonged to the same category as the action of diastase on starch and gave the name *synaptase* to the enzyme. The older name however has continued to be applied to it.

The decomposition of amygdalin which emulsin effects is one of hydrolysis, and can be expressed by the equation

$$C_{20}H_{27}NO_{11} + 2H_2O = C_7H_6O + HCN + 2(C_6H_{12}O_6).$$

| Amygdalin | Benzoic aldehyde | Prussic acid | Glucose |

Besides amygdalin it is able to effect the decomposition of several other glucosides, among which are *salicin, helicin, phlorizin* and *arbutin*. The several reactions may be expressed by the following equations, the reaction being one of hydrolysis in each case :—

$$C_{13}H_{18}O_7 + H_2O = C_7H_8O_2 + C_6H_{12}O_6.$$
Salicin   Saligenin   Glucose

$$C_{13}H_{16}O_7 + H_2O = C_7H_6O_2 + C_6H_{12}O_6.$$
Helicin   Salicylic   Glucose
                       aldehyde

$$C_{21}H_{24}O_{10} + H_2O = C_{15}H_{14}O_5 + C_6H_{12}O_6.$$
Phlorizin   Phloretin   Glucose

$$C_{12}H_{16}O_7 + H_2O = C_6H_6O_2 + C_6H_{12}O_6.$$
Arbutin   Hydro-   Glucose
                       quinone

Emulsin has been found by Fischer to be capable of effecting also the decomposition of certain of the artificial glucosides which he prepared as described in the last chapter.

The occurrence of emulsin has been observed in several of the higher plants and in certain fungi. Of the former, the most striking are the Almond and the Cherry-laurel, both of which belong to the Natural Order *Rosaceae*. Its distribution in these plants is different; in the former it is chiefly found in the seeds; in the latter it occurs in the young stems and leaves. In 1865 Thomé was led to the opinion that the enzyme exists in the seeds of the bitter Almond only, and is localised there in the fibro-vascular bundles of the cotyledons. Portes in 1877 concluded emulsin to be confined to the axis of the embryo and amygdalin to be present in the cotyledons. In 1887, Johansen found emulsin in the seeds of both the sweet and the bitter varieties of the Almond, in the fibro-vascular bundles and the cells abutting on them, particularly in those of the cotyledons. He found amygdalin in the parenchyma of the cotyledons of the bitter variety only. Its distribution in these plants and in a few others closely allied to them has been the subject of a careful study in recent years by Guignard, whose methods have been partly chemical and partly based upon certain reactions leading

to the development of definite coloration in the cells containing the enzyme. The adoption of the latter method of
research is attended by great difficulties, and conclusions based
solely on such reactions must be received with caution. We
may especially allude to the fallacious character of the blue
reaction with tincture of guaiacum said by many workers to be
characteristic of enzymes, but now known to be given by many
other substances usually or frequently present in vegetable cells.

Guignard has been careful not to rely exclusively on
colour-reactions for the identification of the enzyme but to
supplement these by chemical tests. His first experiments
were made with the leaf of the Cherry-laurel. When he
examined under the microscope some of the tissue of this
organ, which he found capable of hydrolysing amygdalin, he
observed that certain cells surrounding the vascular-bundles
were marked by the very finely-granular character of their protoplasm and their freedom from starch and chlorophyll, though
they contained a certain amount of tannin, and were richer
in proteids than the other cells of the leaf. He treated sections
of this tissue with Millon's reagent, which is a mixture of the
nitrates of mercury with 4 cm. of nitric acid, and found that all
the cells of the leaf darkened in colour, becoming almost black,
but the tint was much deepest in the cells in question. On
warming the section the black tint gradually disappeared, and
was replaced by an orange-red in the cells suspected to contain
the enzyme and by a faint pink in the ordinary parenchyma.
In an allied species, *Cerasus lusitanica*, which contains tannin
but no emulsin, this orange-red coloration was not developed
in the corresponding cells by the same treatment. The reaction was not therefore due to the tannin, but to some proteid
constituent with which the enzyme is associated or possibly to
the enzyme itself. Probably the reacting body is a proteid
rather than the enzyme, for proteids give a very similar reaction
when heated with Millon's reagent, the coloration which they
assume being a dull brick-red. Confirmatory tests for proteids
bore out this opinion. Copper-sulphate and caustic potash
coloured these cells a violet-pink, as they do proteids in solution,
while cells containing only tannin did not react to these

reagents more than ordinary parenchyma cells with their lining of protoplasm, which stained a pale pink with no admixture of violet.

The chemical experiments by which Guignard confirmed his conclusion that these cells contain emulsin consisted of a digestion of the tissues with a solution of amygdalin. The cells in question were observed to occur in definite positions in the leaf, and with care Guignard found it possible to separate them from the adjacent tissues. The difficulties of such separation were mainly mechanical but delicate manipulation enabled him to overcome them. In the leaves and young branches of the laurel the cells occur in the endodermis, which is a sheath surrounding the collection of fibro-vascular bundles, but belonging to the cortex, of which it is the innermost layer. When this sheath was carefully dissected out under a microscope and the piece of tissue placed in a solution of amygdalin in a watch-glass, and kept at 50° C., the decomposition of the glucoside was speedily effected, and was recognized by the odour of the benzoic aldehyde and prussic acid which we have seen to result from its hydrolysis. Sections made through the leaf produced the same decomposition when they were cut so as to include part of this endodermal sheath, but not otherwise.

By these methods Guignard determined that the emulsin of the Cherry-laurel exists in the endodermis; that of the Almond was found in the axis of the embryo in the many-layered pericycle which lies immediately under the endodermis and closely surrounds the fibrovascular bundles; in the cotyledons it is in the endodermis as well as in the pericycle. The enzyme exists in both the sweet and bitter Almond, though amygdalin is only present in the latter.

Probably the glucoside and the enzyme do not naturally exist in the same cells, as the decomposition of the former in the seed of the Almond only takes place during germination. The distribution of the amygdalin is not however definitely known. It seems probable that the fluid sap containing it may travel along the cellular tissue of the axis of the plant, and as the ferment which decomposes it is in the immediate neighbourhood of the bast, which is a great conducting tissue for

elaborated products, it is not unlikely that the decomposition
of the glucoside may take place during its transit along the
axis in consequence of its diffusion into the cells containing
the emulsin, so that the latter is charged with the duty of
preparing from the amygdalin certain nutritive products, especially sugar, which may thence easily make their way to the
conducting tissues, and so travel to the actual seats of constructive metabolism. On the other hand it may be that the
amygdalin descends by the conducting tissue of the bast and
undergoes decomposition there as it passes downwards, the
emulsin diffusing into the bast to meet it, though this seems
less likely, as Guignard did not find emulsin at any time apart
from the cells which he describes as the seats of its formation.

The probability that the enzyme and the glucoside are
formed in different cells has been challenged by no less an
authority than Pfeffer, who suggests that they both exist in the
same cells, and that the only degree of separation is that the
ferment is in the protoplasm and the glucoside dissolved in the
cell-sap. This view however does not harmonise with Guignard's experiments.

The plants mentioned so far all belong to the *Rosaceae*.
Emulsin is not however confined to members of this Natural
Order and their near allies. It has been found in plants belonging to the Gymnosperms and the Monocotyledons as well as
other Dicotyledons.

Emulsin has been shown during recent years to have a
somewhat wide distribution among the Fungi. It was observed
almost simultaneously in 1893 by Gérard in *Penicilium glaucum* and by Bourquelot in *Aspergillus niger*. Bourquelot
cultivated the latter fungus in Raulin's nutritive fluid till it
was about to put out its fructification. The liquid was then
decanted off, and the fungus washed several times with distilled water and macerated with water for several days. The
filtrate from the maceration contained a mixture of enzymes,
of which emulsin was one. When a little of it was added to a
solution of amygdalin and kept warm, the characteristic odour
of the essence of bitter almonds could be recognised in the
course of an hour.

Bourquelot investigated its action on several other glucosides and found it capable of hydrolysing some, but not all. He has published the following series of experiments.

10 c.c. of an aqueous solution of *salicin* containing ·2 grm. were mixed with 10 c.c. of the Aspergillus extract and kept for 40 hours at a temperature of 23° C. At the end of that time 55 per cent. of the salicin was decomposed.

To 10 c.c. of the Aspergillus extract ·2 grm. of *coniferin* was added. This glucoside being scarcely soluble in water the liquid remained turbid, the coniferin being only in a state of suspension. The digestion was carried on for 30 hours at 23° C. and then for a further three hours at 45° C. During this exposure the liquid became less and less turbid till it was nearly clear. A little later it became milky, and a new precipitate was formed which settled down in the tube, and increased in quantity till the end of the experiment.

At the end of the digestion the liquid contained ·093 grm. of glucose, which was nearly as much as the ·2 grm. of coniferin was capable of yielding on complete hydrolysis. The final precipitate was the coniferilic alcohol resulting from the decomposition, which like the coniferin itself is almost insoluble in water.

Hérissey has prepared the enzyme from the resting seeds of the Almond, by the following method. The seeds, divested of their outer coatings were powdered in a mortar as finely as possible and extracted for 24 hours with twice their weight of water containing a little chloroform. They were then strained off and a little glacial acetic acid was added to the liquid. This precipitated the bulk of the proteid matter contained in the extract. After filtration a clear limpid liquid was obtained. On the addition of about four volumes of alcohol the enzyme was precipitated. This was collected on a filter, washed with a mixture of alcohol and ether and finally dried in vacuo, over sulphuric acid. The powder thus obtained was soluble in water, and the solution retained the properties of emulsin. It was however not pure but was mixed with a carbohydrate resembling araban.

A more active preparation was obtained from cultures of

Aspergillus. The fungus was cultivated from spores in Raulin's solution till its fructification was mature, generally for about three days. The culture fluid was then syphoned óff and replaced by distilled water. After a few hours this was again replaced by a further quantity, and the process was repeated till the water contained no trace of the salts of the original liquid. It was then macerated in fresh distilled water for about three days at the ordinary temperature, and the liquid filtered off. So prepared the solution was perfectly clear and limpid, and contained only ·2 grm. of residue per litre. It was much more active than the preparation made from the almonds.

Besides *amygdalin, salicin,* and *coniferin,* Bourquelot and Hérissey found the emulsin of Aspergillus to be capable of hydrolysing many other glucosides, including *phloridzin* and *populin.* They obtained negative results however with *solanin, hesperidin, convallarin, digitalin, jalapin,* and *atractylate of potassium.*

Hérissey has found that the emulsin of the almond has no action on either *phloridzin* or *populin.* Either there are two varieties of emulsin or there is another enzyme in Aspergillus, which decomposes the other two glucosides.

Beijerinck states that the emulsin of the almond hydrolyses indican.

Gérard's results with Penicilium were very similar to those of Bourquelot. He prepared his enzyme by macerating the mould in distilled water, and concentrating the extract to a small bulk in vacuo. The emulsin was then precipitated from the concentrated extract by the addition of alcohol. He found it capable of hydrolysing amygdalin and salicin.

Bourquelot has ascertained that emulsin is present in many other fungi than the species mentioned. It is not confined to saprophytic forms, but exists in a large number of parasites, prominent among which may be mentioned the genus *Polyporus.* In all he investigated 43 fungi of parasitic habit, and found 34 to contain emulsin. Most of the latter infest living trees, especially attacking the old wood. It is well known that the cortex, the cambium, and even the woody parts of trees contain glucosides. The distribution of the enzyme in the fungi seems to bear some relationship to the peculiarities of their parasitism.

Fischer has stated that emulsin is capable of decomposing lactose, yielding glucose and galactose. Hérissey made the same observation. It appears however from a consideration of the conditions of the preparation of the enzyme that it is probable that the almond contains also a little lactase. It is a matter of very common occurrence that several enzymes are present in various seeds, and unless careful means of separation are adopted it is rash to assume that the various decompositions set up by an extract of them are all due to a single ferment. The probability that the almond contains lactase is increased by an observation made by Hérissey, that the much more active extract yielded by Aspergillus was altogether without action on lactose.

A similar association of two enzymes in the same tissue was noted by Schunck who found emulsin and erythrozyme together in the root of the Madder plant.

Recently Hérissey has discovered emulsin in several Lichens, among which may be mentioned *Usnea barbata*, *Physcia ciliaris*, *Parmelia caperata* and two species of *Ramalina*. He bruised the lichen, and placed about half a gramme in contact with a solution of amygdalin for two or three days at a temperature of 35° C. At the end of that time both hydrocyanic acid and sugar were present.

It is doubtful whether emulsin exists in the animal organism. Kölliker and Müller have stated that the pancreatic juice is capable of effecting the decomposition of amygdalin, and Moriggia and Ossi, and Gérard, have separately shown that the secretion of the small intestine of certain herbivora has the same power, but the specific enzyme has not been isolated.

Emulsin acts most energetically at temperatures between 40° and 50° C.; above the latter point its power gradually declines, but it is not entirely destroyed till heated to near 80° C. It works best in neutral solutions, but its activity is not materially impeded by the presence of small quantities of either acids or alkalis. It is precipitated by a solution of tannin.

Myrosin.

This body, which is the characteristic enzyme of the Cruciferae, was first discovered by Bussy. Not only is it widely
distributed throughout the Cruciferae but it occurs also in
several closely allied Natural Orders. It acts especially upon
the glucoside *sinigrin* or *myronate of potassium*, effecting its
decomposition into *sulpho-cyanate of allyl*, or essential oil of
Mustard, *glucose* and *hydrogen-potassium sulphate* being simultaneously formed, according to the equation

$$C_{10}H_{18}NKS_2O_{10} = C_3H_5CNS + C_6H_{12}O_6 + KHSO_4$$

Sinigrin	Sulpho- cyanate of Allyl	Glucose	Hydrogen potassium sulphate

It is remarkable that the decomposition of the glucoside is
apparently not associated with the incorporation of a molecule
of water, so that it does not appear to be a process of hydrolysis, as in all the other cases so far examined. Van Rijn has
stated however that sinigrin contains a molecule of water and
that its formula may preferably be given as $C_{10}H_{16}NS_2KO_9 + H_2O$.

When the seed of the black mustard, *Sinapis (Brassica)
nigra*, is bruised and treated with water, the odour of the
sulpho-cyanate of allyl is easily recognisable. Both the myrosin
and the glucoside are contained in the seed, but in separate
cells, and the reaction is the result of their being brought
together by the solvent.

Sinigrin can be prepared from the seeds of the black
mustard by the following method, which is a modification
of that used by Bussy. A kilogramme of the powdered seeds
is extracted with 1·5 litre of alcohol of 82 per cent. strength;
and boiled till the alcohol is reduced to 1250 c.c. The residue
is pressed while hot, and again boiled with a further quantity
of alcohol. After decantation of the spirit it is then pressed
and dried at 100° C., and digested for 12 hours with three
times its volume of cold water. The residue is again separated
by decantation and digested with two volumes of water. The
two aqueous solutions are evaporated to a syrupy consistency,

after the addition of a little carbonate of barium. This syrupy
residue is then exhausted by boiling alcohol, of 85 per cent.
concentration, which dissolves the sinigrin. After filtration the
spirit is distilled off, and the glucoside crystallises out.

The localisation of myrosin has been the object of a very
elaborate research by Guignard, who has investigated its occurrence in a very large number of plants belonging to the
Natural Orders, *Cruciferae, Capparidaceae, Resedaceae, Tropœolaceae, Limnanthaceae* and *Papayaceae*, and has carefully
scrutinised the various regions and organs of typical members
of all these groups. In 1886, Heinricher showed that in many
of the plants of the Cruciferae, special cells, very variously
distributed, could be recognised by the peculiar nature of their
contents. They gave very strongly-marked proteid reactions
and hence he considered them to be reservoirs of albuminoid
material. Guignard has found similar cells widely distributed
in plants belonging to all the Natural Orders mentioned and
by similar tests to those he employed in the cases of the
cherry-laurel and almond, he has identified them as the cells
which contain myrosin. They are recognisable by their finely
granular contents and by their being free from starch, chlorophyll, fatty matter, and aleurone grains, though they are situated
in various regions among other cells which contain one or more
of these constituents. When the tissue in which they lie is
treated with Millon's reagent and warmed, these cells become
orange-red in colour, while the parenchyma in which they are
embedded only takes on a pale pink tinge. They give a violet-
red coloration with cupric sulphate and caustic potash. These
cells contain, associated with their protoplasm, a quantity of
amorphous proteid matter, which is coagulated by alcohol, and
then separates from the peripheral protoplasm in the form of
coarsely-granulated masses, which on subsequent treatment
with Millon's reagent are coloured a more vivid red than
the protoplasm. The cells can be distinguished among the
parenchyma in which they lie by staining with methyl-green
and other anilin dyes. Usually they are slightly larger than
the surrounding cells and are longer and less regular in shape.

The power of decomposing sinigrin possessed by pieces of

different tissues carefully dissected out was found by Guignard
to vary with the number of such cells which the tissue contained. The myrosin seems to be associated in them with
the proteid matter, just as in the case of emulsin as described
above. The demonstration of the presence of myrosin in these
cells was most easily effected in the *Wall-flower*, in which they
form a readily separable layer of the pericycle. When this
was isolated with great care and warmed with a 2 per
cent. solution of sinigrin, the characteristic odour of the
sulpho-cyanide of allyl was perceptible almost immediately.
Guignard found throughout his experiments that any tissue
containing these cells could effect the decomposition, but that
if they were not present the tissue could not act upon the
glucoside.

The results of Guignard's researches into the localisation
of these cells in the plants of the various Natural Orders
mentioned may be briefly summarised as follows :—

Roots. Chiefly in the cortex, but sparsely in the wood. In
fleshy roots the tissue representing the wood is mainly parenchymatous, and contains them. In woody roots they are found
in the secondary bast as well as in the cortex and to a less
extent in the medullary rays. In the Capparidaceae the wood
contains none, but some are found in the pith. In the
Resedaceae they do not occur further inwards than the bast.
In the Papayaceae the root is not very rich in myrosin.

Stems. There is a good deal of variety in the distribution
in different species. Speaking generally, the pericycle and the
tissues derived from it are richest in the enzyme, while the
secondary bast comes next in importance. When the special
cells occur in the region of the wood they are generally in the
medullary rays. The pith also contains some of them. Guignard recognises nine types of distribution in the stems of the
Cruciferae.

(1) The pericycle alone. *Lepidium sativa etc.*
(2) The pericycle and the primary and secondary bast.
     Erysimum cheiranthoïdes.
(3) The cortex chiefly, but the pericycle to a less extent.
     Moricandia hesperidiflora.

(4) The cortex, and the bast under a thin and sclerenchymatous pericycle. *Iberis amara etc.*

(5) The cortex, pericycle and pith. *Nasturtium officinale etc.*

(6) The cortex, pericycle, and secondary bast. *Bunias orientalis etc.*

(7) The cortex, pericycle, secondary bast and pith. *Raphanus sativus etc.*

(8) The cortex and pericycle chiefly; also the primary and secondary bast and pith. *Brassica nigra etc.*

(9) The cortex, pericycle, primary and secondary bast, woody parenchyma and pith. *Cochlearia armoracea etc.*

The distribution in the other families referred to is similar, with the exception of the Papayaceae, in which group the stems contain but little myrosin. In the Limnanthaceae the special cells lie mainly in the lacunar cortex, but a few are to be found in the bast; in the Tropœolaceae they generally form groups or nodules of cells in the hypodermal layer of the cortex, a few being in the bast.

Leaves. Species that contain many of the myrosin-secreting cells in the axis of the plant generally exhibit them in the leaves also, and the relative proportion is often greatest there. They occur throughout the mesophyll but are usually most numerous towards the lower surface. In some leaves they are localised in the mesophyll and the pericycle; in others chiefly in the pericycle and in the bast of the veins. In a few cases they occur in the endodermis of the bundles. In the Capparidaceae they are found two or three together, the groups extending through the parenchyma. In the Limnanthaceae they are principally in the epidermis of the lower surface. They are very long in comparison with their breadth, and appear almost tubular; sometimes two or more lie side by side. In the Papayaceae different species vary as to the abundance of myrosin they contain. A good deal is present in the leaves of *Carica condinamarcensis* and *Vasconcella quercifolia,* the cells being chiefly found in the lamina.

Flowers. In the Capparidaceae the flower contains large

numbers of the secreting cells both in the sepals and petals. They are very numerous in the wall of the ovary. In this order they generally occur in little groups, each of which is developed from a single cell which becomes conspicuous while the tissue is still very young. In *Tropæolum* they are very prominent in the tissue of the spur. In the flowers of the Cruciferae they are met with chiefly in the carpels, where they lie in the neighbourhood of the fibro-vascular bundles.

Seeds. In the Cruciferae the distribution of the secreting cells varies; either the integuments alone, or both embryo and integuments may contain them. Guignard distinguishes five types exhibiting the following regions of localisation :—

(1)   The cotyledonary parenchyma and the cortex of the axis of the embryo.

(2)   The tissue abutting on the back of the cotyledonary fibro-vascular bundles, or the pericycle if that layer is differentiated ; also the cortex of the axis.

(3)   Both regions described under (1) and (2).

(4)   Sometimes they are absent from the radicle and the cotyledons.

(5)   The integuments only.

In the Limnanthaceae these cells can be detected in the parenchyma of the cotyledons when the seed is very young and before reserve-materials are stored in any of the cells. Later, on germination of the seed, they can be recognised in the lower epidermis as soon as the cotyledons become green. The other orders except Papayaceae show a similar distribution. In these the myrosin is almost confined to the outer integument, which forms a thin pellicle swelling considerably when in contact with water. It is doubtful whether any of the enzyme is secreted in the inner layer of the seed-coat.

Jadin has found myrosin in nearly all parts of *Moringa*, a plant closely allied to the Capparidaceae though not a member of that Natural Order.

There is no myrosin in the latex of the Papayaceae.

Besides these Natural Orders, Guignard has discovered myrosin in several species of *Manihot* from Brazil. It is not localised there in the laticiferous system as in emulsin.

Though the action of myrosin has been examined chiefly with regard to the decomposition of sinigrin, it is equally efficacious in splitting up *sinalbin*, the glucoside present in the White Mustard, *Brassica alba*, and there seems no doubt that it can decompose all the glucosides which are so prominent in the Natural Orders mentioned. These are much alike in the products of their decomposition, the aromatic molecule varying in different cases.

Guignard has carried out a research on some of the chemical and physical properties of myrosin, in which he has found it to possess many points of agreement with other enzymes. His material was drawn from two sources, the outer integument of the seed of *Carica papaya*, and the extract of the ground seed of the White Mustard. The enzyme behaved in an almost identical manner in the two cases.

The outer coat of the seed of Carica can be easily separated from the rest of the structure. It swells up considerably in water and assumes an almost mucilaginous consistency. The inner coat is hard and sclerenchymatous, so that the outer envelope can be easily separated from it. The cells of the outer coat contain hardly anything but myrosin, and a fairly pure preparation of the enzyme can be obtained by swelling up this layer, separating it from the inner coat, drying it and grinding it to powder. A very minute quantity of this powder is sufficient to set up decomposition of the glucoside.

Guignard prepared it from the White Mustard by grinding a number of the seeds and soaking them in water at 40° C. for several hours. The filtrate from the paste so formed was then heated to 70° C. to coagulate the larger part of the proteids which were subsequently separated by a further filtration. The myrosin was thrown down from this filtrate by the addition of two volumes of alcohol of 90 per cent. concentration, filtered off and dried over a water-bath at 30° C., being finally washed with ether.

In his experiments on the properties of myrosin, Guignard used both these preparations side by side throughout.

The enzyme was found to show certain peculiarities with regard to its power of resisting destruction by high temperatures.

A series of tubes was prepared, each containing 5 c.c. of the extract of the seeds of Carica or Mustard, mixed with 5 c.c. of water, and they were heated on a water-bath very gradually. When a temperature of 50° C. was reached, the first tube was withdrawn, and another was removed at each increment of 1° C. The heating was continued till the temperature of 85° C. was reached. The tubes were then allowed to cool and ·02 grm. of sinigrin was added to each, and they were digested at the temperature of the laboratory for 18 to 24 hours. The retention of the power of decomposing the glucoside was ascertained by observing whether or no there was a production of the odour of the sulpho-cyanate of allyl in each case. It was found that no deterioration of the power of the myrosin to liberate this body had taken place in those tubes which had not been heated to 81° C. From this point upwards the temperature partially destroyed the enzyme and the destruction was total at 85° C. The curve representing the action at different temperatures was seen to show accordingly a very rapid fall between 81° and 85° C., a much steeper slope than that exhibited by the corresponding curve of any other enzyme so far examined. The maximum temperature or point at which destruction takes place is not very different from that given by Kjeldahl for diastase, which is 86° C., but it is considerably higher than that for invertase, which is 70° C.

Salicylic acid when present in the proportion of less than 1 per cent. does not interfere with the action of myrosin; 1·5 per cent. weakens it very considerably, and the enzyme is quite inoperative in the presence of 2 per cent. Diastase is affected in a very similar way by salicylic acid.

The presence of 1 per cent. of tannin nearly stops the action of both myrosin and emulsin. If the digestion is carried on at a high temperature, such as 80° C. the inhibitory effect of tannin is much greater, ·05 per cent. being sufficient to suspend altogether the decomposition of the glucoside.

Chloral has much less effect on the action of either myrosin or emulsin, which is rather surprising, as this reagent combines very energetically with proteids. Guignard prepared a series of tubes each containing 5 c.c. of extract of the testa of the

seeds of Carica, 20 c.c. of water, and ·02 grm. of sinigrin, and
to the several tubes of the series he added gradually increasing
proportions of chloral, and digested them at 40° C.   He found
that a tube containing 1 per cent. of chloral evolved a strong
odour of allyl sulpho-cyanate in 15 minutes; in the presence
of 2 per cent. much less was produced ; with 3 per cent. the
odour was only just perceptible in that time ; with 4 per cent.
there was no odour till after 30 minutes; and with 5 per cent.
the action could not be detected till the expiration of an
hour.   Subsequent digestion for 12 hours did not increase the
intensity of the odour.   When a tube was heated to 80° C.
for 2 minutes with 1 per cent. of chloral, the enzyme was destroyed.   That temperature alone was without effect upon it.

Alum and borax, which have antiseptic properties, exercised
but little influence upon myrosin, quantities of less than 6 to 8
per cent. being innocuous.

Erythrozyme.

A third enzyme which is capable of splitting up a glucoside
was discovered by Schunck in 1852 in the root of the Madder
plant (*Rubia tinctoria*).   Unlike the others so far described, it
has a very limited distribution and acts upon only a single
glucoside.   The latter, which Schunck named *rubian*, was investigated by him two years earlier.   He found that by soaking
the cortical tissue of the root in water he could extract from it
a body which on the addition of either protochloride of tin, or
acetate of lead, yielded a beautiful purple precipitate.   After
separating this from the liquid by filtration, on suspending it
in water and passing a stream of sulphuretted hydrogen through
it, the sulphide of tin which was formed carried down with it part
of the colouring matter, leaving another part in solution.   The
mixture of these two had been previously described by Kuhl-
mann as a single colouring matter under the name of *Xanthine*.

The precipitated sulphide of tin with its adherent colouring
matter when collected on a filter and well washed with cold
water, gave up to boiling alcohol the colouring matter alone,
forming a yellow solution.   This on evaporation deposited the

pigment, which Schunck named rubian. When it was hydrolysed by dilute mineral acids, the beautiful purple colour
returned and was proved to be due to the formation of *alizarin*,
glucose and other bodies of complex composition being also
present in the solution.

The enzyme, like the glucoside, was prepared from a watery
extract of the madder root. A quantity of the latter was ground
and placed upon a layer of fine canvas and extracted with a
large volume of distilled water, 4 quarts being used for each
pound of ground root. During the process the temperature of
the liquid was maintained at 38° C. The watery extract was
added to an equal bulk of alcohol, when a brown flocculent
precipitate separated out. This was separated by decantation
and washed on a filter with alcohol till the latter failed to extract
any more colouring matter. A reddish-brown mass then remained on the filter.

This crude material contained the enzyme, mixed however
with various pectic bodies and probably some *pectase*, a peculiar
enzyme which will be described in a subsequent chapter.

When a quantity of it was added to a solution of rubian and
left at the ordinary laboratory temperature for a few hours, the
liquid became a jelly of a light brown colour, tasteless, and
insoluble in cold water. After standing for a time, any water
which was passed through it remained uncoloured. Schunck took
this point to be that at which the action of the enzyme ceased.
The liquid remained neutral in reaction, and there was no evolution of any gas. The jelly in all probability resulted from
the action of the pectase on certain constituents of the extract,
as we shall see later. In addition to the antecedent of the
jelly the extract contained certain substances formed from the
rubian. By appropriate treatment Schunck separated from it
alizarin, the purple colouring matter which is formed from
rubian by acids, *verantin* and *rubiretin*, two bodies of resinous
nature, *rubiafin*, *rubiagin*, and *rubiadipin*, the last named being
of a fatty character. The water with which the jelly had been
washed contained glucose and certain pectic bodies.

The decomposition was apparently of a very complex character, for under various modifications of the treatment, the

proportions of the various resulting products were found to vary considerably. Under the conditions described the sugar and the rubiafin and rubiagin were in greatest amount, and the alizarin in smallest, the quantity of rubiretin and verantin being intermediate between the others. When a trace of sulphuric acid was added to a solution in which the action was proceeding, the bulk of the product consisted of rubiretin and verantin, a little alizarin and rubiagin also being formed. When the liquid was made alkaline instead of acid, carbonate of soda being used in small quantities, there was formed a large amount of rubiafin, more than the average quantity of alizarin, a moderate amount of rubiretin and verantin, and no rubiafin. Excess of caustic alkali produced the same result as dilute sulphuric acid. The more the action was retarded by any cause, the more rubiretin and verantin and the less alizarin resulted.

From his numerous experiments under varying conditions Schunck came to the conclusion that the decomposition involved three portions of rubian. The first lost water and gave rise to alizarin. The second also lost water, but produced rubiretin and verantin in equal proportions. The third only was truly hydrolysed, taking up water and producing sugar and rubiafin, or with more water, rubiagin. The rubiadipin he found to be present in very small quantity, and he held both its origin and nature to be very uncertain. Schunck attributed all these actions to the influence of the enzyme, which he named *erythrozyme*.

This body is insoluble in water, but remains in suspension in it, and can be thrown down from such a suspension by the addition of acetate of lead or bichloride of mercury. Its activity is destroyed by heating it to 100° C. in the presence of water.

Erythrozyme is apparently related to emulsin, for the latter enzyme has also a limited power of action on rubian. Schunck used a preparation of emulsin from bitter almonds, which he obtained by precipitating the watery extract of the seeds by alcohol, after getting rid of the oil which they contain. He found that emulsin formed alizarin, verantin, and rubiretin from the rubian, the alizarin being in greater quantity than when erythrozyme was employed. He has made no statement as to the occurrence of either rubiafin, rubiagin, or glucose in his emulsin digestions.

Schunck considered that though the two enzymes were allied, emulsin had a feebler action than erythrozyme.

Rhamnase.

A fourth enzyme also belonging to this group has a still more limited distribution than either of those so far described. It occurs in the seeds of *Rhamnus infectorius*, the so-called *Persian berry*, a species whose fruits yield a brilliant yellow dye. This enzyme, which may be called *rhamnase*, has been investigated by Marshall Ward and Dunlop, and more recently by C. and G. Tanret. The fruits contain a glucoside which is known as *Xanthorhamnin* and which has the formula $C_{48}H_{66}O_{29}$. When decomposed it yields *rhamnetin* or *rhamnin* and two sugars, *rhamnose* and *galactose*. If the pulp of the fruits, or an extract of the pericarp is digested with an extract of the seeds and kept for a short time at a temperature of 35° C. a copious yellow precipitate falls, which consists of the rhamnin, the sugars remaining in solution. Boiling the extract of the seeds destroys its power of producing the precipitate. Very careful histological investigations have proved that the enzyme is confined to the raphe of the seed, which is composed of parenchymatous cells, containing a brilliant oily-looking, colour-less substance. The cells exhibit two or three large vacuoles in which a few brilliant granules can be observed. When Ward and Dunlop prepared a solution of xanthorhamnin from the pericarp of the fruit and floated a small portion of the raphe of the seed upon its surface, the action of the rhamnase became visible in the course of a few minutes. In ten minutes the floating piece of tissue was covered with a golden-yellow precipitate of the colouring matter; in twenty minutes clouds of the same precipitate were sinking through the solution, and in less than an hour the bottom of the test tube which con-tained it was covered by a layer nearly two millimetres deep.

Rhamnase has only been observed in the species alluded to.

It can be precipitated by the addition of alcohol to the cold water extract of the fruit. Its optimum temperature is 70° C. and it is destroyed on heating its solution to 85° C.

Gaultherase (Betulase).

The fifth enzyme of this group which calls for notice has been the subject of investigations by Schneegans and by Bourquelot. It was detected by the former of these observers in the bark of *Betula lenta* and he gave it the name of *betulase*. Bourquelot found it independently in several plants, among which may be mentioned especially *Monotropa hypopythis*, *Gaultheria procumbens*, *Spiræa Ulmaria*, *S. Filipendula*, and several species of *Polygala*. It hydrolyses *gaultherin* with the formation of *methylsalicylic acid* (oil of winter-green) and *glucose*, the following equation representing the reaction:—

$$C_{14}H_{18}O_8 + H_2O = C_6H_4OHCOOCH_3 + C_6H_{12}O_6$$
gaultherin   methylsalicylic acid   glucose

Bourquelot has named it *gaultherase*, from the name of the glucoside.

Gaultherin was first discovered in 1844 by Procter in the bark of *Betula lenta*; he was not successful in preparing it in a pure condition, but he ascertained that on decomposition it yielded methylsalicylic acid, and he stated that the same bark contained a ferment which was capable of splitting it up. Fifty years later Schneegans and Gerock prepared the glucoside from the same source but by a method which enabled them to get it in a crystalline form and to determine its formula.

In the same year Bourquelot found that the oil of wintergreen could be prepared from several species of Polygala (*P. vulgaris*, *P. calcarea*, and *P. depressa*) and from *Monotropa hypopythis*, a parasite which grows upon the roots of several trees, especially some species of Pinus. Monotropa is the typical genus of the Natural Order *Monotropaceae*, which is sometimes considered only a sub-order of *Ericaceae*.

On bruising the roots of the species of Polygala or the stems of Monotropa the characteristic odour of the oil of winter-green is very soon perceptible. Bourquelot states that it arises in consequence of the decomposition of the glucoside under the influence of an enzyme.

Bourquelot prepared both the glucoside and the enzyme from the stems of Monotropa and ascertained that though both are present, they are not found in the same cells, and that consequently the methylsalicylic acid noticed soon after bruising the tissues does not exist in the plant during its intact condition but is formed as a result of the injury.

He prepared the enzyme by the following treatment:— recently gathered Monotropa plants were triturated in a mortar with washed sand ; the mixture was digested for half-an-hour with 95 per cent. alcohol and the latter removed by filtration. The contents of the filter were then washed with alcohol and subsequently with ether, and dried in contact with the air. The powder so produced contained the enzyme.

It is of course apparent that the latter is not in any degree isolated by this method. The powder must contain all the constituents of the plant except such as are soluble in alcohol and ether. As the glucoside however is soluble in the former medium, the method can be relied on to separate the two bodies which take part in the hydrolysis.

Bourquelot found that a little of the powder when added to an aqueous solution of gaultherin very speedily produced the odour of methylsalicylic acid.

Schneegans states that the activity of the enzyme is increased by the presence of small quantities of alkalis or mineral acids.

Gaultherase is present in considerable quantity in the leaves and berries of *Gaultheria procumbens*. The characteristic product of its action, methylsalicylic acid, can also be easily extracted from the same parts of the plant by distilling them with water.

Bourquelot has prepared the enzyme also from the roots of *Polygala calcarea* and *P. vulgaris*, the roots of *Spiræa Ulmaria*, *S. Filipendula*, and *S. salicifolia*, and from the petals and leaves of several varieties of *Azalea*.

Bourquelot has confirmed the early work of Procter and the

researches of Schneegans, showing that the same enzyme exists in the bark of *Betula lenta*. It does not appear to be quite certain that the glucosides of Monotropa, Polygala and the other plants are identical with the gaultherin prepared from Betula by Schneegans and Gerock but it seems probable, as in all cases the same enzyme effects its decomposition and the oil of winter-green appears as one of the products of the hydrolysis.

Tannase.

The fermentation of tannin has been examined by Strecker and by Van Tieghem. The former observer showed it to be a glucoside and to undergo hydrolysis according to the equation

$$C_{27}H_{22}O_{17} + 4H_2O = 3C_7H_6O_5 + C_6H_{12}O_6$$
$$\text{Tannin} \qquad \text{Gallic acid} \quad \text{Glucose}$$

Van Tieghem showed that certain moulds can decompose tannin, gallic acid being one of the products.

Recently Fernbach and Pottevin, working independently and in ignorance of each other's researches, have ascertained the presence of an enzyme in one of these moulds, *Aspergillus niger*, and have proved the hydrolysis to be due to its action. Both prepared the enzyme by cultivating a quantity of the fungus in Raulin's solution, in which tannin in the proportion of about 3 per cent. replaced the sugar of the normal fluid.

Fernbach prepared the enzyme from the crop of fungus by the method adopted by Lintner for the preparation of diastase, which has already been described, obtaining it as a grey powder, soluble in water. Mixing some of this with a 10 per cent. solution of tannin and keeping it at 50° C. all the tannin disappeared and on cooling the vessel a precipitate of needle-shaped crystals of gallic acid rapidly followed.

Pottevin made an extract of the mycelium in chloroform water and sterilised it by filtration through a Chamberland filter. He then prepared two series of sterilised tubes, in each of which he put 10 c.c. of a 30 per cent. solution of tannin which had been similarly filtered. To each he added 10 c.c. of the sterilised extract, and then boiled one series for several minutes. He then sealed them all with the blow-pipe and kept them at 35° C. After a few days the unboiled tubes contained a deposit of gallic acid while the boiled controls remained unaltered.

x] GLUCOSIDE-SPLITTING ENZYMES. LOTASE. 169

Pottevin has found the decomposition to be attended with a formation of glucose, the quantity depending on the purity of the tannin used. It varied from 15 to 98 per cent. of the gallic acid simultaneously formed.

Fernbach also found he could filter his extracts through a Chamberland filter without destroying their activity.

Pottevin says that tannase can be precipitated by alcohol; acts in neutral or faintly acid media, has an optimum temperature of 67° C. and can completely decompose the tannin in a solution.

Tannase attacks not only tannin but the compound of tannin and gelatin, as well as other tannates; also the salicylates of methyl and phenyl.

Pottevin has prepared the enzyme not only from Aspergillus but also from the sumac (*Rhus*).

Lotase.

Another recently discovered enzyme of this group, like rhamnase, has only been observed at present in a single plant. The latter is *Lotus Arabicus*, a member of the Leguminosæ, which abounds in Northern Africa, Egypt, and the Nile Valley. The plant is known to the natives of those parts by the name of *Khuther*. It is one of the few poisonous members of the Natural Order. It has for some time been known that at certain stages of its growth, and particularly just before its time of flowering, its vegetative parts, especially the young leaves and stems, are very deleterious to horses and cattle feeding upon them, fatal results being not infrequent. During the year 1900 an extended investigation of the plant was conducted in the scientific department of the Imperial Institute by Messrs Dunstan and Henry, which led not only to the identification of a particular glucoside, but also to the recognition of a new enzyme, provoking the decomposition of the latter with the production of hydrocyanic (prussic) acid. The name *lotase* has been given to the enzyme by its discoverers.

When the leaves are bruised with water they are found to give off prussic acid with considerable readiness.

The plant contains in certain of its cells the glucoside referred to, which is a yellow crystalline body having the empirical composition $C_{22}H_{19}NO_{10}$.

On hydrolysis by dilute acids after separation from the plant, this body yields hydrocyanic acid as one of the products, together with glucose and a new yellow colouring matter, which has received the name *lotoflavin*. It appears to be isomeric with the pigments which are characteristic of *Reseda luteola*, and *Rhus cotinus*.

Dunstan and Henry have named the new glucoside *lotusin*. They consider the action of the enzyme like that of dilute acids to be one of hydrolysis and they represent it by the following equation:

$$C_{22}H_{19}NO_{10} + 2H_2O = C_{15}H_{10}O_6 + HCN + C_6H_{12}O_6$$

　　　Lotusin　　　　　　　Lotoflavin　　Prussic　　Glucose
　　　　　　　　　　　　　　　　　　　　　　acid

In the plant the glucoside and the enzyme appear to be contained in different and distinct cells of the tissue of the leaves, so that they do not come into contact with each other till the latter are injured. Lotase is rapidly destroyed by contact with alcohol.

Besides these enzymes, several others are known to occur in different plants, but they have not been so completely investigated. One of them is described by Schützenberger as being capable of hydrolysing *phillyrin*, a glucoside existing in the bark of *Phillyrea latifolia*, and *populin*, which is found in that of the *Aspen*. These glucosides are not attacked by yeast nor by the emulsin of the almond but appear to be hydrolised by the lactic Bacterium.

The enzyme does not appear to have been isolated.

Jorissen and Hairs have described some experiments with a species of *Linum* which point to the presence of an enzyme differing somewhat from emulsin but yet able to hydrolyse amygdalin. Linum contains a glucoside to which the name *linamarin* has been given. On hydrolysis it yields hydrocyanic acid, a fermentable sugar, and a body related to the ketones. The hydrolysis cannot be effected by the emulsin of the almond, but is brought about by an extract of the seed of Linum. The latter also hydrolyses amygdalin, as already mentioned.

Bréaudat has discovered another glucoside-splitting ferment in the leaves of *Isatis alpina*, which decomposes indican, giving rise to *indigo-white*, and a peculiar sugar known as indiglucine. It resembles emulsin in being able to split up amygdalin, but the author does not say whether the two are identical. It does not seem improbable that this is the case as Beijerinck has since shown that emulsin can hydrolyse indican.

It has recently been claimed by Sigmund that the enzymes of this group have also the power of splitting up fats into glycerin and free fatty acids. He says that he caused myrosin and emulsin to act upon olive oil in closed glass vessels at a temperature of 38°—40° C. and that free fatty acid was gradually and continuously developed in the mixture, its presence being demonstrated both by litmus and by phenolphthalein.

His mode of preparing the enzymes is however open to criticism. He bruised seeds of the mustard in the one case, and of the almond in the other, and allowed them to macerate in a quantity of water for twelve or fourteen hours. He then decanted the supernatant fluid and added an excess of alcohol, throwing down a precipitate, which he removed by filtration, washed and dried at about 40° C. This method is not likely to isolate either myrosin or emulsin; if any other enzyme, *e.g.* a fat-splitting one, was present in the seeds as well as either of the former, it would certainly have been present in his dried residue. Though hitherto no one has isolated a fat-splitting enzyme from these seeds there seems to be reason for suspecting its presence, as both mustard-seeds and almonds contain oil. Sigmund further states that certain fat-splitting enzymes which he detected in various seeds were able to split up amygdalin and salicin. The same criticism may be applied to this statement. The mode of extraction was similar to that described above, and it is at least possible that his residues contained two enzymes, rather than one as he supposed.

How far these various glucosides are of value in the nutrition of the plant is uncertain. No doubt the sugars which they yield on hydrolysis can be utilised in its metabolism, but whether the other residues can be applied to nutritive purposes has not been determined. Some researches recently published by Treub seem to show that in the case of one plant at least, *Pangium edule*, some compounds of hydrocyanic acid are made use of. Further investigation is necessary on these points.

CHAPTER XI.

PROTEOLYTIC ENZYMES. PROTEOLYSIS.

THE next group of enzymes which must come under consideration comprises those which effect the decomposition of proteids. They occur in both the animal and the vegetable organism and play a very important part in the metabolic processes. Like the diastasic ferments, they may either exert their action in the interior of the cell, as in the case of most vegetable organisms, or they may be extruded in solution in special secretions, as in the alimentary canal of animals.

In examining their action we are at once confronted with greater difficulties than in the case of the diastasic enzymes. The constitution of starch has been at any rate approximately determined and the sequence of changes which it undergoes on zymolysis can be expressed in the form of chemical equations, which certainly appear probable if they cannot be regarded as finally established. But in the case of proteids we meet with similar difficulties to those presented in the case of cellulose. We have no conception at present of their chemical constitution and cannot represent them by any formula. As in the case of cellulose again we are not dealing with a single substance, but with a large group the members of which show considerable differences in behaviour and are no doubt differently related to each other.

We find that certain proteids, such as albumins and globulins, can be split up in various ways by different reagents, and that as a result of such splitting other proteids are formed, which we have reason to think have a simpler composition than

either albumin or globulin. We find further that on very profound decomposition certain of these can give rise to crystalline
bodies which are not proteid, but which belong to the group
of substances known by chemists as *amido-acids*. All these
bodies occur naturally both in the animal and the vegetable
organism. The possibility of a gradual decomposition of a
primary proteid such as an albumin appears thus to be indicated, resulting in the appearance of a certain number of
derived or secondary proteids, which can be broken down finally
into crystalline chemical compounds, the constitution and relationships of which have been fully ascertained.

But this representation of the decomposition appears very
crude when compared with the hydrolysis of starch, in which we
have been able to see group after group of atoms successively
split off and the subsequent fate of each group fairly satisfactorily
traced.

The decomposition of the proteid molecule has been effected
by the action of dilute mineral acids at moderate and at high
temperatures, and different products have been obtained in each
case.

When an albumin is subjected to a prolonged digestion with
·25 per cent. of hydrochloric acid at 40° C. the liquid after a
while is found to contain at least four products of decomposition, known as *antialbumate, antialbumid, hemialbumose* and a
variety of *peptone*. Antialbumate can be separated from the
rest by careful neutralisation of the liquid with a weak alkali.
Gillespie says it falls just before complete neutralisation. It
is soluble only in faintly acid and alkaline solutions and differs
from albumin and globulin by not being coagulated when
its solutions are boiled. Antialbumid appears to result from
the further action of the acid on the antialbumate; it is almost
or quite insoluble in acid liquids, so that it separates out as a
granular-looking residue as the action of the acid proceeds. It
is soluble in a somewhat stronger solution of an alkali, such as
sodic hydrate of 1 per cent. concentration. Hemialbumose
does not appear to be derived from either of the other two. It
is soluble in acid and in neutral solutions, so that it is not thrown
down on neutralisation. It is precipitated on the addition

of a small quantity of nitric acid, or of acetic acid in the presence
of potassic ferrocyanide, and the precipitate dissolves on warming
but reappears on subsequent cooling of the liquid. The peptone,
sometimes called *hemipeptone*, appears to result from the further
decomposition of this body. It is soluble in acid, alkaline, or
neutral solutions, and is not precipitated on addition of nitric
acid or of acetic acid and potassic ferrocyanide, nor is its solution
coagulated on boiling. Further reactions of all these bodies
will be given subsequently.

The decomposition effected by dilute hydrochloric acid at
40° C. is represented by the following scheme by Kühne, to
whose researches much of our knowledge on this point is
due :—

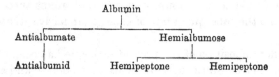

A more complete decomposition is brought about by 3—5
per cent. of sulphuric acid when the digestion is conducted at a
temperature of 100° C. The first splitting into two groups
appears to be the same, but either no antialbumate is formed
or it is at once converted into antialbumid. On the side of the
hemi-groups, both hemialbumose and hemipeptone are formed,
and the latter is still further decomposed, with the formation of
various amido-acids and other bodies, among which *leucin* and
tyrosin are conspicuous. The decomposition is represented by
Kühne as follows :—

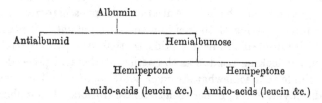

The idea which Kühne has propounded that the proteid
molecule is a double one and may be separated into an anti-
and a hemi-moiety, each of the latter then undergoing further

and different decompositions, is supported by the observations
of Schützenberger on the action of boiling mineral acids on
proteids. As a result of his experiments he comes to the conclusion that only half the proteid molecule is easily decomposed
by the acid, the other remaining in a somewhat modified
condition, but yet comparatively unaltered. This moiety
Schützenberger called *hemiprotein*; it is the same body as
Kühne's antialbumid.

There has been a good deal of controversy about the nature
of the decompositions effected by the proteolytic enzymes. As
we shall see later there are two groups of these bodies, the
types of which are respectively the pepsin of the gastric secretion
of mammals and the trypsin of the pancreas. Each possesses
well marked peculiarities; while they resemble each other in
the general features of their action, they differ in the extent
to which they carry the decomposition of the primary proteid.

Neglecting for the present the earlier researches upon these
enzymes, the first investigations which call for discussion are
those of Meissner and his pupils, carried out during the years
1859—62. While Meissner recognised the power of an extract
or infusion of the pancreas to split up proteids, he carried out
his work chiefly with pepsin in the presence of a small percentage of hydrochloric acid. He digested various primary
proteids with these reagents for varying times at the temperature of the body, about 38° C., and examined the various
products of the several decompositions. On neutralising the
mixture a precipitate settled out just before the neutral point
was reached, which possessed properties resembling those of the
body already described as having been found later by Kühne
and named by him antialbumate as already mentioned. Meissner called this body *parapeptone*. After its removal by filtration,
the addition of a trace of acid sometimes yielded a small amount
of another precipitate, to which he gave the name *metapeptone*.
When casein was the proteid under examination, a small quantity
of a different proteid was formed, which he called *dyspeptone*.
The liquid of the digestion, after being freed from these bodies,
contained three other proteids which he considered to be

varieties of peptone and which he accordingly named a, b, and c peptones.

The reactions of these bodies are described as under:—

(1) Parapeptone.   Soluble in dilute acids or alkalis; solution not coagulated on boiling. Precipitated on neutralisation. Incapable of being converted into peptone by pepsin, but digestible by an extract of the pancreas.

(2) Metapeptone.   Insoluble in very dilute acids ($\cdot$05 to $\cdot$1 per cent.), soluble in $\cdot$2 per cent.

(3) Dyspeptone.   Insoluble in dilute acids; soluble in dilute alkalis.

(4) a-peptone.   Soluble in acid, alkaline or neutral liquids, not coagulated on boiling; precipitated by strong nitric acid, and by potassic ferrocyanide in presence of dilute acetic acid.

(5) b-peptone.   Much like (4) but not precipitated by strong nitric acid. Precipitated by potassic ferrocyanide in presence of an excess of strong acetic acid.

(6) c-peptone.   Like (4) and (5) but not precipitated by either strong nitric acid or potassic ferrocyanide in the presence of any proportion of acetic acid.

Meissner's contemporary opponents failed to confirm his results except with regard to the occurrence of his parapeptone and c-peptone. Meissner claimed that all the bodies he described were due to the specific action of the pepsin upon the primary proteids. The view which came to be generally accepted, largely owing to the influence of Brücke, was that the two bodies mentioned were the sole products of a peptic digestion; that the former, parapeptone, resulted rather from the action of the acid used; and that the pepsin was capable of transforming it into peptone at a subsequent stage of the digestion.

Later work however has shown that Meissner's results come nearer the truth than his opponents would admit.

The subject was again taken up several years later by Kühne and his school, who investigated with great care and thoroughness the decomposition of a primary proteid by both

pepsin and trypsin, the former being as we have seen essentially
an acid digestion, the latter in the animal body being equally
certainly an alkaline one. Kühne found that in the early
stages of both a proteid was produced which was precipitated
on neutralisation of the digesting fluid, falling just before the
neutral point was reached. He found further that the final
products differed when trypsin was employed instead of pepsin,
the amido-acids leucin and tyrosin appearing only in the former
case. However long the digestion was continued the conversion
of the proteid into amido-acids was never complete, the final
products being a mixture of these with peptone and with hemialbumose, the same substances as he obtained under the influence
of acids. The final product of a peptic digestion was similar
except that no amido-acids were produced.

Kühne was led to the view that the decomposition of proteid
is essentially a process of hydrolysis, comparable with that of the
splitting up of cane-sugar under the influence of invertase, and
as the latter decomposition yields two different but similar
sugars, so he suggested that the hydrolysis of proteids leads to
the formation of two forms of peptone which show certain
differences from each other. This idea was supported by the
constant presence of some peptone side by side with the amido-
acids resulting from a long-continued tryptic digestion. Kühne
held that the amido-acids were derived from the decomposition
of one moiety of the peptone formed, and that the other moiety
was incapable of such a splitting up and hence was present unchanged however long the digestion lasted.

Taking a quantity of the peptone formed during a peptic
digestion and submitting it to the action of trypsin Kühne
found his views confirmed by the impossibility of converting the
whole of it into amido-acids.

He ascertained that the hemialbumose which was present
in both digestions was capable of being transformed by pepsin,
partially at least, into a peptone which could be further
converted by trypsin into leucin and tyrosin, while the neutralisation product of the earlier stages of the digestion yielded a
peptone which was not capable of such further decomposition.
Thus he was led to formulate the decomposition of proteid by

the two enzymes severally as under ; the successive stages being
in each case hydrolytic :—

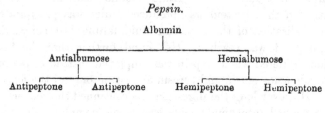

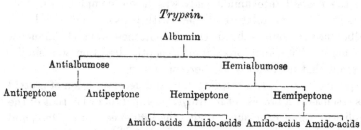

The result of either peptic or tryptic digestion is thus to
split the primary proteid into two residues, each of which gives
rise on further decomposition to a group of compounds. The
first or *anti-residue* is peculiarly resistant to decomposition and
never gives rise under natural conditions to a simpler body than
a peptone; the other or *hemi-residue* behaves differently with
pepsin and with trypsin, the former only converting it into
peptone, the latter decomposing the peptone into leucin, tyrosin,
and other bodies.

The reactions of these products of the cleavage of a primary
proteid may be briefly described as under :

Antialbumose. Soluble in faintly acid or alkaline fluids but
insoluble in neutral ones. Not coagulated on boiling.
Capable of being converted into peptone by either pepsin
or trypsin, the peptone being antipeptone.

Hemialbumose. Soluble in neutral, acid, or alkaline fluids.
Not coagulated on boiling. Precipitated by a small
quantity of nitric acid, by potassic ferrocyanide in
presence of a trace of acetic acid, or by a few drops of a
saturated solution of sodic chloride in presence of strong

acetic acid, the precipitate in each case disappearing on warming the liquid and reappearing as it cools.

Antipeptone. Soluble in neutral, alkaline, or acid solutions which are not coagulated on boiling. Not precipitated by mineral acids nor by acetic acid and potassic ferrocyanide. Incapable of decomposition into amido-acids by trypsin.

Hemipeptone. Differing from antipeptone only in being capable of conversion into leucin, tyrosin &c. by trypsin.

Both peptones can be separated from the albumoses by saturation of a boiling solution with ammonium sulphate, which precipitates all the latter.

A mixture of these two peptones, such as is present in a peptic digestion, is spoken of as *amphopeptone*.

This scheme of hydrolysis though in its general lines accepted to the present time is not quite complete. Further research has shown that it needs modification in some important particulars, the albumoses mentioned not being such definite bodies as Kühne supposed. If antialbumose is prepared and subjected to the influence of either pepsin or trypsin a peculiar insoluble residue always appears during the digestion, which gives the same reactions as the antialbumid resulting from the action of acids. The greater part of it can be converted with some difficulty into antipeptone by trypsin but not by pepsin. The hemialbumose has also been shown to be a mixture of several bodies which are formed successively rather than simultaneously. The decomposition products of the hemipeptone are more varied than Kühne supposed. His researches however mark a very distinct advance in our knowledge of proteolysis, though the changes are more gradual than his theory suggests, as it was first propounded.

It will be interesting here to compare the bodies recognised by Kühne with those described 15 years earlier by Meissner. When we consider their reactions we see that the latter's parapeptone resembles antialbumose in every respect except that Meissner describes his body as incapable of peptonisation by pepsin. Dyspeptone is probably identical with antialbumid, though Meissner's body was most likely mixed with nucleins. Hemialbumose agrees very closely with Meissner's *a*-peptone, though the reactions of the latter have not been very fully described. The *c*-peptone of Meissner was what is now called amphopeptone, a mixture of the two peptones of Kühne's scheme.

Further researches on the hemialbumose carried out by Kühne and Chittenden have shown that it is not a definite body, but a mixture of four closely allied but distinct albumoses. The name *albumose* is being abandoned in favour of *proteose* for the members of the group, the latter term indicating that they are the results of decomposition of primary proteid matter in general and not of albumin especially. The proportions of the different constituents in any given amount of hemialbumose vary according to the conditions of preparation of the latter. The four characteristic proteoses are the following:—

(1) Proto-proteose. Soluble in hot or cold water and dilute neutral saline solutions; precipitated by saturation of the solution with sodium chloride or magnesium sulphate.

(2) Hetero-proteose. Insoluble in water and therefore precipitable by separating the salt from its solutions by dialysis. Soluble in solutions of sodium chloride and precipitated from these on addition of the salt to the point of saturation.

(3) Deutero-proteose. Soluble in water; not precipitated by saturation with sodium chloride unless an acid is added at the same time. Precipitated by saturation of its solution with ammonium sulphate. This body approaches the peptones most nearly.

(4) Dys-proteose. This is very much like hetero-proteose, from which it differs only in being insoluble in salt solutions. It is probably produced from hetero-proteose by a species of coagulation.

All the proteoses can be separated from peptone by saturating a solution containing them with ammonium sulphate, or, according to Bömer, with zinc sulphate. The saturation must be very complete and the liquid must be made successively neutral, acid, and alkaline. Saturation throws down the proteoses.

A modification of Kühne's scheme of hydrolysis has been

advanced by Neumeister. According to him the proteoses do
not all come from the hemi-residue of the early hydrolysis, but
some of them are derived from the anti-group. On this theory
albumin may be considered to be composed of hemi-albumin
and anti-albumin. The first stage of the decomposition is
the splitting of hemi-albumin into proto-proteose and hetero-
proteose, while the anti-albumin gives rise to acid-albumin
and hetero-proteose. The next step is the conversion of the
proto-proteose and both hetero-proteoses into deutero-proteose,
and the final stage is the production of anti- and hemi-peptones
from the members of the separate groups.

Neumeister says that in the hydrolysis brought about by
trypsin proto- and hetero-proteose are not formed, but that
several deutero-proteoses appear as the first stage of the action.

During the past decade very careful and detailed investigations of the transformation of proteids under the influence of
pepsin and hydrochloric acid have been carried out by Chittenden
and his pupils, which point to the process of hydrolysis being
even more gradual than this. Some experiments upon the
digestion of the crystallized globulin or *vitellin* prepared from
hemp seeds show that if the digestion is carried out for different
periods, varieties of proto-vitellose can be detected, which are
exactly alike in their chemical reactions, but which differ in
their percentage composition and perhaps in their specific
rotatory power. The deutero-vitellose obtained under the
varied conditions also showed differences in its percentage composition. In the early stages of all the digestions Chittenden
noted the fôrmation of a body resembling Kühne's antialbumose and having the reactions of an acid-albumin; this
gradually disappeared, leaving an indigestible residue of antialbumid. On filtering this off the liquid contained only
proteoses and peptones. He states emphatically that he finds
no evidence of the production of peptone directly from the
acid-albumin but only from the proteoses.

Pepsin-proteolysis thus leads to the formation of a series of
products by hydration and cleavage, of which four are well
defined, viz. proto-proteose, hetero-proteose, deutero-proteose, and
amphopeptone. The proteoses may each represent a group of

closely allied bodies, so closely related as not to be distinguishable by chemical tests, and the exact composition of each group
may depend upon the length and intensity of the proteolytic
process.

These proteoses represent definite stages in the progress of
the hydrolysis; the proto- and hetero-bodies yielding deutero-
proteose and the latter becoming converted into peptone. This
conversion is not however complete, the amount of proteoses
and peptones existing together at the conclusion of a digestion
varying considerably. In a prolonged digestion of vitellin, of
10 days' duration, there was found to be present 23 per cent.
of proteoses and 77 per cent. of peptone. In another experiment
carried out on egg-albumin, after 17 days' action the proteoses
amounted to 44 per cent. and the peptone to 54 per cent.
The difficulty of the final conversion is not dependent on
an inhibitory influence on the action of the enzyme caused by
the presence of the products of the transformation as in the
case of the hydrolysis of starch into maltose, for the same
results were found in the case of digestions carried out in
dialysing tubes, which permitted the escape of the products
after their formation.

A comparison of proteolysis with the hydrolysis of starch
shows a similarity in another respect. The final product is one
which is capable of passing by dialysis through a moist membrane such as vegetable parchment. Sugar is completely
dialysable, starch not at all. The primary proteids such as
albumin and globulin are also incapable of dialysis—peptone on
the other hand can pass through a membrane with considerable
facility. The proteoses, seen to be intermediate between the two,
show also an intermediate power of dialysis. Hetero-proteose
is very feeble in this respect, deutero-proteose comes next, and
proto-proteose next, the two not showing much difference, and
all being much inferior to peptone. Kühne represents the
relative powers of dialysis as 5·22 for hetero-proteose, 24·1 for
deutero-proteose, 28·3 for proto-proteose and 51 for peptone.

According to Gillespie the comparative rates of dialysis of
proto-albumose, deutero-albumose, and peptone prepared from
egg-albumin are 2·5, 5·4 and 23·8.

The probability that the process of decomposition of proteids by both acids and enzymes is one of hydrolysis is not only supported by analogy with the action of other enzymes and by the general course of the action so far as it can be traced without actual knowledge of the molecular constitution of proteid matter, but also by comparison of the percentage compositions of the various products of the decomposition prepared as pure as possible with our present methods. The proportion of carbon becomes gradually less as we pass from albumin towards peptone, which corresponds to such an increase in the amount of hydrogen as would result from the incorporation of water into the molecule. Too much stress must not however be laid upon this, for our criteria of purity in the case of any proteid are not exact, and with such a large molecule as proteids must possess there is room for considerable error in such a computation.

The decomposition of proteid has still more recently been accomplished by submitting it to the prolonged action of water heated under pressure to temperatures considerably above 100° C. Under such conditions Neumeister has effected the splitting up of fibrin by water at 160° C., with the formation of other proteids, which he calls respectively *atmid-albumin* and *atmid-albumose*, there being at the same time a formation of peptone, and the evolution of a quantity of sulphuretted hydrogen, ammonium sulphide also being distinguishable in the solution. Both the new proteid bodies show resemblances to others of the groups resulting from the action of enzymes and both can be transformed into deutero-albumose by treatment with sulphuric acid. Chittenden and Meara similarly found that egg-albumin could be decomposed by heating it with water in sealed tubes to 150°—160° for several hours. Their resulting products were chiefly "atmid" bodies corresponding to those which Neumeister obtained from fibrin, but they distinguish two forms of atmid-albumose. They also detected the formation of peptone, leucin and tyrosin, while on opening the tubes the evolution of sulphuretted hydrogen and other disagreeable gases could be noticed.

The action of water at such pressures and temperatures as were employed is eminently hydrolytic. The similarity of the products obtained to those yielded on zymolysis greatly supports the view that the latter also is a hydrolytic process.

This view is further supported by the fact that it is possible by the action of dehydrating agents to convert peptone into a body resembling acid-albumin, which we have seen to be an intermediate product formed during the conversion of albumin or globulin into peptone. If 10 parts of dry peptone are mixed with twice their weight of acetic anhydride and the mixture heated for a long time to 80° C., and finally the excess of acetic anhydride distilled off and the residue dialysed, it is found to be changed into a proteid that is not diffusible, is soluble in dilute alkali, is precipitated by acetic acid and potassic ferrocyanide, and by many metallic salts, as ordinary proteids are. According to Hofmeister a similar effect may be produced by heating peptone for a long time to 140° C. The resulting brown mass contains a part soluble in water and another not so, which react after the manner of a globulin and a derived albumin respectively.

Other views of the relationship between ordinary proteids and peptone have been advanced, some observers believing that peptones are polymers and others that they are isomers of primary proteids. Adamkiewicz suggests that they differ in the removal of salts and a rearrangement of the molecule. Schützenberger suggested that peptone is a mixture, which, by treatment with phosphotungstic acid, can be separated into two parts. one containing a little more oxygen than the other, and both being ureïde bodies. Fibrin on the other hand he held to be a kind of compound ether, which is saponified by the enzyme, and on taking up water splits into the two bodies found. The transformation is thus one of hydration, being the result of the decomposition of an ether by saponification.

Kühne's cleavage theory of the hydrolysis of proteids is not now universally accepted. Objections to it are based on the fact that hemipeptone has never been isolated but remains a theoretical substance. The quantities of the various proteoses are not so regular as the theory suggests they should be. The nature of antipeptone is also disputed; Siegfried and Balke deny its proteid character, while Kutscher states that he has found it to be a mixture of hexone bases such as argenin and hystidine, with traces of tyrosin, leucin, glutamic acid &c. If this is confirmed we must consider that both hemi- and anti-groups can be converted into crystalline compounds of nitrogen, or that the hydrolysis of proteid is not a cleavage into two moieties but a series of successive decompositions.

CHAPTER XII.

PROTEOLYTIC ENZYMES (*continued*).

WE have seen in the preceding chapter that the enzymes which can effect the decomposition of primary proteids, such as albumin, fibrin, etc. fall into two categories, represented by the pepsin of the gastric secretion and the trypsin of the pancreas. They are alike in their power of splitting up the proteid molecule, probably inducing hydrolysis or decomposition after incorporation of one or more molecules of water. The first products of the cleavage of the primary proteid are probably the same in both cases, identical or at any rate corresponding secondary products being formed. The two ferments differ however in the extent to which the decomposition is carried out by them, the peptic enzyme never taking it beyond a point at which the final product is still undoubtedly a proteid, *i.e.* peptone; the tryptic ferment splits this up still further, with the formation of amido bodies such as leucin, tyrosin, asparagin, and others. That these bodies are products of the decomposition of proteid is shown further by the fact that in vegetable metabolism, they, or some of them, have been shown to be antecedents of proteid or to take part in its construction, disappearing from the cells as proteid is formed. The tryptic enzymes are accordingly to be regarded as much the more energetic. They are found to be of much wider distribution both in animals and plants, so much so that when only one proteolytic enzyme is present in an organism it is almost invariably a tryptic one. In the metabolic processes concerned in nutrition it seems necessary, or at any rate advantageous, that the proteid molecule should be broken up so far as to produce very much simpler bodies than itself, such indeed as may be compared in complexity with such carbohydrates as the sugars.

Pepsin.

This enzyme can be most conveniently obtained from the stomach of the dog or pig by removing the mucous membrane from the underlying muscular coats, and after finely mincing it allowing it to macerate in a solution of hydrochloric acid of about ·2 per cent. concentration. During the maceration most of the membrane undergoes digestion. After a time the liquid can be filtered from the undigested débris, when it will be found to be strongly proteolytic. Or the membrane may be extracted with glycerine, which has the advantage of not dissolving the proteids to anything like the same extent as the hydrochloric acid solution. As the maceration in glycerine can be conducted in the absence of the acid, digestion of the membrane takes place to only a very slight extent, so that the glycerine extract is much purer than the acid one. Neither method however yields the pepsin free from proteids. Pekelharing says that such an artificial gastric juice can be purified by dialysis, though there is a considerable loss in the process.

The pepsin of commerce is a very crude product; it is generally prepared according to a method first used by Scheffer, which consists of saturating with a neutral salt an acid extract of the mucous membrane of the stomach of the pig. As we have seen, this treatment separates from the liquid the proteoses of digestion; the pepsin adheres to them and can so be withdrawn in large part from the extract. The sticky precipitate is removed and dried at a low temperature, forming a pale yellow or buff-coloured powder.

The enzyme can however be prepared from the stomach in a relatively pure condition by taking advantage of a property it possesses of adhering to any inert precipitate formed in the liquid in which it is dissolved. The method which first yielded it in anything like a pure state was adopted by Brücke in 1861. He extracted the minced gastric mucous membrane with a

considerable quantity of water containing 5 per cent. of phosphoric
acid, and after straining off the débris filtered his extract and
added lime water almost to the point of neutralisation. A
heavy precipitate of calcic phosphate was thus obtained in the
extract and the pepsin was almost entirely carried down with
it, mixed however with a certain amount of proteid matter.
The precipitate was next dissolved carefully in dilute hydrochloric acid, avoiding great excess, and again thrown down by
addition of lime water. A certain portion of the contaminating
proteid escaped precipitation, so that the pepsin clinging to
the lime salt was much purer than before. It was again
dissolved in dilute hydrochloric acid, and a solution of cholesterin in a mixture of four parts of alcohol and one part of
ether was poured through a thistle funnel to the bottom of the
vessel containing the pepsin. Cholesterin is insoluble in water,
and therefore as the light liquid containing it rose towards
the surface, it separated out as a very fine precipitate. The
separation was facilitated by shaking the flask in which the
operation was conducted. The pepsin adhered as before to the
inert precipitate, which was removed by filtration and washed
with very dilute acetic acid and subsequently suspended in a
little water. The moist mass was then agitated with ether,
which dissolved the cholesterin and left the pepsin in solution
in the water. The latter was removed by means of a separating
funnel. It was necessary to repeat the extraction by ether
several times to ensure the removal of the whole of the cholesterin. The separation of the ether from the watery extract was
completed by exposure to the air and the extract filtered and
further purified by dialysis.

A modification of this process has been adopted by Maly,
who instead of using the method of separating the pepsin
by cholesterin, dialysed the hydrochloric acid solution till it
was free from chlorides and phosphates. By this means he got
rid of nearly all the admixed substances.

The methods adopted by von Wittich, Würtz, and Loew,
described in connection with the preparation of diastase, are
also applicable to the extraction of pepsin. Loew's lead method
was formerly used in the preparation of the pepsin of commerce

but was abandoned because it was found very difficult to remove the last traces of lead, and a great proportion of the pepsin was destroyed in the process of manufacture on the large scale.

Pekelharing adopted the plan of dialysing artificial gastric juice into distilled water till the acid and salts had disappeared. The pepsin was thus precipitated upon the dialyser. Collecting it and dissolving it again in dilute hydrochloric acid a purer product was obtained, much insoluble proteid being left behind. Further purification was secured by repeating the operation several times. Ultimately Pekelharing obtained a product sufficiently pure to undergo analysis. He found that it gave proteid reactions and contained phosphorus, the latter comprising less than 1 per cent. of the whole.

Other methods of preparation differing slightly from these have been described by Kühne, Petit, and Sundberg.

The chief seat of the formation of pepsin is the stomach of the Vertebrata, in the mucous membrane of which numerous closely-packed glands occur which secrete it in abundance. The whole of the mucous membrane is furnished with these glands, but its different regions do not supply equally active gastric juice. Most of the pepsin is yielded by the glands which are situated at the cardiac end. As the pylorus is approached by way of the greater curvature the glands produce pepsin in rapidly decreasing amount, and those in the pyloric region itself secrete it in very small quantity.

Pepsin is associated in the gastric secretion with another enzyme which curdles milk.

In some of the lower vertebrates the formation of pepsin is not confined to the glands of the stomach but is produced also by similar structures at the lower end of the œsophagus. A curious distribution of the enzyme has been described by Miss Alcock as occurring in the larva of *Petromyzon*. In this animal the alimentary canal consists of three regions, a pharynx, an anterior intestine, and an intestine proper. The most active preparation of the ferment that is yielded by the walls of the alimentary canal is derived from the pharynx, while those of the intestine contain little, if any, when cleansed from the secretion of the so-called liver, which is poured into that portion of the tube. The glandular structures of the wall correspond in their distribution to this variation of enzymic power. The chief seat of formation of the ferment is however the so-called liver, whose secretion is discharged during life into the intestine. Miss Alcock found that the skin of the larva also secretes the enzyme, the superficial layer of cells being the active ones.

The statement is made both by this author and by Krukenberg that the enzyme of the liver is pepsin. The evidence on which the statement is based is meagre, and room is left for doubt whether it is not trypsin, as is the case with the so-called liver of many invertebrates. Levy says, however, that this organ in *Helix pomatia* secretes pepsin, and Bourquelot claims to have proved the presence of pepsin in the liver of various Cephalopoda, in which however it is associated with trypsin.

Krukenberg found a proteolytic enzyme in the eggs of the fowl, which is active in the presence of ·1 per cent. of hydrochloric acid. He states also that the mesenteric filaments of the Actinozoa can digest fibrin. Hartog has stated recently that pepsin exists in the spawn of the frog. He confirms Krukenberg as to its occurrence in hens' eggs. Dixon and Hartog found it in *Pelomyxa*, an Amœba.

It is very doubtful whether a peptic ferment exists in the vegetable organism, most of the enzymes which have been investigated proving to be tryptic. The most notable case which is still uncertain is that of the secretion of *Drosera*, one of the insectivorous plants. Darwin, who first pointed out the existence of the enzyme, classes it strictly with the enzyme of the mammalian gastric juice, but as he did not examine in detail the products of its action, the matter cannot be regarded as definitely settled. A description of the enzyme of *Drosera* will be found in a subsequent section.

Besides occurring in the gastric secretion, pepsin has been found in the blood, muscles, and urine of the higher animals. Many physiologists consider that this is not normal pepsin, and that its occurrence there indicates the way in which so much of the enzyme as has been made use of is removed from the body. Bechamp on the other hand suggests that the pepsin in the blood, which can be extracted from fibrin, is formed by the leucocytes, and secreted by them into the blood-stream. This view has also been upheld by Leber, who claims for the leucocytes the property of softening and dissolving proteid matter in the tissues by the secretion of an enzyme. This view can however hardly be considered to be proved.

Pepsin exhibits a peculiarity in the conditions of its working by virtue of which it stands almost if not entirely alone among enzymes. It can only carry out its specific decomposition when it is in a state of combination with a weak acid. The pepsin of the stomach of the Vertebrata is always found combined with hydrochloric acid, the latter being found in gastric juice in the proportion of ·2—·5 per cent. So close is the union between the two that many physiologists prefer to speak of the proteolytic fluid as "pepsin-hydrochloric acid." Other mineral acids, especially nitric or phosphoric acid, can replace the hydrochloric. Organic acids can play the same part. Wroblewski has published the results of some very careful researches on this point, using pepsin from dogs, from pigs, and from the human subject. He finds that though acid is essential for the action of all, the different pepsins behave differently in the presence of different acids and are moreover different from each other. He prepared his pepsin by Witte's glycerin method from the mucous membrane of the stomachs in each case. Using pig's pepsin he found it to be most active in the presence of oxalic acid of such strength that one volume of normal alkali solution neutralised 20 volumes of the acid. Then in order came hydrochloric, nitric, phosphoric, tartaric, lactic, citric, malic, paralactic, sulphuric and acetic acids. With human pepsin lactic acid was slightly more efficacious than phosphoric.

The action of pepsin is influenced by several physical conditions, of which the most important is temperature. Klug states that it can act at 0° C., but is then very feeble; as the temperature rises the activity increases to an optimum which Klug puts at from 50° to 60° C.; it then falls off as the digestion is warmed still further, and at 80° C. it ceases and the enzyme is destroyed. Other writers have put the optimum point much lower, at about 38°—40° C., the normal body temperature.

Langley says that pepsin is rapidly destroyed by warming to 55°—57° C.

There appears to be some difference as to the optimum point between the pepsin of warm-blooded and that of cold-blooded animals. Hoppe-Seyler found the pepsin of the pike digested fibrin faster at 15° C. than at 40° C. Many observers have found however that this is not the case with the pepsin of the frog.

Dilution has a conspicuous effect upon the activity of pepsin; the best results are obtained when the digestive fluid contains from ·5 to ·01 per cent. of the enzyme, which is a fairly wide range. With greater or less concentration the activity lessens, though proteolysis will take place with only ·005 per cent. present. The pepsin of the dog is more active than that of the pig or the bullock.

Various neutral salts have been found to have a markedly retarding effect upon the action of the enzyme, especially ammonium sulphate and sodium chloride, the latter if present in greater amount than ·5 per cent. Alkaline salts are also prejudicial in a very high degree, sodium carbonate in such small proportion as ·005 per cent. causing an appreciable. destruction of pepsin in one to two hours at the body temperature. The destruction varies according to the strength of the alkali or alkaline salt, the time during which it is allowed to act, and the amount of proteids present, the latter being preservative up to a certain amount. The protection is however far from complete. Langley and Edkins found that in the presence of 2·5 per cent. of peptone, seven-eighths of the pepsin in an extract of cat's gastric mucous membrane was destroyed at 17° C. by ·5 per cent. of sodium carbonate in 60 seconds. Carbonic acid gas is also deleterious, but much less so than an alkali. Peptone is very efficacious as a protection against this gas.

A great difference may be observed between the effect of alkaline and that of neutral salts. While the former destroy the enzyme entirely, the latter only inhibit its action. If they are removed by dialysis or other means, the pepsin is found to be still capable of carrying out proteolysis.

Loew states that an inhibitory effect is exerted by the presence of 1 per cent. of quinine.

Chittenden and Amerman have found that the removal of the products of the action of pepsin during a digestion has little or no influence upon the course of proteolysis.

The action of pepsin has been chiefly studied upon fibrin or coagulated egg-albumin. The nature of the decomposition which it sets up in these cases has already been discussed. Babcock and Russell state that when it is made to act on casein the products of the digestion consist almost entirely of proteoses, scarcely a trace of peptone being found. Chittenden and Hartwell and later Chittenden and Mendel have studied the proteolysis of the crystalline globulin or vitellin which can be prepared from the seeds of the Hemp, and which may be taken as a representative vegetable proteid. The course of the action was found to be much the same as with fibrin or egg-albumin, syntonin or a form of acid-albumin being formed at the commencement of the digestion, together with a small amount of an insoluble residue corresponding to antialbumid. Later proto- and deutero-vitelloses and peptone could be detected as in the other cases, but only a very small quantity of hetero-vitellose was formed. By varying the length of time during which the digestion was carried on, Chittenden and Mendel identified two proto-vitelloses.

Harlay has recently observed among the products of decomposition produced by pepsin a peculiar chromogen, which, when acted on by tyrosinase, an oxidising enzyme present in the juice of *Russula delica,* gives a red colouration which slowly becomes green. When this chromogen is digested with trypsin it gives rise to tyrosin and tryptophane, a peculiar substance discovered in tryptic digestions by Neumeister, which turns red on treatment with chlorine water. The peptic chromogen is not decomposed by pepsin.

Besides the proteids, gastric juice by virtue of its pepsin can effect the decomposition of *gelatin.* Tiedemann and Gmelin were the first to observe that this body, under the action of the secretion of the stomach becomes liquid, and loses the power of gelatinisation. Metzler, Schweder, Etzinger and

other observers have obtained the same result with a solution
of pepsin.

The process of the decomposition of gelatin has been studied
in detail by Chittenden and Solley, and has been ascertained to
be almost exactly similar to the course of proteolysis by the
same enzyme. The chief difference between the two is the
absence of any neutralisation precipitate corresponding to
acid-albumin. The bulk of the products arising during the
course of the digestion consists of proto-gelatose and deutero-
gelatose, gelatin-peptone also being formed, but only in small
amount. These various bodies show properties very much like
the corresponding ones produced during digestion of albumin.

Dastre and Floresco have arrived at similar conclusions to
those of Chittenden and Solley. They also found similar
changes to those induced by pepsin to follow the action of
water at high temperatures under pressure.

Pepsin also acts upon the so-called *nucleo-albumins*, or
nucleo-proteids, which appear to be compounds of proteids
with a substance called *nuclein*, the latter being the principal
material found in the nuclei of cells. The enzyme appears
to attack chiefly the albumin of the compound peptones and
albumoses being formed, together with a phosphorised residue
which is probably slightly altered nuclein. Pepsin does not act
upon the substance of the nuclei.

The action of pepsin upon the *caseinogen* of milk is
interesting from this point of view. The latter has generally
been considered to be a proteid, much resembling the derived
albumins. Lubavin has found that when digested with pepsin
a phosphorus-containing body closely resembling if not identical
with nuclein is separated out as an insoluble residue. This fact
lends support to Hammarsten's view that caseinogen is rather a
nucleo-albumin than a true proteid.

Pepsin also digests *elastin*, but is apparently without action
on *mucin*.

Trypsin.

This is the characteristic enzyme of the pancreatic secretion of the higher animals and the chief agent in the proteolysis which takes place in plants. It differs from pepsin in two important respects. It effects as we have seen a more complete disruption of the primary proteid, breaking some of it down so completely that crystalline amide bodies result from the decomposition. Further it is not dependent like pepsin upon the presence of an inorganic ally such as hydrochloric acid. The trypsin of the pancreas acts most vigorously in the presence of a small amount of an alkaline salt such as carbonate of soda, but there is no definite relation between the alkali and the enzyme. Digestion by pancreatic juice can go on indeed in a neutral or even a faintly acid medium; Ewald says that the trypsin of the ox can digest fibrin in the presence of ·3 per cent. of hydrochloric acid. Mays obtained the same result. The vegetable trypsins again are most active in a faintly acid medium.

In the pancreatic secretion trypsin is associated with several other ferments, one of which, pancreatic diastase, has already been discussed. Among the other enzymes which it contains are a form of *rennet*, which curdles milk, and *pialyn* or *lipase*, which decomposes fats. The two latter will be considered subsequently. Pancreatic juice as ordinarily obtained contains also a considerable quantity of proteids.

Similar admixture with other enzymes is frequently found to mark the trypsin of vegetable origin.

The preparation of trypsin from the pancreas or its secretion is attended with difficulty, though a proteolytic extract of the organ can be obtained by digesting the minced tissue with a dilute acid or with glycerin, the operation being conducted at 35°—40° C. Such an extract contains however the other enzymes described as well as the trypsin.

The separation of the three especially digestive enzymes, trypsin, diastase, and pialyn or lipase, was first effected by Danilewski in 1862. He took a freshly extracted pancreas and ground it up in a mortar with sand and cold water. Removing

13—2

the débris he saturated the liquid with magnesic oxide, which caused the precipitation of the pialyn, associated with certain other constituents of the extract. The filtrate from this precipitate contained the diastase and the trypsin, and these were separated by shaking it with an alcoholic and ethereal solution of collodion. The precipitate of the collodion carried down the trypsin, leaving the diastase in solution. The two can be separated in another way, by acidulating the extract with phosphoric acid and then neutralising by lime water. The precipitate so caused carries down the diastase, leaving the trypsin dissolved.

Kühne prepared trypsin in what is generally regarded as an approximately pure condition by the following process. The minced tissue of a fresh pancreas was macerated in alcohol and ether and two extracts were prepared from it; the first was acid, the acid used being salicylic acid in the proportion of ·1 per cent. After maceration in this liquid for 4 hours at 40° C. the extract was separated from the tissue by pressure through a linen strainer. The residue was next extracted by a ·25 per cent. solution of sodium carbonate, the digestion as before being conducted at the body temperature; after 12 hours' maceration the liquid was again separated by pressure. The two extracts, which contained the enzymes together with a good deal of proteid matter, were then mixed, and sodium carbonate added to the whole till it contained about ·5 per cent. of the salt. The mixture was then allowed to digest in the presence of some antiseptic till the trypsin it contained had converted the proteids into peptones. A somewhat prolonged digestion was necessary, sometimes extending over a week. The liquid was then allowed to stand in the cold for 24 hours and filtered. It was next made faintly acid with acetic acid and saturated with neutral ammonium sulphate. As this reagent precipitates albumoses, it was very necessary that the preliminary digestion should be prolonged till these were converted into peptones. In the absence of albumoses, the ammonium sulphate precipitated the trypsin in a fairly pure condition and the latter was collected on a filter, washed and dissolved in a little alkali.

The method seems open to criticism to a certain extent.

It has been found by many observers that the total conversion
of the albumoses into peptones is a very difficult process and
probably never takes place, a certain portion of albumose always
remaining in a digestion. When Kühne examined the composition of his product he found it always gave up to boiling
water a certain amount of coagulated proteid and nearly 80 per
cent. of albumose. He attributed this to decomposition of the
enzyme itself, but it appears certainly possible that it may
have arisen from incomplete hydrolysis of the albumoses of the
original extracts.

While trypsin is the most potent proteolytic agent in
vertebrate animals it has also a very wide distribution among
the lower forms. The course of proteolysis has however been
examined only very incompletely among these and in many it
is not certain whether it is peptic or tryptic. Passing by these
for the present and dealing only with those whose secretions or
tissues have been investigated we notice that Fredericq has
prepared an enzyme from certain *sponges* which converts proteids into peptones and partially into leucin and tyrosin. He
has met with the same body in certain *Echinodermata*. In the
pyloric cæca of *Uraster*, which are situated in the rays of the
animal, there is found a digestive fluid which is strikingly like
pancreatic juice. It decomposes fats, saccharifies starch and
dissolves coagulated albumin. Fredericq extracted the enzyme
by the method of macerating the tissue in glycerin after
dehydration by alcohol. So prepared, it digested muscle-fibre
with the formation of leucin and tyrosin. He says it is most
active in alkaline media, less so in neutral fluids and possesses
hardly any power in acid extracts. In the *Cœlenterata* markedly
glandular cells are found in the endoderm, and the body-fluid
possesses digestive properties.

In *Worms* the intestine is partly surrounded by a yellowish
glandular tissue which is said to be pancreatic in function.
Fredericq extracted from the dehydrated tissue of the whole
body of the worm an enzyme which digested fibrin in both acid
and alkaline solutions of various strengths. The most advantageous one was ·6—1·2 per cent. of hydrochloric acid. Though
the extract when neutral was less potent, it was yet capable of

carrying out a slow digestion. Later Fredericq dissected out
the intestine of a large earth-worm and found in the cavity a
slightly alkaline fluid which was capable of digesting fibrin.
This fluid was the secretion of the glandular cells which have
been mentioned as covering the intestine (page 43). Griffiths
has found indol in the products of a proteid digestion effected
by the enzyme prepared from the worm. The evidence thus
points to a tryptic enzyme in the gland.

In *Insecta* there is also evidence of the presence of trypsin;
Griffiths has found the pyloric cæca of *Blatta* to be pancreatic
in function. In the stomach of both larva and imago of certain
Lepidoptera are glandular follicles which secrete a juice corresponding in proteolytic activity to that of a pancreas of the
Vertebrata. Abelous and Heim detected trypsin in the eggs of
certain *Crustacea*, in which diastase was also present, as already
mentioned (page 44).

The so-called *liver* of the *Crustacea* and the *Mollusca* appears
to be in part at all events a digestive gland of the type of
the pancreas. Bourquelot has studied it very carefully in the
Cephalopoda and concludes that it secretes both pepsin and
trypsin. Fredericq who has examined it in many families says
that in them all the fluid it secretes is capable of digesting
starch, splitting up fats, and dissolving fibrin with formation of
leucin and tyrosin. The enzyme can be prepared from the
tissue by dehydrating it with alcohol and subsequently extracting it with glycerin, according to the methods of von
Wittich or Krawkow which have been already described. In
the *Tunicata* the tubules which ramify over the wall of the
intestine are said to constitute a pancreatic gland.

The distribution of trypsin in the vegetable kingdom will
be treated of subsequently.

In the conditions of its action trypsin much resembles
pepsin; it is similarly affected by temperature, and by even
slight excess of acids or alkalis. Pancreatic trypsin is destroyed
by ·1 per cent. of hydrochloric acid and totally inhibited by
·05 per cent. of lactic acid. Neutral salts in excess affect it
much as they do pepsin.

The removal of the products of the digestion as they are

formed has a considerable influence upon the course of proteolysis. Lea has carried out a very careful research upon this point, using trypsin prepared by Kühne's method, and carrying out two operations side by side, the one in a glass flask and the other in a dialysing tube surrounded by a solution containing ·25 per cent. of sodium carbonate and a little thymol. He has found that the typical disintegration of fibrin in its earlier stages is much more rapid in the dialyser than in the flask, and that the digestion is never so complete in the latter case, more antialbumid being left at the conclusion of the experiment. On the other hand the amount of leucin and tyrosin formed in a flask-digestion is always greater than in a dialyser-digestion, other conditions being the same in both cases.

The course of the decomposition of primary proteids effected by trypsin has already been described. A difference between this enzyme and pepsin in the case of the proteolysis of fibrin may be noticed here. When fibrin is digested by pepsin it swells up under the influence of the hydrochloric acid with which the enzyme is associated and is subsequently dissolved, the fluid remaining clear until nearly the end of the digestion, when the resulting antialbumid is left as a granular residue. With trypsin the course of action is different. The fibrin does not become swollen or translucent, but is gradually corroded rather than dissolved, so that the digestive liquid is always turbid and full of a finely granular débris.

It has already been mentioned that the decomposition products arising from the tryptic digestion of hemipeptone are more varied than Kühne supposed. The amido-acids of the fatty series include not only leucin (amido-caproic acid) but *amido-valerianic acid, asparagin, aspartic acid* and *glutamic acid*. Hedin also obtained two basic substances from a pancreatic digestion, *lysin* and *lysatinin*. These were originally prepared from proteid by Drechsel, but he did not prove them to be the result of the action of trypsin. Hedin has since shown lysatinin to be a mixture of lysin and *argenin*, the latter of which can be made to give rise to urea. The amido-acids of the aromatic series are represented by *tyrosin*.

The tryptic digestion of fibrin has been shown by Hirschler

and by Stadelmann to be attended with the liberation of a small amount of ammonia.

Nencki identified among the products of tryptic digestion a certain quantity of *xanthin, guanin, hypoxanthin* and *adenine-hypoxanthin*. Kossell denies that these are due to the action of trypsin on proteids and attributes their presence to the nucleins of leucocytes entangled in the fibrin which was undergoing digestion in Nencki's experiments.

Among the usual products of decomposition set up by trypsin a substance is found which assumes a red colour on the addition of chlorine water and a violet one when bromine water is added to it. This substance was originally observed in an extract of the pancreas by Tiedemann and Gmelin. It was ascertained by Neumeister to be formed simultaneously with tyrosin, so constantly indeed that its presence in a digestion liquid might be taken to be an indication of the formation of tyrosin. He gave it the name *tryptophane*. Stadelmann has suggested the name of *proteinochromogen* for this body and *proteinochrome* for the substance produced from it by bromine water.

According to Nencki the chromogen consists of two substances, from the bromine compound of one of which he prepared *skatol, pyrrolin,* and *indol* by fusing it with caustic potash.

Kossell states that among the products of the action of trypsin on proteids certain bodies occur which he calls *protamines*; by further digestion with the same enzyme a number of basic bodies are formed, to which he has given the name of *hexones*. They include *argenin, histidin,* and others.

Harlay attributes the formation of tryptophane to the action of trypsin upon the chromogen which he says is formed during peptic digestion. This substance undergoes decomposition into tryptophane and tyrosin, which explains their constantly simultaneous appearance, already mentioned as observed by Neumeister.

Trypsin is capable of digesting *mucin*, which resists the action of pepsin; it cannot attack the *collagen* of connective tissue or tendons, though it can digest the gelatin into which this material can be converted by prolonged boiling with water. The products of its action on *gelatin* according to Nencki are gelatin-peptone, leucin, glycin (amido-acetic acid) and ammonia. *Elastin* is more or less rapidly dissolved by it, yielding, according to Wälchli, glycin, leucin, valerianic acid and ammonia.

A curious peculiarity attending the action of pancreatic juice upon milk was first noticed by Roberts and subsequently confirmed by other observers. Milk is an emulsion of fat (cream) in a liquid containing certain proteid, carbohydrate, and mineral constituents. The chief proteid is the one generally called casein, which in its reactions approaches the group of derived albumins. It is so much like them indeed that it was formerly considered to be a form of alkali-albumin. This body differs from a primary albumin or a globulin by not undergoing the process of heat-coagulation as the temperature of its solvent is raised. Roberts has shown that when milk is digested for a short time with pancreatic juice and then boiled, the casein separates out in the form of a scum or coagulum. He suggests that this change is due to the action of the trypsin and indicates a very early stage in the course of proteolysis. Longer digestion with pancreatic juice causes the disappearance of this property and then the ordinary decomposition products of the casein, proteoses, etc. can be detected. Roberts has named this new body, derived from casein and coagulating on boiling "metacasein." His facts have been corroborated, but a different interpretation of them is now suggested. Later writers consider that the action is not connected with proteolysis by the trypsin, but that the metacasein is formed by the rennet enzyme under particular conditions. This point will be further discussed subsequently.

Intracellular proteolytic digestion is not of such general occurrence in the animal kingdom as diastasic. It is however far from unknown, but whether it is brought about by peptic or by tryptic enzymes, and how far the direct action of the protoplasm is involved, is uncertain in most cases. Metschnikoff has described digestion as occurring in the cells of the mesoderm of *Synapta* and *Phyllirhoë*, and has claimed that the leucocytes of the higher Vertebrata possess digestive power. Leidy has described the ingestion of an Amœba by a larger one of another species, and other observers have recorded similar circumstances. There is a general agreement among them that when such ingestion has taken place the absorbed matter often lies for a time in a vacuole in the animal's substance, surrounded by what is apparently a digestive liquid.

Miss Greenwood has described the processes of digestion in several of the lowliest forms. In both *Amœba* and *Actinosphœrium* she has watched the ingestion of solid food particles and the subsequent changes which such particles undergo. The ingested matter is surrounded by fluid, sometimes in not inconsiderable quantity, so that it lies in a vacuole in the animal's substance. If the absorbed matter is innutritious the vacuole of ingestion soon disappears; if nutritious it undergoes change which is not effected by direct contact with the living substance, but by something passing out of the protoplasm into the vacuole, which thus becomes a digestive cavity, and receives a secretion from the living substance. This secretion is probably not acid, but it exerts a digestive action on the imprisoned food material, which after a time disappears. If the ingested matter is an Alga surrounded by a cellulose membrane, the latter is not dissolved but its proteid contents are digested. This fact clearly shows that in these cases the digestive action is exerted by a secretion which can diffuse through the membrane, and not by contact with the protoplasm itself. The fact that innutritious substances do not provoke this formation of a digestive vacuole points to the absorption of food-material being a stimulus setting up the secretion of a digestive juice.

In *Actinosphœrium* soon after the ingestion of food-material there is a marked appearance of granularity in the region of the living substance surrounding the prey. These granules are not formed round indigestible matter that may be absorbed. Miss Greenwood states that this granular gathering is significant of secretory activity.

It is worthy of notice that various observers have indicated that the vacuoles round ingested matters frequently contain an

acid liquid. Among others, Engelmann states that litmus
particles turn red when ingested by *Paramœcium, Stylonichia*,
and some species of *Amœba*.

In *Carchesium*, one of the *Vorticellidœ*, the digestive changes
can be seen within even greater clearness. Miss Greenwood
describes the appearances which can be made out when the
animal is fed with coagulated white of egg in a fine state of
division. A number of particles suspended in fluid are ingested
together and a vacuole containing them can be seen in the
animal's sarcode. They pass from the mouth towards the more
deeply-seated substance of the animal, becoming aggregated
together and losing the fluid which surrounds them. They are
then stored for a time, which sometimes lasts for more than an
hour, when vacuoles again form around them, containing freshly-
secreted fluid which has a solvent action on proteids. The
included particles are gradually dissolved, the vacuole containing them moving slowly towards that region of the animal
from which the finally undigested portions are discharged.

The process of secretion of the proteolytic enzymes in the
cells of the mammalian glands follows the same course as that of
the formation of diastase in the salivary glands. The general
structure of the pancreas and of the glands of the stomach is
similar to that already described in the case of the latter
organs. When the glands are at rest there is a conspicuous
formation of peculiar granules in them, which may accumulate
so as to make the whole of the cell opaque, or which may be
stored more or less on the side which abuts upon the lumen
of the gland. In the pancreas we have thus two well-marked
zones, the outer being small but transparent, while the inner is
opaque. When secretion takes place the granularity gradually
becomes less evident, the granules retreating from the outer
parts of the cells towards the lumen of the gland and the whole
outlines of the cells being well-defined. At the conclusion of
the process the cells are nearly free from granularity. The
granules are then gradually reformed till the resting condition
is again attained, the cells at the same time becoming somewhat larger. Comparison of cells taken from the gland during
different conditions of secretion lead to the view that the act of

secretion consists of three well-marked stages. (1) The growth
of the cell at the expense of material brought to it by the
lymph. (2) The construction of the granules from the protoplasm of the cell. (3) The extrusion of these with a certain
amount of fluid in which they are more or less completely dissolved. In some glands these stages are apparently more or
less consecutive. In others they appear to go on simultaneously
but at relatively different rates. In such cells the clear and
granular zones cannot be clearly recognised, though the amount
of granularity varies from time to time.

The granules do not appear to be the actual enzymes but
to be composed of substances which are readily and rapidly
converted into them. If a fresh pancreas, taken from the body
immediately after death, is extracted with water or glycerin, the
extract is almost without action on proteids. If instead of
being extracted at once it is kept warm for 24 hours, or if it
is warmed for a shorter time in dilute acetic acid, and then
extracted, the extract has well-marked proteolytic powers. The
inert extract prepared from the fresh pancreas can also be made
active, by slightly acidulating it with acetic acid and keeping it
for some time at 40° C. The cells contain the enzyme in the
condition of a *zymogen*, which is readily converted into the
actual ferment by the acid used. The zymogen of the pancreas
can also be converted into the enzyme by passing a current of
oxygen for about 15 minutes through the neutral extract.
That the granules really contain this zymogen appears evident
from the fact that the amount of enzyme which can be formed
is proportional to the granularity.

There is little doubt that these changes are controlled by
nervous action as in the case of the salivary glands, but the
paths of the impulses are not so clearly understood. Stimulation of the medulla oblongata increases the secretion of the
pancreas, but it is not dependent altogether upon nervous supply
from the central nervous system. When all the nerves going
to the gland are cut, the secretion is not materially affected.
In some animals there is probably a reflex nervous mechanism
influencing it, for secretion commences immediately food is
taken into the stomach.

The secretion of both pepsin and trypsin is affected if not altogether provoked by the absorption of nutritive products. Schiff found that when an insoluble proteid such as fibrin or coagulated albumin was introduced into the stomach of a fasting animal no pepsin was secreted and the proteid remained un-digested. If with the latter certain soluble matters, to which he gave the name of *peptogens*, were introduced, pepsin began to be secreted immediately on their absorption. Among these pep-togens Schiff mentions solutions of dextrin, extract of meat, gelatin and peptone. He found that they were just as effective when introduced into the blood by injection, or when absorbed in any other way. The action they set up is clearly due, therefore, not to their mere presence in the stomach, and they must act chemically and not mechanically upon the gland cells. They may be of value as setting up the necessary nutritive changes in the gland cells, acting directly upon the protoplasm; or they may work indirectly by stimulating some nervous mechanism in the gland or in the stomach wall, so causing secretion. Some physiologists hold the view that there are in the pancreas as in the salivary glands two sets of nerve fibres, the one influencing the formation of the enzymes, and the other their discharge during the activity of the gland. This view rests not so much upon experimental proof as upon analogy with the salivary gland, in which there is evidence of such nerve supply.

There is a marked difference between the two enzymes with regard to the way in which they are affected by acids, alkalis, and neutral salts. Pepsin can only effect proteolysis in the presence of a dilute acid; in gastric juice it is closely associated with hydrochloric acid in quantity varying from ·2 to ·6 per cent. Pancreatic trypsin on the other hand is destroyed by a relatively small quantity of free acid. Trypsin is aided in its activity by most salts in small amount, particularly alkaline salts; of these sodium carbonate is the most favourable, the best results being obtained when from ·9 to 1·2 per cent. of the salt is present. This body is most injurious to pepsin, ·5 per cent. destroying almost the whole of the enzyme in a solution in as short a time as 15 seconds. The presence of soluble proteids lessens the rate of destruction, but not very materially, even 2·5 per cent. of peptone not preventing it. The pepsin of the frog is less easily destroyed than that of a mammal.

The presence of an excess of neutral salts in a solution of either enzyme is deleterious, but pepsin is more readily affected than trypsin. The action of the latter, according to Edkins, is assisted by the presence of 1—2 per cent. of sodium chloride, but very greatly retarded by 8 per cent.

A still more marked difference between pepsin and trypsin lies in their relative powers of effecting the decomposition of proteid. Pepsin can only carry the hydrolysis up to the stage at which peptone is formed, while trypsin can split up one form of peptone so far that the proteid is completely decomposed. The initial stages of the proteolysis do not differ greatly in the two cases, and the chief cause of such difference as there is seems to be closely related to the fact that the one acts in an acid, and the other in an alkaline medium.

The process of peptonisation by the aid of pepsin is a slow and incomplete process, the relative amount of proteoses and peptones found when such a digestion approaches its completion being roughly 65 and 30. Chittenden and Amerman believe that complete peptonization is not a property of gastric digestion either in the natural or artificial process. The action of pepsin is rather a preliminary stage in proteolytic digestion, a preparation for the more vigorous transformations brought about in the intestine under the action of trypsin. This view is not a new one, having been advanced originally by Claud Bernard. It is supported by the consideration that among the lower animals pepsin is of rare occurrence, the only proteolytic ferment secreted by them being a tryptic one. Trypsin acts more energetically than pepsin, converting primary proteids rapidly into proteoses, and these more or less completely into true peptone. The only rapid transformation effected by pepsin is the first change into primary proteoses, the subsequent action being slow and incomplete.

Galactase.

An enzyme which belongs to the tryptic group has been discovered by Babcock and Russell in the course of some researches they carried out on the ripening of cheese. They have given it the name of *galactase*.

They found in the course of experiments on the changes which take place in milk that is kept for a long time under antiseptic conditions that there was a gradual increase in the nitrogenous constituents which are soluble in water, which are continuously formed at the expense of the caseinogen. In a milk 12 days old the nitrogen existing in this condition was 30 per cent. of the total nitrogen; in one 240 days old it had become 63 per cent. In another sample, 12 hours after milking it was less than 10 per cent., but after three weeks standing it had risen to 30 per cent. During all this time bacteria were found to be entirely absent.

These changes in the nitrogenous constituents of the milk did not take place in samples that were boiled at the outset.

Similar results were obtained with a cheese which was kept under a bell-jar in the presence of the vapour of ether and chloroform.

The experiments clearly pointed to the presence of a proteolytic enzyme, and this after considerable trouble they succeeded in extracting from fresh milk. Taking advantage of the property of enzymes to attach themselves to finely divided material in suspension, they passed the milk through a centrifugal separator and collected the slime which remained adhering to the instrument. They mixed the slime with an equal weight of 40 per cent. alcohol and added an antiseptic. After 24 hours it was filtered and the filtrate concentrated by evaporation at 25° C. to one-tenth its original volume, more antiseptic being added. The solution was faintly acid, and was therefore neutralised by adding a little sodic carbonate, when a precipitate of syntonin and a little calcic phosphate fell. The filtrate from this precipitate was found capable of curdling milk and subsequently dissolving the curd. It contained the proteolytic ferment together with a certain quantity of another enzyme, *rennet*, which will be discussed in a subsequent chapter.

More complete examination of the proteolytic enzyme showed it to differ in some important respects from trypsin, with which its discoverers were at first inclined to identify it. The general course of digestion of caseinogen by its activity resembles that brought about by trypsin in that it gives rise to amido-acids as well as to albumoses and peptones. It differs however in carrying the decomposition of the proteid further, a certain quantity of ammonia being always found. Trypsin from pancreatic juice was found not to be capable of doing this, though it is stated by Hirschler and by Stadelmann, the formation of ammonia takes place during the tryptic digestion of fibrin.

The enzyme has been named *galactase* by its discoverers. Its principal features seem to be the following:—

It is capable of hydrolising proteids, particularly caseinogen, carrying the decomposition to the stage of the liberation of ammonia.

It also liquefies gelatin.

It is active in neutral, faintly alkaline and faintly acid solutions, but very slight excess of acidity rapidly inhibits it.

It decomposes peroxide of hydrogen with liberation of oxygen.

Its optimum point is 37°—42° C.; its activity at 52° C. is about equal to that it exhibits at 28° C.; its proteolytic power is destroyed at 76° C., and its action on gelatin at 65° C.

It is inhibited by mercuric chloride, formalin, phenol and carbon-disulphide, but not by chloroform, ether, benzol or toluol.

CHAPTER XIII.

PROTEOLYTIC ENZYMES (*continued*). *VEGETABLE TRYPSINS.*

IT is uncertain whether pepsin is represented in the vegetable kingdom. All the proteolytic enzymes which have been fully investigated have been found capable of carrying the hydrolysis beyond the stage of peptone. The work of the earlier observers did not include a careful examination of the products of the decomposition, and hence for the present it remains uncertain whether or no some of the ferments belong to the peptic category.

The chief enzymes of the proteolytic class that have been examined with any minuteness are *papaïn*, obtained from the juice of *Carica papaya*; *bromelin*, from the fruit of the Pineapple; the trypsin of germinating seeds; and the enzyme of the secretion of *Nepenthes*, one of the so-called pitcher-plants. Several others are known to occur in different plants, but our information respecting them is not very complete.

Bromelin.

Though this enzyme was not the earliest to be investigated, it is the one about which we have at present the fullest information, and as it seems typical of most of them it is advisable that its consideration should be undertaken first. It occurs in great quantity in the juice of the *Pine-apple* (*Ananassa sativa*) and is becoming of commercial importance. Attention was first called to it by Marcano in 1891, and it has been examined very fully since then by Chittenden and his pupils in America, who have given it the name of *bromelin*, from the Natural Order to which the pine-apple belongs.

The juice of the fruit is extremely abundant, nearly a litre being yielded by one of average size. It is distinctly acid, the acidity being equal to that of a solution of hydrochloric acid of about ·45 per cent. strength. The proteolytic enzyme is associated in the juice with a rennet or milk-curdling ferment, just as pepsin is in the gastric secretion of animals.

The juice contains two proteids, one of which appears to be a protoproteose and the other either a heteroproteose or a globulin. The former is associated chiefly with the bromelin, and coagulates when the juice is heated to 75° C. If the latter is neutralized before warming the coagulating point is raised to 82° C. The second proteid is associated chiefly with the rennet and is only coagulated when the juice is boiled.

Bromelin can be extracted from pine-apple juice by saturating the latter with crystals of either sodium chloride, magnesium sulphate, or ammonium sulphate. The first of these salts gives the best result, the digestive activities of the three preparations having been ascertained by submitting 10 grams of moist coagulated egg-albumin suspended in 100 c.c. of water to the action of ·05 gm. of each for 6 hours at 40° C. The sodium chloride preparation under these conditions digested 27·6 per cent. of the proteid, the magnesium sulphate one 14·1 per cent., and the ammonium sulphate one 12·3 per cent. only. The ferment loses some of its activity during the process of extraction, the neutralised juice always having greater proteolytic powers than solutions of the different salt precipitates made up to the original bulk.

The action of the enzyme shows slight variations when it acts in neutral and acid solution respectively, and its course is not quite the same when acting on different proteids. As the sodium chloride precipitate is the most potent, it was used by Chittenden and his pupils in most of their experiments.

When fibrin was used as the proteid to be digested the course of the action was shown in the following experiment:—

300 grams of moist fibrin were warmed at 40° C. with 2500 c.c. of ·025 per cent. hydrochloric acid and ·25 gram of

bromelin for 5 hours: it had then been all dissolved except a small flocculent residue. The filtrate from this residue when neutralised deposited a small amount of precipitate. After filtration the neutralised fluid was boiled for a short time, and a heavy coagulum was formed, which dissolved in dilute acids and alkalis. A further precipitation occurred on adding alcohol to the concentrated filtrate from the heat-coagulated proteid, and from the alcoholic filtrate leucin and tyrosin separated out in large quantities under appropriate treatment. The original undissolved residue was found to consist of an antialbumid mixed with a proteose which resisted further action of the enzyme. The neutralisation precipitate was probably not acid-albumin but a form of hetero-proteose, as it became converted into dysproteose on standing. The heat-coagulated proteid was a heteroproteose which was not precipitated by neutralisation. In different experiments the quantity of the neutralisation precipitate and that thrown down on boiling varied inversely, pointing therefore to their both being the same proteid, which did not always behave in the same way with regard to the reaction of the liquid. The precipitate given by the alcohol was a mixture of proteoses and peptone. When dissolved in water the former were separated from the latter by saturation of the solution with ammonium sulphate, and were found to consist of a mixture of proto- hetero- and deutero-proteose. The peptone was almost if not entirely antipeptone. The course of action was thus found to be almost identical with that of pancreatic trypsin, but there was no hemipeptone. The amount of crystalline amides indicated that the latter had been entirely decomposed.

When the digestion was carried out in neutral solution, in the presence of 1 per cent. sodium chloride, the course of action was similar, but the heat-precipitate was in greater quantity and consisted chiefly of dysproteose, which yielded deutero-proteose on further digestion. The greater part of the peptone was antipeptone, but some hemipeptone appeared to be present as well. The proteoses mixed with the peptone contained no heteroproteose, and the leucin and tyrosin were in smaller quantities than in the acid digestion.

14—2

Fibrin was not digested so easily in neutral as in acid solution of the enzyme, and the presence of such a salt as sodium chloride was almost essential to the activity of the bromelin in the former case.

When coagulated egg-albumin was substituted for fibrin several differences were manifested. The coagulated albumin from 3 dozen eggs was digested with 2 litres of distilled water and ·8 gram of bromelin for 40 hours at 40° C. under antiseptic precautions. There was then still some insoluble residue, consisting of antialbumid with a little undigested albumin. The liquid had become slightly acid but neutralisation gave no precipitate, nor was there any coagulum formed on boiling. Alcohol threw down a mixture of proteoses and peptone, and the filtrate from this yielded leucin and tyrosin. There was very little proto- or hetero-proteose, these having probably been further hydrolysed during the somewhat prolonged period of digestion; abundant deutero-proteose was present.

The action of bromelin on myosin was also examined. It digested this proteid best in ·025 per cent. hydrochloric acid solution. The course of digestion showed a difference in one respect from what was observed in the other cases. Shortly after it had begun the solution became thick and semigelatinous from the separation of what appeared to be acid-albumin. This disappeared however as the action proceeded. In other respects myosin behaved like the other proteids, yielding proteoses, peptone, and amides.

Careful quantitative analyses of the bodies formed during the digestions showed a gradual decrease in the nitrogen they contained, suggesting successive separations of nitrogen-containing radicals. Thus while myosin contains 16·86 per cent. of nitrogen, the neutralisation precipitate only contained 15·8, and the deutero-albumose 13·9 per cent. Somewhat singularly the peptone prepared from albumin contained 14·5 per cent. and that from myosin 15·7 per cent. of nitrogen.

The effect of various quantities of different acids on the action of bromelin on fibrin is expressed in the following table.

Acid used	With NaCl preparation	With MgSO$_4$ preparation	With (NH$_4$)$_2$SO$_4$ preparation
None	Enzyme almost inactive	As with NaCl preparation	As with NaCl preparation
HCl	With ·012% active: activity increases with increase of acid up to ·025%; then is not altered up to ·05%; then decreases, and ceases with ·1%	Increases up to ·05%; above this concentration activity declines	As with MgSO$_4$ preparation
Acetic	Almost inactive up to ·25%; then is active, and remains so up to 1%	——	As with NaCl preparation
Citric	Activity begins with ·03%; remains good up to ·12%; then declines, is inactive with 1%	——	Maximum activity with ·5—1·5%
Tartaric	Slight action with ·06%; maximum with ·15—1%	——	——
Oxalic	Active only with ·12—·25%; inactive with 1%	——	Maximum activity with ·25—·5%

With egg-albumin coagulated by boiling, bromelin was most active in neutral solutions; slight amounts of hydrochloric acid (·012—·025 %) did not inhibit it, nor did ·025 % of sodium carbonate; organic acids interfered with the action; ·1 % of citric acid reduced it in the proportion of 31 : 12·5.

With raw egg-albumin different acids showed further variations. Hydrochloric acid in a series of experiments gave the following results when 10 c.c. of albumin solution, ·05 gm. bromelin, and 90 c.c. water, were digested for 16 hours at 40° C. in the presence of thymol:

Neutral (standard)   HCl ·012%   HCl ·025%   HCl ·1%
digested 16·6%,   digested 17·3%   digested 21·9%   digested 6·2%

Sodium carbonate was prejudicial to a greater extent; in the presence of ·05 % only ·8 % of the albumin was digested.

In another series of experiments, carried out with the same relative proportion of bromelin and proteid, the effect of other acids was examined, with the results shown in the following table:

	Neutral (standard)	Tartaric 0·25 %	Tartaric 0·5%	Oxalic 0·1 %	Citric 0·1 %	Citric 0·2%
Percentage of proteid digested	35·5	23·3	16·6	19·0	18·6	18·7.

The effect of neutral salts on the action of bromelin varied again with the preparation of the enzyme used, with the particular proteid digested, and with the reaction of the digesting fluid. With *neutral* preparations digesting fibrin, the NaCl preparation was assisted by the presence of 1—3 % of neutral salts, the other two by proportions ranging from 1—5 %, but larger quantities than these were prejudicial to the action. Neutral salts did not affect *acid* digestions unless they were present in large amounts.

They did not facilitate neutral digestions of coagulated egg-albumin.

The neutralised pine-apple juice works most powerfully at a temperature ranging from 50°—60° C., and still shows considerable action even when heated to 70° C.

With the precipitated enzyme the effect of temperature upon the working is shown in the following table, the conditions of the digestion being 10 grams of coagulated egg-albumin, ·05 gram of the precipitate, and 100 c.c. of water, digested at the temperatures quoted for 5 hours:

Temperature	Percentage of proteid digested
40° C.	7·8
45° C.	9·6
50° C.	11·4
55° C.	11·9
60° C.	12·5
65° C.	9·4
70° C.	8·3.

The optimum point here is about the same as for the neutralised juice.

Papaïn.

The fruit of the Papaw tree has long had traditionally the property of rendering meat tender when cooked with it. During recent years it has been ascertained that the juice of the fruit contains a proteolytic enzyme of considerable power. It was first investigated by Würtz in 1879; he discovered that the enzyme is not confined to the juice of the fruit, but is present also in the sap obtainable from the stem and leaves. It can be prepared in a very crude condition by expressing the juice from the tissues of the plant, adding alcohol till a precipitate falls, collecting the latter by filtration, washing it with absolute alcohol, and drying it at a low temperature. The sap extracted from the plant is neutral in reaction and contains proteid matter in solution. Würtz considered that the enzyme was itself a proteid, and gave as its reactions that it is not precipitated on boiling, gives no precipitate with corrosive sublimate, gives a precipitate with nitric or hydrochloric acid, which is soluble in excess of the precipitant; also gives a precipitate with acetic acid and potassic-ferrocyanide. Wurtz says further that in a neutral solution it dissolves animal proteids, with formation of peptones and leucin. At the time when he wrote, the differences between peptone and intermediate products, particularly albumoses, were not understood, and many bodies of the latter class were incorrectly considered to be peptones.

A more complete examination of the papaw juice and of the enzymes it contains was made by Martin in 1883 and 1884. He found that the proteids in the neutral juice consisted of an albumin, a globulin, and two forms of albumose, which he named a- and β-phytalbumose. Comparison of these bodies with those more recently investigated points to the former being a form of proto-proteose and the latter related to the hetero-proteoses. As we have seen, comparatively little was known of the character and behaviour of the proteoses, or albumoses until they received careful study in the laboratory of Professor Chittenden, of Yale University. A comparison of Martin's two bodies with the several classes of proteoses described by Chittenden leaves little doubt that they may be considered forms of proto-proteose and hetero-proteose respectively. Martin found that the papaw juice contains two enzymes, one capable of curdling milk (rennet), and the other the proteolytic one, now known as *papaïn*. The latter is associated closely with the a-phytalbumose, or proto-proteose.

Martin agrees with Wurtz that the action of papaïn takes place best in a neutral medium like that of the fresh juice.

When fairly pure it is easily destroyed by hydrochloric acid of greater concentration than ·05 °/₀, but in a faintly alkaline medium, such as a ·25 per cent. solution of sodium carbonate, it is almost if not quite as active as in a neutral one. A higher percentage of alkali destroys it.

Martin has stated that the optimum temperature for papaïn digestion is 35°—40° C., though it is active at lower temperatures, such as 15° C.

When fibrin is subjected to the action of papaïn in neutral solution it is affected as it is by pancreatic trypsin, being corroded rather than dissolved, and converted into a pultaceous mass, a good deal of turbidity accompanying the action. The products of the digestion Martin determined to be a globulin-like body, apparently intermediate between a true globulin and alkali-albumin, some peptone, and leucin and tyrosin, the leucin being in the greater proportion of the two.

Martin also examined with considerable minuteness the action of papaïn on the proteids of the papaw juice. When digested with the globulin or the albumin a small quantity of an insoluble residue was left, which was probably a form of antialbumid, such as Chittenden obtained with bromelin. At the end of the digestion the most prominent body found in solution was the hetero-proteose which Martin called β-phytalbumose, while traces of peptone and both leucin and tyrosin were also present.

Digestion of the a-phytalbumose or proto-proteose yielded a little antialbumid, a quantity of a deutero-proteose, no peptone, and a certain amount of both leucin and tyrosin. The β-phytalbumose showed the same decomposition.

It seems probable from Martin's experiments that the course of proteolysis in the papaw is the same as that brought about by bromelin. The juice as it is extracted from the plant contains the two albumoses or proteoses, but it seems quite likely that these are formed by the action of the enzyme before the juice is extracted and are not the primary proteids of the plant. Both yield the deutero-albumose when subjected to the action of the separated papaïn.

The native albumin and globulin are both capable of giving rise to the hetero-proteose in the course of laboratory experiments, and it is not improbable that the proto-proteose may be formed at the same time, but may undergo further hydrolysis as fast as it appears. If this is the case the course of proteolysis proceeds on exactly the same lines as that set up by bromelin. The primary proteids, albumin and globulin, yield at the outset the two proteoses, proto- and hetero-proteose, which may come from both the hemi- and anti-residues of the primary cleavage. The anti-residue also gives rise to the small amount of anti-albumid, which occurs in this case as it does in bromelin digestions. The proto- and hetero-proteoses on further digestion yield deutero-proteose, which Chittenden, as already mentioned, has shown to be a stage nearer to peptone. The deutero-proteose is then hydrolysed to peptone. Being itself derived from the proteoses of both the anti- and the hemi-group, the peptone as formed is a mixture of antipeptone and hemipeptone. The latter is split up into leucin and tyrosin, while the former remains as peptone at the end of the digestion.

Martin's results have since been confirmed by other observers. Davis has found further that the action of the enzyme is slightly improved by the presence of ·005 per cent. of hydrochloric acid, or ·25 per cent. of sodium carbonate, but is inhibited entirely by ·05 per cent. of hydrochloric acid.

Sharp agrees with Martin that proto- hetero- and deutero-proteoses are formed during the digestion of albumin, but he doubts the production of peptone. He says that a little dys-proteose also occurs.

Halliburton says that papaïn converts animal proteids into proteoses and peptone, but that when acting on vegetable proteids it stops short at the proteoses, no peptone being formed.

It is difficult to reconcile these statements as to the non-formation of peptone with the appearance of leucin and tyrosin. The latter have not been determined to arise from the direct decomposition of proteoses. We must conclude that in the experiments of these writers the peptone was altogether decomposed. Martin notes the same disappearance of the peptone when digesting some of the albumoses.

Helbing and Passmore found peptones in the papaïn diges-

tion of both myosin and egg-albumin, and separated them from
the other proteids by saturation of the liquid with ammonium
sulphate according to the method adopted by Chittenden.

Rideal also claims to have shown the existence of peptone
in papaïn digestions by determining the rate of dialysis of the
proteids through a parchment membrane. When experimenting
with egg-albumin he placed the whole product of the digestion
in a dialyser and found evidence of proteid in the water outside
it in 10 minutes. Though the proteoses are capable of dialysing
through a parchment membrane the rate of their diffusion is
very slow compared with that of peptone. Rideal concludes
that the rapidity of the passage points unmistakeably to the
presence of peptone as well as proteose. He confirmed this
result by the method of saturating the liquid with ammonium
sulphate.

Recent work carried out in Chittenden's laboratory fully
confirms the statement of previous writers that true peptone is
produced by papaïn. Neumeister makes the same assertion.

Rideal calls attention to the fact that papaïn is most
active when working in a relatively small amount of fluid.
Dilution of the digestive solution exercises a very marked
retarding influence on the progress of the hydrolysis. He says
the most advantageous quantity of liquid to use is from $1\frac{1}{2}$
to 3 times the weight of the proteid submitted to the action of
the enzyme.

He finds the optimum temperature for the working of
papaïn to be 40° C.

Harlay has stated recently that the digestive action of
papaïn points to its being intermediate between pepsin and
trypsin. It forms much proteose but very little tyrosin. In
papaïn digestions he observed the occurrence of the same
chromogen as he found produced by pepsin, the body, that is,
which with tyrosinase gives a red colouration slowly turning
green. Papaïn has a slight action on this body, producing from
it a small quantity of tyrosin. According to Harlay papaïn is
not destroyed by heating unless the temperature of its solution
is raised to 80° C.

Other Vegetable Trypsins.

The occurrence of proteolytic enzymes in those seeds whose reserve stores consist largely of proteid materials was suspected almost as soon as the existence of diastase was demonstrated in them. The search for them was first undertaken by von Gorup-Besanez in 1874. He detected such an enzyme in the seeds of the Vetch, and subsequently in those of Hemp, Flax, and Barley. He describes it as having power to convert fibrin into peptone, but apparently he did not investigate its action upon the proteids with which it is associated in the seeds. Beyond pointing out its existence his work does not give us very much information about it. Krauch writing in 1878 opposed the views of von Gorup-Besanez and denied the existence of such an enzyme. Von Gorup-Besanez says that the body he prepared from the vetch seed dissolved the fibrin, and that the solution when filtered gave a good biuret reaction, which he attributed to peptone. Krauch insisted that the biuret reaction was due to something present in the extract of the seeds and was readily yielded by the latter alone. He attributed the diminution of the fibrin to shrinkage of its substance and not to solution. Krauch's work however appears untrustworthy, for von Gorup-Besanez says the diminution of the fibrin went on to the point of disappearance, and this Krauch does not explain. Neither does he show that there was no further formation of a body giving the biuret reaction, though the extract itself may have shown the same. Krauch's own control experiments were somewhat scanty. Von Gorup-Besanez did not determine that the decomposition of the fibrin was carried beyond the stage of peptone, but he points out that under certain conditions large quantities of amido-acids, such as leucin and asparagin, could be detected in the shoots of very young vetch plants.

A series of investigations was carried out by the writer in 1886 upon the seed of a species of Lupin (*Lupinus hirsutus*). The experiments were made with germinating seeds, germination being allowed to proceed for four days, when the radicles were nearly 3 inches long. The cotyledons were separated, ground in a mill, and extracted with glycerin. The extract was strained and dialysed till free from crystalline amido-acids. It contained such of the proteids of the germinating seeds as were soluble in neutral fluids.

This extract was found capable of digesting fibrin in a faintly acid solution. The fibrin used was boiled in water and subsequently in ·2 per cent. hydrochloric acid solution, in which it swelled up and became transparent. Control experiments, in which the fibrin was suspended in dilute acid of the same degree of concentration as was used with the extract, showed that no change took place during the digestion unless the latter was present. In the acid extract of the seeds the fibrin became corroded, causing the liquid to become very turbid. It was then gradually dissolved, with the exception of a small amount of granular residue. Further control experiments were conducted in which the extract was boiled.

The digestions were generally carried out in parchment dialysing tubes and the liquid outside them was frequently changed. This liquid contained the same percentage of acid as that inside the tube. After a short time the dialysates showed the presence of a peptone, or at any rate of a body giving the biuret reaction, and appropriate treatment separated from them crystals of leucin and tyrosin, the latter in small amount only. When the digestion had proceeded for some time the liquid inside the dialyser contained a proteid which was precipitated upon neutralisation, and was soluble in dilute acids and alkalis; a considerable amount of proteoses, chiefly hetero-proteose, was also present. The course of digestion thus appeared to be the same as that brought about by papaïn.

The boiled controls showed the fibrin intact at the end of the experiment.

The action of the enzyme was further tested upon the proteids of the resting seed. These were three, a globulin, and two albumoses or proteoses, apparently proto- and hetero-proteose. The action of the glycerin extract on these was to transform them into what appeared to be peptone, and to cause the appearance of amido-acids, chiefly asparagin, but a certain amount of leucin as well.

Comparing the digestive effects upon animal and vegetable proteids the course of the action seemed to be the decomposition of a primary proteid, such as fibrin, into primary proteoses and acid-albumin, with a small quantity of antialbumid; the primary proto- and hetero-proteoses, whether formed from the fibrin or preexisting in the seed, were then changed into what was taken to be peptone, but which was probably a mixture of the latter with deutero-proteose. This seems probable in the light of Chittenden's work on bromelin, but no certain statement can be made, as his methods of separating the two were at that time unknown. Part of the peptone was then decomposed, yielding the amido-acids mentioned.

The enzyme was found to work most advantageously in a liquid containing ·2 per cent. of hydrochloric acid, which corresponded approximately to the natural reaction of the germinating seeds. When the acid was increased the digestion became much slower. The ferment was totally inactive in weak alkalis; indeed exposure to the presence of ·5 per cent. of sodium carbonate destroyed it entirely. Neutral salts such as sodium chloride impeded the action but did not destroy the enzyme. The optimum temperature for the activity of the ferment was about 40° C.; it then worked about twice as quickly as at the usual temperature of the soil.

Examination of the *resting* seed showed that it did not yield any active enzyme to glycerin. The glycerin extract became active however on warming it for some time with a dilute acid, in which particular it agreed with the extract of the pancreas. The experiments therefore indicate the existence of a zymogen in the seed which is converted into an enzyme in the presence of a feeble acid. It may be noted that just such a change in the reaction of the parenchyma of the seed takes place when germination commences. It is difficult, though not impossible, to prove the presence of the zymogen, as the digestion of the proteid must take place in the presence of such an acid. It is possible however to confirm its existence by a method adopted by Langley and Edkins in their study of the zymogen of pepsin. A description of this method will be given in a subsequent chapter.

The writer subsequently detected the existence of this vegetable trypsin in the germinating seed of the Castor-oil plant (*Ricinus communis*).

The presence of this enzyme in the Lupin seed has recently been observed by Butkewitsch.

Neumeister has shown that a similar enzyme is not uncommon in seedlings. He has extracted it from those of the Barley, Poppy, Wheat, Maize and Rape. It does not occur in the early stages of germination but is developed as the plantlet grows, and is plentiful when it has attained a length of from 15 to 20 centimetres. He prepared it by a somewhat novel method. Having obtained an extract of the seedlings he soaked moist fibrin in it. This proteid appears to have the property of absorbing the ferment from the solution. Fibrin so impregnated with the enzyme, was removed and placed in acid or alkaline liquids and the results observed, control experiments being carried out in which fibrin that had not been in contact with the extract was placed in similar fluids. Neumeister says that the enzyme is active only in acid liquids, and that the acid must be organic, oxalic being the best. Mineral acids such as hydrochloric destroy it.

Fernbach and Hubert have shown that a similar enzyme exists in malt and can be precipitated by alcohol with the other enzymes which have already been described. It shows a similar behaviour to that of diastase in the presence of phosphates. Laszcynski states that kiln-dried malt contains the enzyme but that it is absent from barley seedlings. He could not detect its action upon fibrin till it had been digested with it for six hours at 40° C.

Few observations have been made on the digestion of proteids in the living plant. Schulze has stated that during germination a mixture of nitrogenous compounds results from the decomposition of the proteids, among which aromatic and fatty amido-acids and arginin are probably invariably present.

The course of proteolysis in the cereals has not been fully investigated at present, but it appears to present some differences from that already described. Osborne and Campbell have examined the proteids of barley and of malt with a view to

seeing what changes can be traced during the preparation of
the latter. Barley contains nearly 11 per cent. of proteid
matter; of this a globulin, *edestin*, with a small inseparable
amount of proteose, constitutes about 2 per cent.; an albumin,
leucosin, is present to the extent of ·3 per cent.; and a peculiar
proteid, *hordein*, amounts to 4 per cent. The latter differs from
all proteids of animal origin by being freely soluble in 75 per
cent. alcohol, while it is insoluble in water. The remainder of
the barley proteids, about 4·5 per cent., is insoluble in either
water, potash, or alcohol. On examining malt, hordein was found
to have disappeared, being partly replaced by another proteid,
also soluble in alcohol, to which the investigators gave the name
bynin. Another globulin, *bynedestin*, replaced the edestin. Two
proteoses were present, amounting together to about 1·3 per
cent. of the total. A great deal of the insoluble proteid also was
present, though nearly 1 per cent. less than the corresponding
body in the barley. The total amount of proteid matter in
malt was a little less than 8 per cent., so that during germination about 3 per cent., or nearly one-third of the whole, had
disappeared. The authors conclude that the proteids change
extensively during germination before acquiring the properties
of proteoses. The proteids replacing hordein and edestin are
richer in carbon and poorer in nitrogen. The authors did not
discover any enzyme capable of effecting these changes.

It was mentioned above that the proteolytic power of
pine-apple juice was first observed by Marcano. Some years
earlier he found that the expressed sap of the leaves of certain
species of *Agave* was capable of digesting meat, if the latter
was soaked in it and exposed to a temperature of 35°—40° C.
The digestion was somewhat slow, but after it had continued for
26 hours Marcano ascertained that about 20 per cent. of the
meat used had been converted into "peptone," which was
probably a mixture of peptone and proteoses. The results of
the experiments show therefore that the Agave contains a
proteolytic enzyme. The digestion proceeded equally well in
the presence of chloroform, so that it was not due to putrefactive changes. Marcano says he has found this proteolytic
power in the juices of a great number of fruits.

In 1880 Bouchut carried out some researches upon the juice of the common fig-tree (*Ficus Carica*), and found it to contain a powerful enzyme capable of dissolving proteid substances. Hansen investigated the same plant in 1883 and 1884 and confirmed Bouchut's results. He described the enzyme as working most advantageously in an acid medium, but as not being without proteolytic power in an alkaline liquid. The enzyme was the subject of a more extended research in 1890 by Mussi. He took the juice expressed from the branches, leaves, and fruit of the fig-tree, and after filtration to remove the débris accompanying the liquid he precipitated certain of its constituents by the addition of absolute alcohol. The resulting precipitate was insoluble in water but dissolved easily on the addition of a trace of acid or alkali. When either solution was placed in contact with moist fibrin it dissolved it readily. Mussi gave the name *Cradina* to the enzyme. It is inactive in neutral fluids. Mussi gives several reactions for it, but as by the method of preparation used it must have been very impure these have no particular importance.

Another Indian fruit, the Kachree gourd (*Cucumis utilissimus*), was examined by the writer in 1892 with a similar result. The fruit is in appearance much like a small vegetable-marrow, about 6 inches in length. It is yellow in colour, and when cut has an aroma similar to that of the melon. Its pulp is extremely succulent and the expressed juice is faintly acid in reaction. Both the juice and the pulp of the pericarp contain an enzyme which is associated with a globulin-like proteid. The ferment is most completely extracted from the pulp by a dilute salt solution, containing 3 per cent. of sodium chloride. Such an extract acts slowly on coagulated egg-albumin, yielding proteoses, peptone, and leucin. It is most effective in a faintly alkaline medium, less so in a neutral one, and acts still more feebly in the presence of acid.

A proteolytic ferment was described in 1892 by Daccomo and Tommasi as obtainable from *Anagallis arvensis*. It can be prepared as a white amorphous substance, easily soluble in water. The authors say that if the fresh plant is reduced to powder and kept in contact with fresh meat or moist fibrin for

4 or 5 hours at a temperature of 60° C., the proteid is considerably softened, though complete disintegration is not effected in less than thirty-six hours. The enzyme is stated to have the property of destroying fleshy growths and horny warts.

The normal action of the proteolytic enzymes so far described is presumably intracellular. This is certainly the case with the trypsin of germinating seeds, which indeed is formed in the cells in which the reserve-proteids lie. We find however in the vegetable kingdom cases in which a proteolytic secretion is formed and poured out upon the surface of the plant, or into particular receptacles, to carry on there a digestive process remarkably similar to that of animals. The plants in question are the so-called insectivorous plants, which by various methods capture, kill, and digest various insects which alight upon them. The most conspicuous of them are the pitcher-plants, *Nepenthes*, *Sarracenia*, *Darlingtonia*, and others, certain of the leaves of which are in part or altogether transformed into large receptacles, containing in life a considerable amount of fluid. Insects attracted to the plants are enticed into entering the pitchers and are drowned in the liquid they contain. Some of these plants, particularly Sarracenia and Darlingtonia, have nothing but water in the pitchers and the insects drowned therein undergo ordinary putrefaction, the products of which are absorbed by the plant. Nepenthes stands out conspicuously from these in possessing a series of glandular structures in the lower portion of the pitcher. The glands secrete into the pitcher a proteolytic enzyme by whose agency the organisms which are captured undergo a process of true digestion, in the absence of putrefactive changes. The liquid of the pitcher indeed has antiseptic properties. The enzyme has generally been compared to pepsin, chiefly because during digestion the reaction of the liquid is acid. It has recently been shown by Vines to have features which associate it preferably with trypsin.

The existence of the proteolytic property in the liquid of Nepenthes was first ascertained in 1874 by Hooker, who however did little more than determine that it is capable of dissolving boiled white of egg.

Lawson Tait in the following year prepared the enzyme in a crude form and spoke of it as resembling pepsin. A little later von Gorup-Besanez made a detailed study of its action on fibrin. He found it to be capable of dissolving fibrin with the formation of a soluble body giving the reactions of peptone so far as they were then known. He ascertained that for the formation of the enzyme it was necessary that the pitchers should be stimulated by the absorption of some digested matter, much as is the case with the gastric glands of a mammal. Such stimulation was followed by the secretion of an acid liquid which had well-marked proteolytic powers. The latter were only manifested while the reaction remained acid and were considerably greater at a temperature of 40° C. than at 20° C. These results of von Gorup-Besanez were confirmed by other observers; Vines in 1877 showed that the pitchers yield a proteolytic extract when pounded up and mixed with glycerin. Such an extract is not quite so active as the actual secretion of the glands as it is poured into the pitcher. If an unstimulated pitcher is extracted with glycerin the resulting preparation is inactive, but it develops proteolytic powers if warmed for a time with a dilute acid. The pitcher wall therefore contains a zymogen, just as do the gastric and pancreatic glands.

Recently a certain controversy has arisen as to these proteolytic powers, Dubois and Tischutkin separately asserting that they have entirely failed to obtain digestion of animal matter by the aid of liquid in the pitchers when the experiments have been carried out under antiseptic precautions. They attribute the digestion observed by von Gorup-Besanez, Vines and others to the agency of bacteria. On the other hand Goebel maintains that he found the liquid from a pitcher, when made acid with hydrochloric acid, capable of readily digesting fibrin, and that when a gelatin film was inoculated with some of the same liquid no micro-organisms of any kind made their appearance upon it.

It is difficult to explain the discrepancy between these statements; possibly the age of the pitcher, the condition of the plant, or some other condition may affect the formation of the enzyme. At the same time it is impossible to avoid laying greater stress on positive than on negative results.

More recently Vines has re-investigated the subject, with the result that he has proved the enzyme to exist and to be tryptic in character, breaking down the digested proteid partially to the state of amide crystalline bodies. His experiments were carried out under strict antiseptic precautions. A typical experiment carried out with the secretion of *Nepenthes Mastersiania*, may be quoted here:

"Two test-tubes were prepared, each containing 5 c.c. of neutral pitcher-liquid and a shred of fibrin; to the one (A), 5 c.c. of ·25 % HCl were added; to the other (B), 5 c.c. of distilled water: the tubes were placed in the incubator (about 35° C.) at 11.30 a.m. At 2.30 p.m. the fibrin in tube A was completely dissolved, the liquid giving a good biuret-reaction. The fibrin in tube B was still undissolved at 9 a.m. on the following morning; 5 c.c. of ·25 % HCl were then added to it, with the result that the fibrin had completely disappeared by 11.30 a.m."

In considering this experiment it is difficult to avoid admitting the presence of the enzyme; fibrin is not dissolved by ·125 % HCl except with extreme slowness; bacterial action would not be sufficiently rapid to dissolve it in anything like the time in which it disappeared.

Vines quotes several other experiments in which the digestions were carried out under the influence of such antiseptics as potassium cyanide, thymol, chloroform, and corrosive sublimate. In all these cases digestion of the fibrin was complete in a few hours. Egg-albumin was digested also, but not quite so rapidly. The two proteids behave similarly under the action of the animal enzymes, the digestion of fibrin being easier than that of white of egg.

The writer has also examined the question of bacterial agency in the process. Some faintly acid pitcher-liquid was put into a small bottle and a little coagulated egg-albumin added. It was then tightly corked and set aside. The proteid was gradually digested, leaving a little granular residue. The bottle with its contents was kept under observation for

12 months, and at the expiration of that time the liquid it contained was perfectly limpid and clear, showing not a trace of bacterial contamination. Had the digestion been due to micro-organisms it would have become turbid and shown evidence of considerable growth in a few days, as the amount of proteid in it was fairly large.

Clautriau has observed the digestion of proteid matter in the liquid of the pitcher while the latter remained attached to the plant. He denies that bacteria were present, for he found the liquid to remain limpid and to possess no odour of putrefaction.

It has been noticed in connection with many of the enzymes we have discussed that they exist in the various juices in association with some form of proteid. So close is the connection that many observers have held the view that the enzyme and the proteid are identical. It is interesting to note in the case of Nepenthes that Vines says many very active liquids taken from pitchers contained hardly a trace of proteid matter; they gave only a very faint xanthoproteic reaction, no precipitate with nitric acid, or with acetic acid and potassic ferrocyanide, and no turbidity when boiled.

The course of proteolysis under the influence of the Nepenthes enzyme appears to be similar to that effected by bromelin and the other trypsins described. There is always a little insoluble residue of the nature of an antialbumid, a neutralisation precipitate which is probably hetero-albumose, a quantity of deutero-albumose, not precipitated by nitric acid, a variable but slight amount of peptone, and some leucin. Some observers question the existence of the peptone, but the balance of evidence is in favour of its presence.

Clautriau did not find any amido-acids and he consequently held the enzyme to be a form of pepsin.

Besides the pitcher-plants, there are others which capture and digest small insects. The chief of these are *Drosera*, *Dionœa*, and *Pinguicula*. Our knowledge of the nature of the enzymes which these form is very much less complete than that which we possess in the case of Nepenthes, being chiefly confined to observations on the fate of small pieces of nitrogenous matter placed upon the leaves and subjected there to the action of the secretion which exuded from the latter.

Darwin investigated the behaviour of the leaves of Drosera in great detail. These structures are provided with long-stalked glands which when stimulated pour out upon the surface of the leaf a peculiar viscid secretion having an acid reaction. When an insect, or a small piece of nitrogenous matter, is placed upon the leaf, the glands or tentacles bend over slowly and enclose it, at the same time pouring out the viscid fluid. The surface and the margins of the leaves are alike provided with the tentacles, but the secretion of the central ones is more acid than that of those at the periphery. The imprisoned matter, living or dead, is slowly dissolved by the secretion, and the resulting products are absorbed by the leaf-surface. Darwin found that the secretion could dissolve not only proteid matters, but also connective tissue, cartilage, and gelatin, while it had no action on mucin. The secretion of the enzyme like that of the animal glands is dependent upon the absorption of nitrogenous matter by the leaf, and the acid is only developed under the same condition. If the leaf is stimulated by putting upon it a piece of indigestible matter, less secretion takes place and the liquid poured out has no proteolytic powers.

Darwin discovered that the same enzyme exists also in the leaves of Dionæa. These differ in the arrangement of their glands from those of Drosera; the leaves have their upper surfaces covered with small, almost sessile, secreting glands of a purplish colour. Like the leaves of Drosera those of Dionæa do not secrete any enzyme until they are excited by the absorption of nitrogenous matter. Then they pour out a fluid which is colourless and slightly mucilaginous. It is more acid than that of Drosera and like the latter dissolves coagulated egg-albumin. Pinguicula also secretes a similar digestive fluid on the edges of the upper surface of the leaf, which folds over to enclose its captives.

No examination has been made at present of the products of digestion by these enzymes. It is consequently uncertain whether they are peptic or tryptic. From analogy with the pitcher-plants it seems probable that they belong to the latter category, but the point must for the present be left undecided.

Similar considerations affect certain enzymes which have

been found to be secreted by several of the Fungi. One of the earliest known of these is the ferment which Krukenberg found to be procurable from the plasmodium of *Æthalium*, one of the Myxomycetes. A glycerin extract of the plasmodium was found to have very marked proteolytic powers in the presence of lactic or hydrochloric acid. Krukenberg's statement has been confirmed by Miss Greenwood, who has stated that the plasmodium of another member of the same group yielded to ·4 per cent. hydrochloric acid an extract which showed marked solvent action on fibrin.

Bourquelot states that an extract of the mycelium of *Aspergillus niger* has the property of dissolving both fibrin and coagulated egg-albumin when digested with them for 2 hours at 40° C. There is an unmistakeable formation of peptone during the digestion, which takes place best in a neutral medium. The extract of Aspergillus has also the power of liquefying gelatin. Malfitano has found that the secretion of the enzyme is constant during the life of the fungus, but is chiefly associated with the death of the cells which form it.

Bourquelot has also obtained a feebly proteolytic enzyme from *Polyporus sulfureus*, and Zopf and Bourquelot and Hérissey state that ferments of this kind exist in several other fungi. Bourquelot's results have been confirmed by Hjort.

Descending still lower in the scale we find evidence of the probable existence of proteolytic enzymes in *Yeast*. If this organism is kept in a medium which supplies it with no nourishment, such for instance as a liquid containing no sugar, and if it is at the same time deprived of oxygen, a process of true digestion of its own proteid reserve-material takes place. If yeast is pressed till it is dry, and the exuding liquor collected, it will be found to contain a great deal of proteid material, which forms a bulky coagulum when the fluid is heated to 45° C. After standing a few days under antiseptic precautions, this form of proteid will have disappeared, so that the liquid gives but a slight coagulum even on boiling. These facts point to the presence of a proteolytic enzyme in the yeast cells. Hahn has recently prepared it by the following method. He pressed yeast till it was dry and then carefully ground it up

with kieselguhr and sand till the cells were disintegrated,
when he prepared an extract by making the fine powder into
a paste with water. On pressing this till the liquid was
squeezed out, a yellowish solution resulted, which contained
a quantity of proteid material. Hahn treated a small volume
of this extract with chloroform and added to it some solid
gelatin containing a trace of phenol. He found that an
appreciable amount of gelatin dissolved in 24 hours and that
the process of solution went on till the whole was liquefied.
The absence of either yeast-cells or micro-organisms was ensured
by the antiseptics used.

The same author in conjunction with Geret proved the
presence of the enzyme by another method. It was stated
above that the extract of the yeast contains a quantity of
proteid matter. The authors added chloroform to some of this
extract and kept it at a temperature of 37° C. for several weeks.
The chloroform served two purposes; it kept the liquid free from
microbes, and it slowly precipitated the proteids. The bulky
precipitate so formed slowly but gradually diminished, and at
the end of several days the extract was almost clear. After this
stage was reached the liquid again became turbid and the
turbidity increased for several days. The second precipitate
was found to consist chiefly of crystals of tyrosin, while some
leucin was ascertained to be in solution in the mother liquor.

The course of action has thus been shown to be comparable
with that taken by pancreatic trypsin.

Geret and Hahn have shown more recently that the enzyme
is capable of digesting added albumin as well as the yeast
proteids. They say that hypoxanthin occurs as a product of
the digestion.

Various forms of Bacteria also are known to form proteolytic
enzymes. In 1887 Bitter showed by sterilising a culture at a
temperature of 60° C. that one of these organisms excretes the
enzyme into the medium in which it is growing. He found
this treatment killed the micro-organisms, but did not destroy
the enzymes, which continued able to liquefy gelatin and to
peptonise albumin. Hankin has extracted from the bacillus
of anthrax (*Bacillus anthracis*) an enzyme which is capable

of forming albumoses from fibrin. Several toxic bodies of this class have been traced to similar agency, an extract prepared from the organisms being capable of effecting the proteolysis in the absence of the cells. The ordinary putrefactive bacteria may excrete or yield an enzyme resembling trypsin in its action on proteids. Sirotinin has shown that culture fluids in which certain microbes have been growing have been able to liquefy gelatin after they have been filtered through porcelain. The conditions of life of these bacteria frequently influence the secretion of the enzyme. Lauder Brunton and MacFadyen found that a particular microbe secreted a proteolytic enzyme when it was cultivated in meat broth, but not otherwise. Indeed the secretion showed a very definite relationship to the culture medium, having diastasic powers when the latter consisted of starch-paste. The two enzymes were quite distinct, the proteolytic one being most easily extracted when present. Acids favoured and alkalis impeded its activity.

The relative amounts of trypsin and diastase in the pancreatic juice have also been shown to be largely influenced by diet. Vassilief established this fact in the case of dogs.

Among the microbes which secrete proteolytic enzymes may be named four which are well known. These are Koch's *cholera bacillus*, Deneke's *cheese-bacillus*, Finkler's *cholera-nostras-bacillus*, and Miller's bacillus. The enzymes have been extracted by Wood from culture fluids in which the microbes had grown, and which had subsequently been sterilised by antiseptics. Wood found that the enzymes from the different bacilli varied a good deal in their power of resisting the influence of acid media, those from Koch's bacillus being destroyed by very little acidity, while those from Finkler's and Miller's bacilli could act in distinctly acid solutions. Wood noticed that the bacilli themselves showed a varying susceptibility to acids exactly corresponding to that of the enzymes. Vignal discovered that *B. mesentericus vulgatus* secretes a peptonising ferment, in addition to those already mentioned.

Babcock and Russell have shown that *Bacillus subtilis* and

two other species of Bacillus carry the proteolytic decomposition
as far as the liberation of ammonia.

The proteolytic enzyme in all these bacteria is associated
with at least one other, generally rennet.

Fermi has ascertained that the proteolytic enzymes secreted
by several Schizomycetes have the power of liquefying gelatin,
and that this corresponds to their ability to form peptone from
fibrin. He has found such enzymes in cultures of *Bacillus
subtilis, B. anthracis, B. megatherium, B. pyocyaneus, Vibrio
cholerae asiaticae, V. Finkler-Prior, Micrococcus prodigiosus,
M. ascoformis, M. ramosus,* and a few others. Fibrin is dissolved by these enzymes as well as gelatin, though with a little
greater difficulty. Egg-albumin and coagulated serum-albumin
are still more resistant, so that the enzymes hardly correspond
to either pepsin or trypsin. Indeed Fermi gives reasons for
supposing that several enzymes exist in these plants, pointing
out that they are destroyed at different temperatures. Thus
the enzyme of *Micrococcus prodigiosus* is unable to work at
a temperature of 55° C., while those of *Bacillus pyocyaneus,
B. anthracis* and *Vibrio Finkler-Prior* are not rendered inactive
till heated to 60° C., 65° C. and 70° C. respectively. These
enzymes resemble trypsin in working most advantageously in
faintly alkaline solutions, though they will attack a solution of
gelatin containing ·5 per cent. of hydrochloric acid.

Another proteolytic enzyme has been described by Duclaux
which is associated with the digestion of casein or tyrein both
in milk, and in cheese during the operation of ripening. He
gave it the name of *casease.* It is secreted by several species
of bacteria belonging to the genus *Tyrothix,* which also contain
a rennet ferment such as will be described in a subsequent
chapter.

This proteolytic enzyme can be precipitated from a culture
of any of these species by the addition of a large excess of
alcohol. It acts not only on the clotted proteid produced by the
rennet, but on the unchanged casein of the milk, and its action
can be traced by the disappearance of the opalescence of the
liquid, a fat-splitting enzyme being associated with it. This
enzyme occurs also in the hepato-pancreatic secretion of *Sepia,*

which has the same effect on milk. Casease is a tryptic ferment, leucin and tyrosin being formed among the products of its activity. When it is working in the cheese it produces also the amido-bodies already mentioned which occur together with other products that are traceable to the metabolic activity of the micro-organisms. Weigmann states that casease can be prepared from bacterial cultures, and that when added to fresh cheese it accelerates its ripening just as do the organisms themselves.

On a review of all these vegetable proteolytic enzymes it will be seen that our knowledge is not at present sufficiently definite for us to say whether we have to do with one or many. Some of them may be peptic only, though it seems probable that they are all tryptic. Those which have been at all exhaustively examined undoubtedly carry the proteolysis to the stage of crystalline amido-acids, some of those secreted by bacteria even liberating a little ammonia. We do not yet know again whether there is one enzyme only, varying somewhat in its features according to the conditions of its secretion, or whether the different plants discussed yield different varieties of trypsin. Bromelin and papaïn certainly show very little difference in their behaviour, and one is tempted to pronounce them identical. For the present however it is perhaps advisable to leave this question undecided.

The identity of the trypsins of animal and vegetable origin is not established. From a consideration of the composition of the proteoses and peptone formed during their respective action Chittenden has concluded them to be different bodies. Speaking of bromelin he says, " We are impressed with the exceptionally low percentage of nitrogen in the deutero-albumose and peptone. In fact this is so contrary to our general experience with digestion products formed by animal ferments that we are forced to consider it as something peculiar to the vegetable ferment. As is well known, secondary proteoses and peptone formed by the action of pepsin and trypsin usually show a much lower content of carbon than the proteid undergoing digestion, as in the bromelin products, but the percentage of nitrogen is ordinarily increased in proportion to the decrease of carbon....

The vegetable ferment bromelin is peculiar in giving rise to secondary proteoses and peptones with a much lower percentage of nitrogen than is contained in the mother proteid, thus implying a cleavage of a nitrogen-containing radical as part of the proteolysis. This, if true, would constitute a good ground of distinction between the animal and vegetable proteolytic ferments."

CHAPTER XIV.

FAT-SPLITTING ENZYMES. LIPASE. (*PIALYN*, *STEAPSIN*.)

THE transformations which attend the digestion of fat in the mammalian alimentary canal are twofold. A variable quantity of the fat of a meal—considered by some physiologists the greater part—appears suspended in the liquid contents of the intestine in the form of very fine globules which show no tendency to run together. Such a condition constitutes what is known as an *emulsion*. It is apparently a purely physical condition, and the unaltered fat can be separated from the liquid by shaking the latter with ether. This is the condition in which the fat remains permanently in milk. But beside this physical change a chemical alteration also occurs, the molecule of the fat being hydrolysed, with the result that it is decomposed into fatty acid and glycerin. How much of the fat undergoes hydrolysis is at present chiefly a matter of speculation, many observers thinking that only a small percentage is so changed and that the greater amount is only emulsified and absorbed by the intestinal epithelium otherwise unaltered. The two processes occur together in the duodenum and there is little doubt that the emulsification is largely helped by the coincident hydrolysis. A small amount of soap, which is formed when free fatty acid combines with an alkali, materially accelerates the process of emulsion. If a *neutral* fat is shaken with a dilute alkali no emulsion occurs, but if the fat is *rancid*, that is if a little fatty acid is present, an emulsion is very readily produced.

Claud Bernard was the first physiologist to point out that in the digestion of fat the pancreatic secretion plays a most important part. He called attention to the fact that both the phenomena mentioned occur under its influence, showing that when neutral oil and pancreatic juice are shaken up together an emulsion rapidly results, and further that the action of pancreatic juice on oil produces free fatty acid. In his opinion the chemical change took much longer to set up than the physical one. By many physiologists these have both been held to proceed from the action of enzymes, which have been called *emulsive* and *saponifying* respectively. Twenty years after Bernard wrote it was established that there is no emulsive enzyme but that the emulsion is due to the free fatty acid which results from the action of the saponifying one. Among the workers upon whose investigations this view is based may be mentioned Brücke and Gad. The enzyme has been variously named by different writers, having been called *Pialyn, Steapsin*, and more recently *Lipase*. The decomposition it sets up is expressed in the following equation

$$C_3H_5(C_{18}H_{35}O_2)_3 + 3H_2O = C_3H_5(HO)_3 + 3(C_{18}H_{35}OHO)$$

stearin   glycerin   stearic acid

which represents the hydrolysis of stearin, one of the neutral fats.

The enzyme which, adopting the most recent terminology, we may call *lipase*, can be detected in the secretion of the pancreas when this is collected by means of a cannula inserted into the duct of the gland, or it can be prepared from a fresh gland by extracting it with glycerin or water. It is important to use a neutral solvent, as the enzyme is very easily injured by acids. Lipase has also been found in the vegetable kingdom, where it plays an important part in the utilisation of the fatty reserve-products which are stored in many seeds. During the germination of such seeds the fats undergo decomposition with the liberation of free fatty acids. The process of emulsion does not appear to occur as it does in animals.

Lipase has a fairly wide distribution in the animal kingdom. In mammalia the great seat of its formation is the pancreas, from which it is easy to extract it by either of the methods already mentioned. It is said to occur also in the stomach, but in much smaller quantity.

Hanriot, who has detected its occurrence in the blood of several vertebrates, has tested its action on monobutyrin. He says this body is readily saponified by blood serum in neutral or slightly alkaline solution, but a cautious addition of alkali is necessary as the action proceeds, for it is inhibited by the free fatty acid produced as soon as the latter reaches a certain percentage. The serum loses its power of hydrolysing monobutyrin if it is heated to 90° C.

Cohnstein and Michaelis have also observed that the blood has the power of causing the disappearance of the fat of the chyle, which they attribute to a lipolytic substance in the red corpuscles.

The power of hydrolysing fats is found also among the invertebrata. Fredericq extracted from sponges a substance which formed an emulsion with neutral fats and finally decomposed them into fatty acids and glycerin. The same body is present in *Uraster*, one of the Echinodermata; also in the so-called liver of the spider, and of the edible snail, the latter only containing it in summer. The "liver" of several of the Crustacea yields an extract which appears to have a similar action, as it renders milk transparent. The corresponding organ in certain of the Cephalopoda has the same action, and it has been shown to hydrolyse neutral fats. Lipase has been prepared from the "liver" of *Sepia*. Biedemann has found a lipase in the secretion of the intestine of the larva of *Tenebrio molitor*, the Mealworm. The eggs of several of the Crustacea have also been shown by Abelous and Heim to contain this as well as other enzymes.

The presence of lipase in a solution, secretion, or extract can be tested by preparing an emulsion of a neutral oil, with the aid of a little powdered gum arabic and a very little water. The emulsion should then be mixed with the extract and the whole neutralised with care, as the gum has sometimes an acid reaction. The whole may then be digested at 40° C. together with a minute quantity of neutral solution of litmus. If the enzyme is present the fatty acid it liberates from the oil causes the liquid to turn pink or red as it affects the litmus. The rapidity of the change of colour gives some idea of the amount of lipase present.

Another method which may be used to test smaller quantities of solution is based upon Gad's experiments on the spontaneous emulsions set up when free fatty acids and oil are added in appropriate quantities to dilute solutions of carbonate of sodium. In this case no shaking is necessary, but a beautiful emulsion results as soon as the liquids are brought into contact. Gad's method may be described as follows:

Carbonate of sodium solution of ·25 per cent. strength is placed in a series of watch-glasses, and drops of oil containing different percentages of fatty acids placed carefully upon the surface of the alkaline solution. With a certain percentage of free fatty acid in the oil, an emulsion at once results. If the optimum percentage is present, this emulsion is instantaneously complete with less or more than this amount, the emulsion is more or less imperfect. With the percentage of sodium carbonate mentioned and at a temperature of about 18° C. the oil must contain 5·5 per cent. of free fatty acid to give the best results. The emulsifying agent in this method is the soap produced by the combination of the free fatty acid with the alkali. In the presence of this no gum arabic is necessary.

The formation of a spontaneous emulsion is therefore a very satisfactory test for the presence of free fatty acid. If perfectly neutral oil is taken and digested for a time with a solution containing lipase the enzyme will set free sufficient acid to make the emulsion, while the completeness of the latter will depend upon the proportion so liberated.

This method has been employed by Rachford in a series of researches upon the lipase of pancreatic juice, and incidentally upon certain problems connected with the digestion of fat in the small intestine. Rachford obtained his pancreatic juice by inserting a cannula into the pancreatic duct of a rabbit and collecting the secretion which exuded therefrom. He mixed a small quantity of the juice with twice its volume of neutral olive oil in a small test-tube and after shaking the mixture he allowed it to settle. Separation of the two fluids was almost immediate. Then taking a drop of the floating oil he placed it upon the surface of a small quantity of ·25 per cent. solution of carbonate of sodium, and noticed whether any emulsion resulted. After a few minutes' interval he repeated the operation and continued to experiment in the same manner for some time. By taking a series of such observations, with definite intervals of a few minutes after each, he was able to see that the power of forming an emulsion was gradually developed in the oil. It follows necessarily that lipase was present in the pancreatic juice and that it gradually decomposed part of the neutral oil used in the experiment. Rachford says that the alkalinity of the juice employed was not sufficient to absorb the free fatty acid liberated, so that the mixture did not form an emulsion in the test-tube.

The lipase of the pancreas does not appear to be inhibited by the free fatty acid as Hanriot says is the case with the enzyme in blood. It will decompose most of the neutral fats, but its action on castor-oil is comparatively feeble. It hydrolyses not only the fats which are fluid at the temperature of the body, but also solid ones, such as spermaceti, though its action on these is very slow.

In the process of the digestion of fats in the intestine the action of lipase is complicated by the free acid which the chyme contains as it leaves the stomach and by the bile which is poured into the duodenum coincidently with the pancreatic juice. Rachford found that the effect of the addition of an equal volume of bile to the extracted juice and the oil in a test-tube is to increase the action of the lipase in the proportion of three and one-fifth to one. If ·25 per cent. solution of hydrochloric acid is substituted for the bile the action is materially retarded. If however both are added simultaneously the hydrolysis of the oil is increased in the proportion of four to one. The influence of the bile is chiefly associated with the glycocholate of sodium which it contains. Rachford found the action of lipase was considerably impeded by excess of alkali.

The extent to which lipase takes part in the process of the digestion of fat appears now to be more considerable than was formerly supposed. It has been held by many observers that fat is absorbed by the intestinal epithelium in the form of an emulsion, and that the work of lipase is completed when enough fat has been decomposed to form the quantity of soap required to set up such an emulsion. Recent investigations have led many physiologists to question the accuracy of this view and to hold that the fate of the great bulk of the fat is to be hydrolysed, and absorbed either as free fatty acid or as soap. The emulsion may then be regarded as helping the further hydrolysis and saponification. It is however outside the scope of the present work to enter upon a discussion of this point.

The activity of lipase like that of other enzymes is remarkably influenced by temperature. The optimum point for its working is not that of the body, as would be expected, but 55° C. It has not so wide a range as some ferments, being very materially slowed by increasing the temperature to 60° C., and it ceases to be active at 72° C. When the conditions are so arranged that the action is rapid, the amount of change is proportional to the amount of lipase present.

Hanriot has published the results of experiments made with the view of comparing the activity of lipase prepared from different sources. The most interesting comparison is that between the enzyme of the serum and that of the pancreatic juice of the dog. In his researches he prepared solutions of lipase from the two sources in such a way that under standard conditions they had the same hydrolysing action on monobutyrin in the presence of a little sodium carbonate. When his solutions were neutralised and left to work for 20 minutes the quantity of fatty acid liberated by the serum was nearly twice as much as that set free by the pancreatic solution. It appears therefore that serum lipase is more energetic than pancreatic in the presence of the products of the action. Solutions of the two lipases that showed equal hydrolytic power at 15° C. were examined at different temperatures. Serum lipase liberated one and a half times as much fatty acid at 30° C. and twice as much at 42° C. Pancreatic lipase showed no change of activity through that range of temperature. Serum lipase retained its activity for several months under antiseptic precautions, while the pancreatic enzyme was inert after a few days.

Hanriot says that the lipase prepared from the serum of

the horse is materially assisted by weak alkali. In one case he found that the presence of ·2 per cent. of sodium carbonate increased the production of fatty acids fourfold.

Lipase exists in the vegetable as well as in the animal kingdom. The very frequent occurrence of oils in seeds suggests that a good deal of carbonaceous matter is stored in this form for the nutrition of the young embryo. It is noteworthy that oil and starch seldom occur together in seeds. The transformations of the oily reserves in many endosperms during germination are however very quickly followed by the appearance of starch in the young plant, to which the products of the germinative changes have made their way. The view that this indicated an actual transformation of oil into starch was advanced by many observers, based of course upon the elementary facts of the disappearance of the oil and the formation of the starch. This view was held by Sachs in 1859 and by other writers who followed him. More accurate views were introduced in 1871 by Müntz, who pointed out that during germination a fatty acid appears in the seed, pointing to a process of hydrolysis of the oil. Schützenberger in 1876 showed that when an oily seed is bruised in water, an emulsion is obtained in which careful observation will soon show the presence of glycerin as well as fatty acid. He pointed to the evident hydrolysis going on and suggested that it is due to an enzyme. Detmer in 1880 attempted to reconcile the new theory with the old one by the suggestion that the fatty acid arising during the hydrolysis is the immediate antecedent of the starch, and put forward the following theoretical equation to explain the process:

$$C_{18}H_{34}O_2 + 27O = 2C_6H_{10}O_5 + 6CO_2 + 7H_2O.$$

It is evident however that no suggestion involving the direct transformation of oil into starch is at all adequate to explain the course of events, for neither oil nor starch is diffusible, and as the two bodies appear at a considerable distance from each other, separated by a number of cell-membranes, there can be no very direct connection between the disappearance of the one and the formation of the other.

The enzyme whose presence was suggested by Schützenberger was discovered by the writer in 1889 in the germinating
seeds of *Ricinus*, the castor-oil plant. Some of these seeds
were germinated for about 5 days, until the embryo was of some
considerable size, its hypocotyledonary portion being $2\frac{1}{2}$ inches
long and a fair root system developed. The endosperm was
swollen and semi-mucilaginous in appearance where it was in
contact with the cotyledons. Such endosperms were ground up
in a mortar with a solution containing 5 per cent. of sodium
chloride and ·2 per cent. of potassium cyanide, the latter being
used as an antiseptic. After standing 24 hours the liquid was
filtered till it was nearly clear, retaining only a slight opalescence.
A thick emulsion of castor-oil was then prepared and a little
of the extract carefully stirred into it, a little neutral litmus
solution being added at the same time. A boiled control was
prepared in the same way, and the two were placed in an incubator at 35° C. In about half-an-hour the litmus in the
unboiled preparation began to redden, indicating the liberation
of fatty acid. This was extracted by shaking the digestion with
dilute soda solution, when the acid formed a soluble soap with
the alkali. On decomposing the soap solution with a mineral
acid a quantity of fatty acid soon rose as a scum to the surface.

In a subsequent experiment a larger quantity of the
emulsion treated in the same way was allowed to digest in a
dialysing tube, suspended in distilled water. The digestion was
carried on for a week; during this time the reaction in the
dialyser became more and more acid, while that of the surrounding fluid remained unchanged. At the end of the week
the dialysate was concentrated and examined for the presence
of glycerin, which was detected in it by the acrolein test. A
control experiment with boiled extract, which was carried on
side by side with the other, showed no change of reaction and
no glycerin was present in the dialysate. The experiments
showed that lipase was present, that it hydrolysed the oil,
forming fatty acid and glycerin, the latter passing into the
dialysate, while the former was not able to do so. Many
subsequent experiments confirmed these results.

An examination of the properties of the lipase showed that

16—2

it worked most advantageously in neutral solution, but was also active in dilute alkalis, and to a much less extent in dilute acids. The activity was tested by noting the amount of a standard alkaline solution which was necessary to neutralise the fatty acid formed. Comparing three digestions, the first of which was neutral, the second containing ·066 per cent. of sodium carbonate, and the third ·066 per cent. of hydrochloric acid, the relative activities were as 78 : 69 : 21. Hydrochloric acid when present in the proportion of ·13 per cent. stopped the action almost entirely; sodium carbonate needed to be present in the proportion of ·6 per cent. to lessen its power one-half. The effect of prolonged action of both acid and alkali was the destruction of the lipase and not merely an inhibition of its working. Neutral salt such as sodium chloride impeded the working of the enzyme, but did not destroy it.

Lipase is present only in the endosperm of Ricinus, the embryo containing none. This is what would be expected, as the oil does not pass from the one to the other and the carbonaceous food material temporarily appearing in the embryo is in great part starch.

There appears to be no lipase present in the resting seed of Ricinus, but ground seeds when kept at 35° C. for a few hours in the presence of very dilute acetic acid develop the power of hydrolysing oil. An extract of the resting seeds, made with salt solution, and then faintly acidulated with acetic acid and kept warm, also undergoes the same change. At first quite inactive, it gradually develops the enzyme, just as the pancreas under the same treatment develops trypsin. The lipase therefore may be regarded as existing in the form of a zymogen in the resting seeds, the latter being converted into the enzyme when germination begins. The same transformation takes place without acid, if an extract of resting seeds is allowed to stand some days under adequate antiseptic precautions.

The further changes which the products of hydrolysis undergo may be briefly alluded to here. The fatty acid is not further affected by lipase; after a prolonged exposure of the latter with ricinoleic acid no change can be detected. During germination however the fatty acid does disappear from the cells in which it is formed, and as it cannot dialyse through a membrane it is evidently decomposed. As the germination proceeds, other acids, crystalline in character, make their appearance in the cells, the quantity being approximately proportional to the diminution of the fatty acid. There is every probability that the ordinary oxidative processes going on in the cells under the influence of the protoplasm transform the heavy fatty acids into others which can leave the cells by dialysis. There is a continuous production of sugar during the germination which must come from the oil. It is most probable that its immediate antecedent is the glycerin which is the other product of the hydrolysis. This travelling through the tissue is absorbed by the embryo, and is the forerunner of the starch which the latter soon exhibits in its cells.

The formation of sugar from the fat or oil has been shown recently by Mazé. He ground up germinating seeds of Ricinus with clean sand and made a friable paste, which was easily penetrable by the air. This paste was spread out in a thin layer over a water bath kept at a temperature of 53° C. A control was prepared and the paste heated to 100° C. for 10 minutes. They were allowed to stand for several hours, at the end of which time the liquid was drained off from both. The control gave an emulsion, the other a limpid liquid which filtered clear. The latter contained sugar in abundance, the former scarcely a trace. Mazé determined the quantity of sugar present twice, after 7 hours and again after 22 hours. The amount was nearly doubled in the interval. Mazé attributes the formation of the sugar to an enzyme, but he does not indicate its course of action. In his experiments he found he was able to obtain a quantity of sugar equal to about 3·5 per cent. of the dry weight of the ungerminated seeds or about 7 per cent. of the weight of its oil.

The existence of lipase was demonstrated two years later by Sigmund, who found it in both resting and germinating seeds of the *Rape*, the *Opium Poppy, Hemp, Flax* and *Maize*. His mode of experiment was to crush the seeds with water, and estimate the free fatty acid in the resulting emulsion. Comparing the amount of standard alkali needed to neutralise part of the emulsion immediately on crushing and a further part after allowing it to stand for 24 hours at 30° C., he found in all cases a larger quantity was required after the interval. Free fatty acid was accordingly developed on standing, and this was due no doubt to the presence of lipase. Sigmund says that the resting seeds he examined contained a certain amount of enzyme, and that this was increased at the onset of germination.

It has already been mentioned that Sigmund attributes the power of hydrolysing oils to myrosin and to emulsin. Causing what he took to be a preparation of these enzymes to act upon olive oil in closed glass vessels at a temperature of 38° to 40° C., he found that there was a gradual and continuous formation of free fatty acid in the mixture, its presence being demonstrated by both litmus and phenol-phthalein.

On looking over his experiments it does not appear to be at all certain that the results were due to emulsin or to myrosin; it seems far more likely that he was dealing with lipase present in the seeds in addition to the other enzymes. The occurrence of more than one enzyme in a tissue is very common, both in vegetable and animal organisms, and many instances of it have already been given. Sigmund prepared his two glucoside-splitting enzymes by bruising in water seeds of the mustard in one case and of the almond in the other, and allowing them to digest with excess of water for 12—14 hours. He then decanted the supernatant fluid and added to it an excess of alcohol. This caused the formation of a bulky precipitate; removing the latter by filtration he washed it with alcohol and dried it at about 40° C. This method would certainly not prepare any enzyme pure; his precipitate would contain all the soluble proteids of the seed and all the enzymes that might be present. The action of a solution of this powder at once suggests that it contained lipase, and the natural deduction from the experiment would be that mustard seeds and almonds must be added to the list of those in which the existence of this enzyme has been established, rather than that glucoside-splitting enzymes possess the power of hydrolysing oils. The constitution of oil on the one hand and of glucosides on the other opposes very strongly the view that there is one hydrolytic agent which is capable of decomposing them both.

It is interesting to note that Sigmund attributes to his lipase the power of splitting up amygdalin and salicin. On his hypothesis therefore lipase and emulsin are practically the same body. This appears equally unlikely.

The germination of several oily seeds has been examined also by Leclerc du Sablon, whose work has embraced the *Castor-oil* plant, *Hemp*, *Colza* and *Flax*, besides one or two less conspicuously of an oily nature. Du Sablon found the oil of these seeds to diminish markedly during germination, and the diminution to be accompanied by the formation of sugar of both the cane-sugar and glucose types. In his study of the processes of germination he relied chiefly upon analyses of the contents of the seeds from time to time. He pronounces against the hydrolysis of the oil on the ground that he has not been able to detect the presence of glycerin. This however is not very surprising, as most other observers have shown that no appreciable quantity of it is present in the tissue of the seeds during the progress of the germination. It should not however be hastily inferred that none is ever formed; it is quite conceivable that it may undergo further decomposition as fast as it is set free. Du Sablon did not attempt by any method to secure an accumulation of it, and hence his denial of its existence cannot be accepted as decisive. He did not attempt either to bring about the decomposition of the oil by any extract containing lipase, but relied entirely on his analyses.

In his discussion of his results he suggests the existence of a modified lipase which sets free fatty acid from the oil, but does not liberate glycerin; such fatty acid however represents but a small part of the oil decomposed. The glycerin radical of the whole of the oil together with so much of the fatty acid as is not set free goes directly to form sugars, of which those of the cane-sugar type appear first and give place subsequently to those resembling glucose, in which latter form they are absorbed. Transitory starch also appears in the seeds during these changes; and ultimately this also is converted to dextrin and sugar.

Both the cane-sugar and the starch he is inclined to consider as indications that the oil is decomposed more rapidly than it is required by the young plantlet, and hence he regards them as secondary and temporary reserve-stores. He identifies both invertase and diastase as present in the watery extracts of his seeds. In some cases, especially in the seed of Colza (*Brassica oleracea*), he thinks the dextrins occur as inter-mediate bodies between the oil and the non-reducing sugar.

One difficulty in the way of accepting du Sablon's views on the course of the transformation is the great unlikelihood of fatty acid serving directly as an antecedent of carbohydrates. No instance of a similar transformation is known, and no laboratory method can at present bring it about. The great stress he lays on the presumed absence of glycerin is also noteworthy, for it is not difficult with proper methods to prove its occurrence, as the writer has already shown in the course of his own work on Ricinus. Its presence or absence appears very important as indicating whether or no the enzyme which he predicates is identical with lipase.

It has already been noticed that *Penicilium glaucum* is the source of several enzymes. To those already mentioned, lipase must be added. Gérard has found that the watery extract of this fungus is capable of hydrolysing *monobutyrin*. The best means of showing its power is the process of cultivating it in Raulin's fluid to which a little monobutyrin has been added. Butyric acid is very quickly liberated, just as with Hanriot's lipase from blood. It may be noticed in view of Sigmund's work discussed above, that Gérard finds emulsin does not hydrolyse monobutyrin.

The conclusions of Gérard have been confirmed inde-pendently by Camus, who has also found lipase in another fungus, *Aspergillus niger*.

Lipase has been extracted by Biffin from a fungus which sometimes attacks Coco-nuts during germination. The fungus in question has not been completely identified but it appears to be a member of the Hypocreaceæ, a division of the Pyrenomy-cetes. It grows freely on vegetable tissue which is rich in oil.

In his experiments Biffin cultivated the fungus in sterilised coco-nut milk till the liquid became acid to litmus. The mycelium was then taken out of the flasks, washed rapidly in distilled water and ground to a thin paste with kieselguhr and water in a bacterium-mill. On filtering the paste under pressure through several thicknesses of filter-paper a faintly brown opalescent fluid was obtained which gave a slight acid reaction. It was neutralised and made sterile by the addition of potassium cyanide.

Sections of coco-nut endosperm containing oil, and cover-slips coated with a thin layer of oil, were placed in some of the extract and in all cases the oil was decomposed. Control experiments in which the extract had been boiled gave negative results. The extract also decomposed monobutyrin liberating butyric acid.

Biffin found that the lipase could be precipitated by alcohol from the extract of the fungus. It was thrown down as a flocculent whitish grey substance, which powdered readily after being dried over calcium chloride. A solution of the powder had the same power of decomposing fats as was exhibited by the original extract.

CHAPTER XV.

THE CLOTTING ENZYMES. RENNET.

THE enzymes which next present themselves for examination form a group which at first sight appears well marked and sharply separated from those we have so far discussed. The prominent feature of their action is that it results in the formation of a semi-gelatinous clot or jelly, which soon after its production undergoes a species of contraction or shrinkage and ultimately becomes semi-fibrous in character. Instead therefore of producing from a primary body, often insoluble and indiffusible, a more freely soluble and diffusible material, they appear to invert this order. In two cases however the clot or coagulum is accompanied by another product more soluble than the original one, or at any rate not less so. Another feature which is common to three of them is that the action of the enzyme is peculiarly associated with certain compounds of Calcium, in the absence of which the clotting does not occur. It is however not certain that the resemblance in this particular is more than superficial, for the action of the inorganic salt cannot at present be said to have been proved the same in all cases.

One of these enzymes, generally known as *rennet*, occurs in both the animal and vegetable kingdoms; the so-called *fibrin ferment*, and the ferment which is instrumental in causing the formation of myosin in muscles, are only found in the former. Another, *pectase*, is of purely vegetable origin. The three first mentioned may be regarded as related to the proteolytic enzymes: pectase is associated with certain changes in the materials which constitute the cell-membranes of plants.

Rennet (Lab, Chymosin, Presure, Rennin).

The first of these enzymes which we shall discuss is the so-called Rennet, which is so largely employed in the production of cheese from milk. When a small quantity of this body is added to a relatively large volume of milk, the latter rapidly changes to a stiff jelly; after standing a shorter or longer time, the jelly shrinks somewhat, and a watery liquid known as *whey* oozes out of it, and after a little while a firm clot is found floating in a quantity of the watery liquid.

The formation of this clot, which is really the crude cheese, or curd, is due to an alteration of one of the constituents of the milk. This is a body which is known as *caseinogen*; it is a member of the group of nucleo-albumins, but is generally spoken of as the principal proteid of the milk.

As it exists in milk, caseinogen can be readily precipitated from solution by the cautious addition of a dilute acid: it then separates in a loose flocculent form, and can be easily separated from the liquid by filtration. When so separated it can be redissolved in dilute alkalis and again reprecipitated.

When caseinogen either in milk, or in solution in dilute alkalis, is acted upon by rennet in the presence of a small quantity of calcium phosphate, it is at once chemically altered. It gives rise to the formation of a body which has been named *casein*, which is the proteid constituent of the clot. When the whey has separated out, it is found to contain a by-product having properties resembling those of a soluble albumin, except that it is not made insoluble by boiling. It is consequently distinct from the serum albumin, or lact-albumin, present in normal milk. The formation of this by-product when solutions of pure caseinogen are treated with rennet shows that it results from the alteration of the caseinogen. The clotting of milk by rennet is consequently not to be confused with the precipitation of caseinogen by acid. In the latter case the separated body can be redissolved and is found to be unchanged in constitution; in the former case it has been decomposed, with the formation of an insoluble curd (consisting chiefly of casein) and a certain quantity of a variety of albumin. The curd differs in composition from caseinogen by containing a relatively large amount of calcium phosphate, which on ignition of the proteid remains as ash. If this calcium salt is removed from it and it is then dissolved in dilute alkalis it cannot again be clotted by rennet as precipitated caseinogen can. Halliburton suggests that it is caseate of lime.

The chief source of rennet in the animal body is the mucous membrane of the stomach, and it is present in largest quantity in the stomach of young animals, particularly the calf. It can be obtained however from the stomach of almost any animal and may be regarded indeed as one of the normal constituents of gastric juice. It exists in the pancreas of several animals, especially the pig, ox, sheep, horse, dog and cat; also in the human pancreas. Halliburton and Brodie found it present in pancreatic juice obtained from the dog by means of a cannula inserted in the pancreatic duct. Edmunds prepared it from various tissues of the mammalian body, including the testis, liver, lung, kidney, spleen, thymus, thyroid, brain, intestine and ovary, but it was only in very small amount in any of these organs. He also obtained evidence of the presence of a small quantity in the blood. In animals belonging to groups below the Mammalia, Benger has prepared it from the stomach of the cod-fish, and appropriate treatment has shown that it is formed in the stomach of the pike. Roberts obtained it from the digestive organs of the fowl, and Harris and Gow extracted it from the pancreas of the eagle. It is stated that a small amount may be present in urine, but this is probably an indication of its excretion from the body. Babcock and Russell have found that milk itself contains a very small quantity. It has a wide distribution in the vegetable kingdom, but this will be treated of in greater detail in a subsequent section.

The preparation of rennet in an impure form from the stomach or pancreas is very easy. All that is necessary is to mince the organ finely, either in the fresh state, or after dehydration by alcohol, and then to extract the pulpy mass with glycerin, water, or solutions of neutral salts. If the reaction of the solvent is made faintly acid, a more active preparation is obtained than if it is perfectly neutral. The most efficient extractive is a solution of sodium chloride, containing about 5—10 per cent. of the salt.

Many attempts have been made to prepare the enzyme free from admixture with other ferments or with proteids. The first experiments in this direction were made by Deschamps in 1840, but they were subsequently shown to be unsuccessful. The isolation was however carried out by Hammarsten, who published his method in 1872. It consists in the first place of a fractional precipitation of an acid aqueous extract of the stomach by means of magnesium carbonate or a solution of acetate of lead. Both pepsin and rennet are thrown down by these reagents, but pepsin is precipitated with the greater readiness. It is possible therefore to free the liquid from pepsin while the greater part of the rennet remains in solution. After filtering this off, further addition of the lead acetate throws down the rennet, the process being facilitated by the simultaneous addition of a little ammonium hydrate. The precipitate is then filtered off, suspended in water and decomposed by very dilute sulphuric acid. This acid solution, which contains mere traces of albumin, is mechanically precipitated according to Brücke's method for the separation of pepsin, with the aid of cholesterin, as already described.

This method is useful in cases where the rennet exists in considerable quantity, but it always involves the loss of a good deal of material.

A few years later the rennet and the trypsin of pancreatic extract were separated by Roberts by the following method. A salt-solution extract of a pancreas was slightly acidulated with hydrochloric acid and kept for 3 hours at 40° C. When neutralised it was found that the trypsin had been destroyed while the rennet was unharmed. Roberts says that he found that when some of the same extract was filtered through porcelain under pressure, the rennet passed through the porcelain and the trypsin remained behind.

In 1886 a different method was introduced in America by Blumenthal. In his process, the stomach of the calf is cut into small pieces and macerated for 24 hours in a solution of common salt of ·5 per cent. concentration, kept at about 30° C.

The solution is then filtered and ·1 per cent. of mineral acid is added. This produces a thick precipitate of mucous matter which can be separated mechanically, leaving the enzymes in solution. The acidity is next raised to ·5 per cent., and the liquid saturated with powdered sodium chloride. It is then kept for 2 or 3 days at 25°—30° C. with constant stirring, and the temperature is gradually raised to 30°—35° C. After standing for a day or two, a white flocculent scum separates out which can easily be removed by skimming or filtration and dried at about 28° C. This is an amorphous white gelatinous substance, greatly resembling aluminium hydrate in appearance. It is without taste or smell, and dissolves readily in water, forming a clear solution. This precipitate is nearly pure rennet, a small portion of it readily inducing coagulation in milk. The mother liquid from which it is removed has no curdling action but is possessed of very considerable peptic powers.

Blumenthal says that the same treatment will separate the two enzymes from the so-called *pepsin essence* of commerce, which is very impure.

Friedburg found on repeating Blumenthal's experiments that the prolonged standing after saturation with salt was not always necessary, but that if comparatively small amounts of material were used, the rennet separated almost immediately. He recommends its immediate removal, for on standing it shows a tendency to sink in the liquid.

Hammarsten quotes the following reactions as characteristic of his purified product:—

(1)   it does not show the xanthoproteic reaction.

(2)   its watery solution does not coagulate on boiling.

(3)   it is not precipitated by alcohol, nitric acid, tannin, iodine, or normal lead acetate.

(4)   it is precipitated by basic lead acetate.

Friedburg found his product obtained by Blumenthal's method gave exactly the same reactions.

The effect of heat on the rennet enzyme is to some extent dependent on the reaction of the solution. Its optimum point is about 40° C. Neutral preparations will remain active after a short exposure to 70° C., but if kept long at that temperature

they are rendered inert. A momentary exposure to a higher
point than this is destructive of the enzyme. If the preparation
contains ·3 per cent. of hydrochloric acid it will not survive
heating to 63° C. Prolonged exposure of such an acid solution
to only 40° C. is very deleterious. The presence of a little
alkali has a similar influence.

Rennet is active in neutral, or in faintly acid or alkaline
solutions, but the presence of only a little more than the
optimum quantity of either acid or alkali speedily destroys the
enzyme at the ordinary temperature. Hammarsten found that
clotting occurred most rapidly when a faintly acid extract was
used, and Ringer has confirmed the observation, care being
taken that the acid was added in insufficient quantity to precipitate the casein. The clotting was not due to the action of
the acid on the caseinogen, as the addition of the same amount
of acid without the rennet produced no effect upon the milk.
Roberts says that pancreatic rennet differs from gastric in being
active in the presence of larger amounts of alkali. Alcohol in
small quantities does not injure the enzyme, but large quantities or prolonged exposure destroy it.

The effect of neutral salts such as sodium chloride, or
magnesium sulphate in greater concentration than 4 per cent.,
is to impede the action, but not to destroy the rennet. Small
quantities of less than 1 per cent. are beneficial.

The action of rennet is materially impeded by the presence
of peptone or some kinds of proteose. Edmunds has published
an account of some experiments on this point which are very
interesting in the light of the effect of the same proteids on the
process of the coagulation of the blood. He mixed in several
tubes 10 c.c. of milk, a varying quantity of Witte's "peptone,"
containing a certain amount of proteoses, and 150 cubic millimetres of a preparation of rennet and exposed them to a
temperature of 40° C. In the absence of any "peptone"
clotting took place in 10 minutes; with ·625 per cent. present,
it did not set in for 20 minutes; with 2·5 per cent. no clot was
formed till after 45 minutes, and in the presence of 5 per cent.
coagulation was delayed for several hours.

It has already been mentioned that the salts of calcium

play an important part in the action of this enzyme. The most pronounced effect is produced by calcic phosphate; indeed, according to most observers, the clotting of the milk will not take place in the absence of this salt. The experiment can be made by dialysing the milk till all the salts have disappeared, when rennet is inoperative upon it. The same effect can be seen when experiments are carried out on a pure solution of caseinogen prepared by precipitating it from milk with dilute acetic acid, filtering and washing quickly with distilled water. The solution of this in water gives no clot with rennet. If however it is dissolved in lime water and the solution carefully neutralised with dilute phosphoric acid, a milky-looking liquid is obtained which behaves exactly like milk itself, curdling indeed even more readily on the addition of rennet. It is noteworthy in this connection that cheese always contains a certain amount of calcic phosphate. Ringer states that other salts of calcium, especially the chloride, can replace the phosphate.

Hammarsten, to whom our knowledge of this peculiar behaviour is due, explains the action by saying that the enzyme induces the conversion of caseinogen into casein, and the calcic phosphate makes the latter separate out in the condition of the clot or curd. He quotes in support of his view the following experiment.

Some pure caseinogen prepared as described above, and ascertained to be ash-free, was dissolved in dilute hydric-disodic-phosphate and the solution divided into two parts, A and B. A was treated with rennet and both were digested for half-an-hour at 30° C. A was then boiled to destroy the enzyme, and B was boiled also that the two might be capable of comparison. A quantity of the rennet solution equal to what had been added to A was then boiled and added to B. After both had been cooled the same amount of calcium chloride was added to each. A, in which the rennet had acted for thirty minutes without producing a visible change, at once gave a thick precipitate of casein. B, in which rennet though present had been prevented from acting, remained unchanged. The curd appears to be therefore not casein, but a compound of casein and the calcic salt. Ringer has published an account

of similar experiments which completely confirm those of
Hammarsten.

The part played by calcium salts may perhaps explain the
inhibitory action of peptone. This body can combine with such
salts, and it may be that no clot is formed owing to their being
taken out of the sphere of action by means of such a combination. The subject needs however further research before definite
opinions can be formed upon the point.

The proportion of water to caseinogen in the milk must be
taken into account in considering the curdling. Mayer found
that milk which gave a normal clot in 25 minutes took 30
minutes when 5 per cent. of water was added; further addition
of water retarded it longer, and when 20 per cent. had been
added the time taken was 73 minutes, the proportion of rennet
and the temperature of the experiment being alike in all cases.
The clot produced in diluted milk is much more flocculent than
a normal one.

The dilution of the rennet extract itself has still greater
effect in the same direction. Soxhlet found that when one
part of a particular preparation of rennet was made to act upon
ten thousand parts of milk, curdling took place in 40 minutes;
but when the same quantity of rennet was mixed with five
hundred parts of milk, the clot was formed in two minutes and
six seconds. From his experiments he inferred that the time
taken to form the clot was inversely proportional to the amount
of rennet used. When however smaller quantities of the
enzyme are employed, this relation does not hold. The writer
found that with a particular preparation of the enzyme from the
germinating seeds of *Ricinus*, while 1 c.c. clotted 5 c.c. of milk
in two minutes, ·2 c.c. did not produce its effect till after four
hours. In another series of experiments in which the extract
was diluted so that the total amount of fluid added to the milk
in each case was the same, the same result was obtained. 1 c.c.
clotted 5 c.c. of milk in five minutes; ·125 c.c. added to ·875 c.c.
of water, did not clot 5 c.c. of milk till the expiration of $5\frac{1}{2}$ hours.

The character of the clot depends also to a large extent
upon the source of the milk. When cows' milk is coagulated
by rennet, or precipitated by acids, the curd, or the precipitated

caseinogen, is more solid than when human milk is used. This is particularly noticeable in the case of true coagulation, the curd of human milk always being of a loose, flocculent consistency. From what has already been said we can see that the solidity or compact condition of the curd formed by the rennet of the stomach is mainly determined by three factors:—

(1) The concentration of the caseinogen solution.

(2) The amount of soluble calcium phosphate.

(3) The acidity of the milk.

Comparing the milk of the cow with the human secretion we find it contains about twice as much caseinogen, and six times as much calcium, while it has about three times the acidity of human milk. It is natural therefore that the coagulum should be flocculent in the case of the latter, while in cows' milk it is more compact and leathery. By dilution with water and appropriate adjustment of reaction, cows' milk can be so altered that it will coagulate very nearly in the same way as human milk. There will always be some difference between the two, on account of the preponderance of calcium phosphate in cows' milk.

Rennet will induce the formation of a clot in boiled milk, but the curd is very flocculent and a considerable quantity of the enzyme is needed for the coagulation. The alteration which causes this difference is greater in proportion to the temperature to which the heating has been carried and the time the milk has been exposed to it.

Before leaving the subject of the action of rennet on milk it seems not inappropriate to allude to some theories of its action which have been advanced, but which are now disproved. The most prominent of these has been that the rennet acted upon the sugar of the milk, converting it into lactic acid, and that the separation of the curd was due to the presence of the latter. Liebig in 1865 advanced this view and suggested that the lactic acid neutralised the alkali which keeps the caseinogen dissolved, thus precipitating the latter.

Soxhlet had a slightly different conception. He thought that the difference between the coagulation of milk by rennet and the precipitation of the caseinogen following upon its turning sour on standing, was only that the former took place most rapidly. According to him rennet transformed sugar of milk into lactic acid, and this reacting on the alkaline phosphate existing in milk, converted it into the acid salt. This in turn precipitated the caseinogen. Hallier imputed the process to the micro-organisms present in the stomach of the calf.

The question was finally decided by Hammarsten. Not only did he show that casein was essentially different from caseinogen, but he also proved that the curdling was quite independent of the presence of sugar in the milk, by preparing a solution of caseinogen in the manner already described and seeing it clot in the presence of rennet. In the method of preparing caseinogen which he adopted, the sugar of the milk was left behind in the mother liquor from which the proteid was precipitated. Hammarsten's pure rennet, or *lab*, was further found to be quite without action on milk sugar, while it clotted milk or casein solutions with great rapidity.

The theory of the action of the acid is negatived also by the fact that curdling can be brought about quite easily in faintly alkaline solutions.

Certain peculiarities of the rennet prepared from pancreatic sources have been noticed by several observers. It has already been mentioned that Roberts when examining the proteolytic effect of pancreatic juice or extract upon milk noted a curious change in its proteids which seemed to be different from the ordinary proteolysis brought about by trypsin. When milk was exposed to the action of pancreatic juice for a short time and then boiled, the boiling caused a precipitation of the caseinogen in a curd-like form. Roberts considered this change to be brought about by the trypsin. Subsequent observers have however established the fact that it is due to the rennet which is secreted with the trypsin in the pancreas. The product differs from casein in not separating out as a curd at the ordinary temperature. Roberts called it *metacasein*, and the name has been retained by subsequent observers. It can be separated from milk that has been subjected to the action of a small quantity of pancreatic rennet, being thrown down by the

17—2

addition of an equal volume of a saturated solution of sodium chloride. It differs from caseinogen in turn, in that it will not give a clot when subjected to the further action of rennet.

The conditions that lead to its formation are a large preponderance of milk over the rennet solution. Too little of the latter must be added to bring about the ordinary clotting and then metacasein appears to be formed instead of casein. The by-product, or whey-proteid already mentioned, is formed simultaneously with the metacasein.

Halliburton and Brodie draw a distinction between gastric and pancreatic rennet which has not been noted by previous observers. According to them the clotting indicated by pancreatic rennet is not a true coagulation, but a precipitation which takes place in the warm bath at 35°—40° C. The precipitate is finely granular, and cannot be detected by the naked eye. On cooling it to the temperature of the air, it sets into a coherent curd which can be again broken up by warming to 35° C., when the granular condition returns and the milk appears fluid. This may be repeated several times. Halliburton and Brodie call the proteid in this condition "pancreatic casein," and they say it can be converted into casein by gastric rennet.

Rennet does not appear to be stored in the cells of either gastric or pancreatic glands in a condition in which it will effect coagulation, but rather to occur in the form of an antecedent body or zymogen. This has been established by Hammarsten and subsequently by Langley. The latter physiologist examined the gastric glands of the dog, cat, rabbit, mole, and frog; also the œsophageal glands of the latter; none contained rennet, but the zymogen was present in them all. The zymogen, like that of trypsin or pepsin, can be converted into enzyme by warming it for fifteen minutes with a dilute acid. Langley found that the zymogen, like the rennet itself, is destroyed by keeping it for a few minutes at 38° C. in the presence of 0·5—1·0 per cent. of sodium carbonate.

Lörcher has recently confirmed these observations of the existence of rennet zymogen.

In addition to mammalian sources Hammarsten found that

rennet zymogen exists in the gastric glands of certain fishes, among which the Pike is conspicuous.

The coagulating action of rennet can be inhibited by some substance which is normally present in the serum of blood. Our knowledge of this subject is due to the researches of Camus and Gley and of Biot. The latter observer has stated that when the serum of the horse is added to milk at the same time as a solution of rennet the coagulation is retarded and that there is a quantitative relation between the serum and the rennet. The substance in the serum which has this property is incapable of dialysing through a moist membrane, is destroyed by heating, and can be precipitated from solution by sulphate of ammonia or by alcohol. Biot suggests that it is itself an enzyme. This seems on the whole unlikely as egg albumin has been found to have the same effect, if added in sufficient quantity.

Vegetable Rennet.

The existence of rennet in the vegetable kingdom has been noted by many observers from the sixteenth century to the present day. Linnæus stated that the power of curdling milk was known in his day to exist in the leaves of *Pinguicula vulgaris* and that certain of the Lapland tribes with whom he became acquainted always used them for that purpose. Prior in his *Popular Names of British Plants* speaks of a curious property of *Galium verum*, which was noted by Matthioli in the sixteenth century, who wrote of it, " Galium inde nomen sortitum est suum quod lac coagulet." In West Somersetshire and certain parts of Herefordshire this property of Galium verum is still recognised and it is customary for dairymen to put the plant into milk to set the curd ready for cheese-making. The active principle seems to be located in the flowers, though the whole plant is used. It is somewhat singular that our other native species of Galium do not appear to contain the enzyme.

Many other observations of a similar character have been made by various investigators. Pfeffer says that *Pinguicula vulgaris* is employed in the Italian Alps by the peasantry to

prepare curd from milk. Darwin noticed that the secretion of
the glands of *Drosera* not only digested proteids, but curdled
milk as well. The association of rennet with a proteolytic
enzyme is very often met with. Baginsky found that it was
present in the juice of the papaw (*Carica Papaya*) which we
have seen is especially rich in vegetable trypsin. Martin confirmed Baginsky's observation in the course of the researches
which we have already discussed. Chittenden has shown that
rennet exists side by side with bromelin in the juice of the
pine-apple, but is associated with a different proteid, as already
mentioned.

Rennet has a very curious distribution in some plants. It
is found in the bast of the stem bundles of *Clematis Vitalba*,
and in the petals of the artichoke (*Cynara scolymus*), neither
of which localities would appear likely ones for an enzyme.

The most constant situation for vegetable rennet appears
to be the fruit and seed of the plant; at any rate it is these
regions which furnish most instances of its occurrence. It is
more frequently present in the seed than the fruit.

A very active preparation can be made from the pericarp of
the fruit (pepo) of the singular plant known as " Naras," which
is met with in very dry desert places in Namaqua Land and
Whalfish Bay in South Africa. Naras (*Acanthosicyos horrida*)
is a typical creeping desert plant, of very wiry habit and well
furnished with spines. It has squamiform leaves which are
very quickly deciduous, and while they are attached to the stem
they are closely adpressed to it, so that at a little distance it
appears to be leafless. There are two sharp spines at the base
of each, which do not fall off. The flowers are developed
without stalks between these spines and therefore in the axils
of the frequently abortive leaves. The fruit is an almost
globular pepo with a hard tough skin enclosing a very pleasant
acid pulp something like that of an orange, in which a large
number of seeds are embedded. Though the pulp is very
palatable, strangers eating it are often very singularly and
painfully affected by it. It is said to produce an almost in-
tolerable itching of the anus. The rennet is confined to the
succulent parts of the fruit, the seeds strangely being without

it. The enzyme is not destroyed by drying but can be preserved almost indefinitely by exposing the pulp to the sun. Like other enzymes it is destroyed by boiling. According to Marloth it is soluble in alcohol of 60 per cent. strength. It is developed only as the fruit ripens.

In seeds rennet seems to be of very common occurrence. Its preparation from this source was first successfully carried out by Lea, who found it in the resting seeds of *Withania coagulans*, a shrub belonging to the Solanaceae, which grows freely in Afghanistan and Northern India. Withania has a capsular fruit containing a large number of small seeds of a dark brown colour. It can be extracted from the ground seeds by means of glycerin or a moderately strong solution of common salt. It is however much contaminated by the colouring matter of the seed-coats, which dissolves in the extracting liquid, causing it to have a dark brown hue. Its activity is about the same as that of most commercial samples of animal rennet and the details of its behaviour are undistinguishable from those of the latter. It is similarly affected by temperature, but it can withstand a moderately prolonged exposure to alcohol without destruction.

The writer has met with rennet in the seeds of *Datura Stramonium, Pisum sativum, Lupinus hirsutus*, and *Ricinus communis*; in the two former in the resting, and in the two latter in the germinating condition. Though in Ricinus it does not exist in the resting state, the seed contains a zymogen which is soluble in water, and which readily develops the milk-curdling property on being warmed with dilute acids. The rennet itself can be extracted from the germinating seeds by either salt-solution or glycerin. It exists in the endosperm, in close association with both lipase and trypsin. There is generally a considerable quantity present, or it has very energetic powers; in one experiment a glycerin extract curdled two and a half times its volume of milk in 5 minutes. The salt-solution extract acts more slowly than the glycerin one, sodium chloride being a hindrance to the activity of rennet as it is to trypsin. Different seeds however contain very varying quantities of the enzyme.

Ricinus rennet is capable of acting in either acid, neutral, or alkaline solutions. Too great a degree of acidity obscures the action, as the acid itself tends to precipitate the caseinogen of the milk.

In the germinating lupin seed rennet exists side by side with trypsin, but there is much less of it present.

The enzyme exists also in many of the more lowly forms of plants. Duclaux has demonstrated that it is produced by several of the filamentous fungi, and many micro-organisms secrete it. The bacilli examined by Wood which we have seen produce proteolytic enzymes also furnish rennet. That two enzymes are present in the secretion from these plants is shown by their different powers of resisting different destructive reagents. When they are exposed to the action of gradually increasing quantities of mineral or organic acids the power of peptonisation disappears before the destruction of the rennet, while when carbolic acid is used the reverse is the case. Vignal has shown that rennet exists in *Bacillus mesentericus vulgatus,* side by side with four other enzymes. Conn separated the rennet from the proteolytic enzyme in some of these Bacilli by a method resembling those already described. He mixed the secretion of the microbes with milk and allowed coagulation to take place. The clot was then broken up, shaken well with distilled water and the whole filtered through porcelain. Sulphuric acid was added till the liquid contained 1 per cent., and it was then saturated with sodium chloride. A white granular looking scum floated to the surface, which on removal was found to be almost pure rennet, the proteolytic enzyme being left in solution in the brine. Conn found that the rennet enzyme was secreted most plentifully by the microbes when the culture of the bacillus was carried out at about 20° C., while the proteolytic one preponderated if the temperature was maintained at about 35° C.

The action of rennet is most probably proteolytic, but the details of its action are still very obscure. That it splits up the caseinogen seems probable when we consider the coincident appearance of the whey proteid and the casein. Neither of these bodies however falls into line with the course of proteolysis carried out by pepsin or trypsin, and the relation of the two groups of enzymes must at present be left undecided. Nothing at all is known at present as to its action in the course of vegetable metabolism. Its constant occurrence in the stomachs of young animals whose chief food is milk points to its being of use in the preliminary digestion of proteid, but why the clot of rennet should be more beneficial from this point of view than a precipitation of the caseinogen by the acid of the gastric juice is obscure. It may be however that the combination of proteids with small quantities of free acid which is known to take place in the stomach would delay the acid precipitation so long that a certain quantity of the milk might make its way out of the stomach into the intestine unchanged.

CHAPTER XVI.

THE phenomena which are noticeable when mammalian blood, freshly drawn from an animal, is collected in suitable vessels, have received the name of coagulation. The blood, preferably drawn from an artery, leaves the blood-vessel in a perfectly fluid condition; after an exposure of a minute or two to contact with the sides of the collecting receptacle it becomes viscid and gradually sets into a firm jelly. After some hours a yellow liquid exudes in drops from the surface, and gradually increases in quantity till the jelly, now become much firmer, is floating in a considerable bulk of the liquid. The exudation of this fluid, known as *serum*, is due to a shrinkage of a certain constituent of the jelly which thus squeezes out of itself the liquid part, leaving a firm clot of a leathery consistency. This clot is of a deep red colour, owing to the presence in it of the red corpuscles of the blood, which are mechanically entangled in it owing to the rapidity with which the change sets in.

It was originally supposed that the coagulation or formation of the clot was due to a mechanical coherence of the corpuscles, which can be seen to run together and form rouleaux when a drop of blood diluted with water or a weak solution of common salt is examined under the microscope. It was first shown by Hewson in 1772 that the process is independent of the presence of the red corpuscles and is due to the separation of a peculiar fibrous-looking substance from the plasma or *liquor sanguinis*.

This substance is now called *fibrin,* and coagulation is known to result from its formation from some antecedent substance which exists in the fluid portion of the blood. The separation of this antecedent of fibrin was attempted in 1859 by Denis. He saturated the liquor sanguinis with sodium chloride and obtained a sticky precipitate of a proteid nature. When this was separated from the liquid and redissolved in a dilute solution of the same salt, the solution underwent coagulation, in the same way as normal blood. Denis called his precipitated proteid-matter *plasmine.* A little later Schmidt showed that certain serous fluids that would not clot spontaneously could be made to do so by adding serum to them. From one of these, hydrocele fluid, he prepared by Denis' method a proteid which he called *fibrinogen.* The serum which set up the coagulation in the hydrocele fluid yielded another proteid when similarly saturated. Schmidt called this *fibrino-plastin.* Neither serum nor hydrocele fluid will clot spontaneously, nor will solutions of the precipitates obtained from them in this manner. The mixed liquids or the mixed solutions of their saturation precipitates will however give a true coagulum. Schmidt found that the formation of the clot depended on the presence of a third factor which was present with his fibrino-plastin when prepared by saturation of the serum with salt. When this proteid was prepared by passing a stream of CO_2 through considerably diluted serum, its solution did not always cause clotting when added to a solution of fibrinogen. Schmidt held that the presence of this third body was necessary for the interaction or union of the other two. He sought for it in the serum, and in 1872 separated it in a crude condition by precipitating the proteids of the serum with a large excess of alcohol. The precipitate was allowed to stand under alcohol for a long time till the proteids were rendered almost entirely insoluble; it was then separated by decantation or filtration and dried. A watery extract of the powder so produced was found to be capable of bringing about coagulation in such of his preparations of fibrinogen as would not clot spontaneously. Schmidt gave this third factor the name of *fibrin-ferment.* His fibrino-plastin was subsequently named *paraglobulin* or *serum-globulin,* its reactions showing that like fibrinogen it was a member of the globulin class of proteids. Schmidt explained coagulation by saying that these two globulins interacted with each other under the influence of the fibrin-ferment to form fibrin.

Some years later, Hammarsten showed that the presence of serum-globulin is not necessary for the formation of fibrin, but that the part which it takes in the process can be sustained by other substances, such as calcium chloride and impure casein. Serum-globulin itself when pure is without effect. He was thus led to the view that the formation of fibrin is due to the action of the fibrin-ferment upon fibrinogen only, under certain favourable conditions, which are usually realised in the presence of serum-globulin.

The process of the coagulation of shed blood is thus found to be a ferment action, and the fibrin-ferment has within the past twenty years been the subject of much careful investigation. Its behaviour indicates that it belongs to the great group of enzymes, more closely approaching rennet than any other so far discovered. It shows the same dependence upon temperature as the other enzymes, having a minimum activity at a very low point, rising thence to an optimum at about 38°—40° C., and being destroyed at about 50° C. It does not apparently enter into the reaction by uniting with fibrinogen, nor does it undergo destruction in the course of its activity. The decomposition it sets up is not certainly known, but it appears to behave after the manner of rennet, decomposing the fibrinogen with the formation of an insoluble proteid, *fibrin,* and a soluble one of globulin character which remains in solution. This decomposition recalls the formation of casein and the whey proteid.

We have seen that the fibrin ferment, which has recently been named *thrombase,* or *thrombin,* was originally prepared by Alexander Schmidt from defibrinated blood or from the serum which separates out from a normal clot. This is not however its only possible source, for it may be obtained with equal ease and in much greater amount from the fibrin which is formed. An indication of this fact was first obtained in 1835 by Buchanan, who found that both hydrocele and pericardial fluids, which are not spontaneously coagulable, would produce a clot on the addition of some fibrin. Buchanan diluted blood with a considerable quantity of water, thereby delaying the process of coagulation, and causing the clot when formed to be of a loose fibrous character. The fibrin could be removed from the blood as it was formed by stirring the latter constantly with a bundle of twigs to which the fibrin adhered. Buchanan called his preparation "washed blood clot." Many years later Gamgee ascertained that the washed blood clot when extracted with a solution of sodium chloride containing about 8 per cent. of the salt, yielded a very active ferment. Subsequently other observers showed that normal fibrin, obtained from blood without dilution, yielded an equally active extract. The thrombase can consequently be prepared from the fibrin of the clot or from the serum which exudes from it after standing.

Thrombase does not exist in normal blood before coagulation has taken place. If blood is shed from a vein or artery directly into alcohol and subsequently treated in the way Schmidt treated his serum or defibrinated blood, the solution or extract of the precipitated proteids will not cause coagulation of serous fluids. The appearance of the thrombase appears to be almost coincident with the act of coagulation. Schmidt stated that just at this moment there occurs a disintegration of the colourless corpuscles or leucocytes, and suggested that this was the decomposition which gives rise to the enzyme. This is supported by an earlier observation of Buchanan that when coagulation is retarded to such an extent that the red corpuscles sink to the bottom of the liquid before it takes place, the supernatant plasma, known as the *buffy coat*, yields a more active "washed-blood-clot" than the rest of the blood. This buffy coat contains the greatest abundance of the leucocytes, which sink much more slowly than the red corpuscles. He therefore concluded that the power of the washed clot to set up coagulation depended on its cellular elements.

The nature of the fibrin-ferment as obtained from the two sources mentioned does not appear at first sight to be the same. When prepared by Schmidt's method it seems improbable that it is of a proteid nature, as all the proteids of the blood he used may be supposed to be coagulated and rendered insoluble by a prolonged exposure to the action of the alcohol. As extracted from fibrin however it appears likely to belong to the class of globulins, proteids which are insoluble in water, but soluble in solutions of sodium chloride or other neutral salts. There is reason to think however that the difference is more apparent than real, for Schmidt says he never obtained his ferment free from proteids, and he found it most active when the admixture with proteid was greatest. It is not difficult to suppose that his proteids were not completely rendered insoluble by the alcohol. He says moreover that he was never able to obtain his fibrino-plastin (serum globulin) free from ferment.

These considerations led most observers to entertain the opinion that the ferment was proteid in nature and was closely associated with the serum-globulin.

On the other hand, the globulins are always precipitated or coagulated by being heated. Heat coagulation of a proteid must not be confused with the coagulation of the blood or of milk, as it only indicates a conversion of the proteid into a peculiarly insoluble form and not the formation of a clot. It is a little unfortunate that the same term is applied to two different processes. When an albumin or a globulin in solution is heated to a temperature usually about 70° C., it is converted into the insoluble body in question, which is known as *coagulated proteid*. Schmidt says that when aqueous solutions of his ferment were boiled they did not throw down any of this proteid, though the specific action of the enzyme was destroyed at a temperature of 70° C. The same observation was made by Sheridan Lea, in conjunction with the writer, in the course of some experiments made upon the preparation of thrombase from fibrin.

The view that the enzyme is a proteid was very strongly urged by Halliburton, who prepared it from the leucocytes of the lymphatic glands. He associated it with a particular proteid, much resembling serum-globulin, but differing from the latter in a few particulars, one of which is its heat-coagulation point. The change on heating in 5—10 per cent. sodium chloride solutions takes place at 60—65° C., while serum-

globulin is not affected below 75° C. Halliburton originally
called this proteid "cell-globulin."

The probability that the serum-globulin and the ferment
are distinct is confirmed by Hammarsten's statement that he
obtained from hydrocele fluid, which is not spontaneouly coagulable, a pure paraglobulin, (serum-globulin) which was free from
ferment and which had no fibrino-plastic activity.

In serum obtained from a normal clot these two globulins
exist together, and the presence of two rather than one escaped
detection until Halliburton's researches.

More recently Halliburton has found reason to doubt that
his cell-globulin is a true proteid, as it leaves on gastric
digestion a residue of nuclein, a body containing phosphorus.
In his latest work he includes it in a group of bodies allied to
proteids and known as *nucleo-proteids*. Pekelharing has come
independently to the same view. Halliburton says further that
when thrombase is prepared according to Schmidt's method,
it gives reactions which show that it belongs to this group.
Thrombase is therefore a nucleo-proteid, or is associated very
closely with such a body.

The story of the coagulation of shed blood is however much
more complicated than would appear from what has so far
been advanced. It is found that by many methods it may be
very greatly delayed or altogether inhibited. It may be hindered considerably by keeping the shed blood at a very low
temperature, the blood of the horse being peculiarly susceptible
to this influence. If the blood when shed is at once mixed
with strong solutions of various neutral salts it will remain
fluid for a very long time, the most efficient salts being the
sulphates of magnesium or sodium. On copious dilution, so as
to make the percentage of salt very small, coagulation slowly
takes place. If a certain amount of commercial peptone, which
is very largely mixed with proteoses, is injected into the veins
of a dog previously narcotised with chloroform, and the blood
then withdrawn from the animal, it remains fluid indefinitely.
In some cases the same result follows when solution of peptone
is added to the blood as it is being shed. If an extract prepared from the head and pharynx of the common leech is mixed

with shed blood, the process of clotting does not take place. This extract contains a peculiar proteose. Other methods of maintaining its fluidity when shed will be referred to subsequently.

Blood kept fluid by such methods yields thus different plasmas which are not spontaneously coagulable but which by different treatment can be made to yield a clot. These are generally referred to as "salted plasma," "peptone-plasma," "leech-extract plasma," etc.

The influence of the inorganic salts present in the liquor sanguinis must be taken into account in the study of the phenomena of coagulation. If blood is kept fluid by any of the methods in vogue and dialysed till quite free from salts, no coagulation can be induced. Of all the salts present in the plasma, the compounds of calcium invite particular attention. It was first demonstrated by Brücke that the ash of fibrin always contains this metal in some combination. The writer found in 1884 that fibrin prepared by whipping blood always contains a certain quantity of calcium sulphate which can be extracted by appropriate means. As already stated Hammarsten showed that calcium chloride could take the place of paraglobulin in aiding the action of thrombase.

The writer found in 1885 that when a salted plasma was diluted so that its normal coagulation took place in about an hour, it could be made to clot in a few minutes by the addition of a few drops of calcium sulphate solution. The salted plasma itself of course contained some of the enzyme, the action of which was temporarily inhibited by the magnesium sulphate used. Addition of calcium sulphate to a serous fluid such as hydrocele which contains no ferment caused no coagulation.

The conclusion was obvious, that in some way or other the thrombase was dependent upon a calcium salt, possibly the sulphate, for its manifestation.

This conclusion was borne out by depriving horse's blood of its salts by a long process of dialysis, which was carried out in tubes surrounded by ice. When the blood of the horse is maintained at a temperature of $0°$ C. as already stated, coagulation may in favourable cases be indefinitely postponed. After a long period of dialysis all the salts were removed except sodium chloride, the presence of which is necessary to keep the globulins from precipitation. This was therefore kept in the blood by causing the dialysis to take place into ·6 per cent. sodium chloride solution. Such plasma removed from the ice and subjected to the temperature of the body refused to clot, either with or without dilution. On the addition of a little solution of calcium sulphate solution clotting speedily took place. Precipitation of calcium sulphate from the blood by the addition of barium chloride materially hindered coagulation but did not altogether prevent it. It seemed from these experiments probable that calcium sulphate was a necessary factor in the process. Its removal by barium chloride was not complete.

In 1888 Arthus and Pagès discovered that coagulation may be almost entirely prevented by mixing the blood with a 0·2 per cent. solution of potassic oxalate. This reagent forms an insoluble compound with calcium salts and thereby removes nearly all that metal from the liquid. It is much the best precipitant, as no other throws the calcium out of solution to so great an extent.

It is stated by Schäfer that even oxalated plasma slowly coagulates, so that even rendering the greater part of the calcium insoluble does not entirely prevent its interaction with fibrinogen.

Other observers, especially Ringer, have found that other salts of calcium than the sulphate are able to cause coagulation in salted plasmas. Of these may be mentioned the chloride and the phosphate. Neither acts however so rapidly as the sulphate. Whether the latter is the actually interacting body must be left at present uncertain, but it is not impossible, for the blood contains sulphates of other metals, which would cause the formation of calcic sulphate on addition of any other soluble calcium salt. The occurrence of calcium sulphate in fibrin may be remembered in this connection.

The theory of coagulation already stated needs amplifying in the light of these facts; the conversion of fibrinogen into fibrin occurs only under the influence of certain salts of calcium.

The part played by the salt of calcium is at present still obscure. When we consider that calcium is always present in the ash of fibrin it seems at first sight very possible that the latter is produced by the union of the fibrinogen and either calcium or some compound of it, and that the work of the thrombase is to secure their combination. But experiment and analysis alike negative this supposition. With a considerable quantity of fibrinogen in a solution the weight of fibrin formed does not vary with the amount of calcium salt introduced. A very small quantity induces coagulation, and a larger quantity does no more. Hammarsten has shown recently that fibrinogen and fibrin both contain the same amount of calcium, so that the latter cannot be a calcium compound of the former.

From the experiments already referred to the writer was led to advance the view that the thrombase might exist in the blood in the form of a zymogen, and that under the influence of the calcium salt it might be converted into the active enzyme. He was unable however to obtain any evidence in support of this hypothesis. More recently the same view has been put forward by Pekelharing, who has been able to bring forward some experiments which tend to establish its accuracy. He has prepared from various forms of extra-vascular plasma a substance which possesses no fibrino-plastic properties, but which by treatment with a salt of calcium, acquires the power of setting up coagulation in solutions of fibrinogen. This substance appears to be identical with Halliburton's so-called " cell-globulin," which is really a nucleo-proteid, as already shown. Hammarsten also has shown that the nucleo-proteid can be prepared from plasma, and that it is incapable of acting as a ferment until it has been exposed to the action of soluble salts of calcium. Pekelharing has suggested that the interaction of the calcium salt with the nucleo-proteid zymogen, which has been called *prothrombin*, is really a union of the two, but this view has not at present met with general acceptance by physiologists.

Thrombase is thus seen to possess the properties of an enzyme. Its activity shows the same dependence on temperature that has been noted in the cases of the other soluble ferments which have so far been discussed. A very small quantity will induce clotting in a very large amount of blood, the time it takes to bring about the change being approximately in inverse ratio to the amount present. It exists in the living body in the state or condition of a zymogen, but it is peculiar in its mode of conversion from the latter, calcium in one or other of its combinations being necessary for the transformation. The reason why the interaction of the two does not normally take place in the living blood-vessels has not been satisfactorily ascertained.

Thrombase is peculiar in many ways. It shows a very great susceptibility to the influence of certain neutral salts. Its action is impeded by the sulphates of magnesium and sodium, and to a less degree by the chloride of the latter metal. Coagulation is almost entirely inhibited by soluble oxalates, but on the other hand it is aided conspicuously by the presence of salts of calcium, especially the sulphate; indeed the action of the enzyme is only possible in their presence. Whether all these affect the formation of thrombase from its zymogen, or whether they influence rather its activity, are points which are still matters of controversy. We have seen reason to believe that the salts of calcium are concerned in the formation rather than the working of the enzyme.

According to Dastre a neutral reaction is the most favourable for the exhibition of its activity. Indeed he says that when working with any artificial plasma, neutralisation always accelerates coagulation. Hammarsten found that the neutralisation of many serous transudations not spontaneously coagulable, often led to the formation of a clot.

The action of thrombase is inhibited by certain organic bodies, of which the chief are certain albumoses or peptones. The secretion of the buccal glands of the leech also prevents its action.

The reason of the inhibitory effect of some of these bodies has been ascertained. The soluble oxalates form a compound with the calcium which is insoluble and which consequently withdraws this metal in great part from the solution. Schäfer has shown that this removal does not in all cases prevent the

18—2

ultimate clotting of the plasma. The interaction however between the oxalates and the soluble calcium salts does not effect
the total removal of the latter. When two soluble salts which
can exchange their metals with one another are mixed in a solution, the interchange between them is never complete, but takes
place up to a certain point at which a dynamical equilibrium is
reached between all four of the possible combinations. This
point will vary according to the relative amounts of the
originally interacting bodies, but the interchange is never
complete unless one of the new products is withdrawn as
fast as it is formed. Even on the addition therefore of considerable excess of the oxalates, some soluble salt of calcium will
remain. This agrees with Schäfer's observation that the onset
of coagulation is very materially postponed but not altogether
prevented.

The action of peptone and leech-extract is considered by
some physiologists to be of the same order. They hold that
these bodies enter into combination with the calcium salts, and
so prevent the formation of the enzyme. It is noteworthy that
if a stream of carbon-dioxide is passed through a peptone-
plasma which is not spontaneously coagulable, it soon clots.
The carbon-dioxide appears to break up the union between the
peptone and the calcium salt, and so restores the conditions
necessary for clotting. Addition of more calcium salt to the
peptone-plasma will, in like manner, enable coagulation to take
place.

Other physiologists, with less probability, attribute to the
peptone or albumose a power of destroying the enzyme altogether. Dickinson claims that this is undoubtedly the effect
of the leech-extract.

It is not possible at present to explain the inhibitory effect
of magnesium and sodium sulphates. It seems probable that
they interfere in some way with the enzyme and not with the
calcium salt. Sodium chloride in excess may be supposed to
act like the oxalates, but much less completely, calcium chloride
being soluble.

A further condition of the action of thrombase is that the
blood shall be withdrawn from the body. In some way or

other the blood in the living vessels, or the vessels themselves, either destroy it or altogether inhibit its action. Most observers have found that the injection of a powerful ferment-extract into the veins of a living animal does not produce any intravascular clotting, even if considerable quantities are used and if soluble salts of calcium are simultaneously introduced. Halliburton says that if very large quantities of thrombase are injected into the vessels, the blood coagulates more rapidly than normal blood when it is subsequently withdrawn. There appears to be a destruction of the enzyme on its introduction into the body. Such a destruction moreover may be normally taking place in the blood. We have already seen that one source of the enzyme is the leucocytes of blood or lymph. Schmidt was of opinion that the cause of its formation was the disintegration of these leucocytes, which he found to take place as soon as blood is shed. As these cells are continually being decomposed in the living body, we must suppose that they do not under these conditions form thrombase, or that it is destroyed as soon as formed, or that the living blood or the vessels in which it flows prevent the action of the enzyme on the fibrinogen. The absence of coagulation cannot be due to a deficiency of calcium salts, for the mere shedding of the blood cannot affect the amount of these which it contains. The observation of Schmidt that when normal blood is allowed to flow into alcohol, no enzyme can be extracted from the precipitate which is formed in the spirit, suggests that an actual destruction of thrombase takes place in the blood-vessels.

The action of thrombase on fibrinogen may be compared with that of rennet on caseinogen. In both cases we have the formation of a very insoluble body, which constitutes the solid matter of the clot, together with the appearance of a soluble proteid which appears to be formed simultaneously. Denis found that when a solution of his "plasmine" coagulated, a new proteid substance was present in the solution. He gave it the name of "fibrine soluble," to distinguish it from the normal fibrin of the clot. Hammarsten has observed that the decomposition is attended by the appearance of this new body, which he says is a globulin coagulating when heated to 64° C. More recently Arthus has confirmed the observation. He finds that in an oxalated plasma the weight of fibrin produced by the addition of a soluble salt of calcium is less than the weight of the fibrinogen which can be obtained by heating an equal volume of the plasma to the point at which this proteid is precipitated, and is therefore considerably less than the total weight of the fibrinogen present, for many observers have noticed that this proteid does not all separate out when its solutions are heated to 56° C. If this observation is correct, it follows that in the formation of fibrin from fibrinogen there must be a splitting or decomposition of the latter. Coagulation cannot therefore be due to a combination of fibrinogen with calcium; a conclusion which we have seen already to be reached by a different method of experiment.

Hammarsten holds that the change in the fibrinogen is not one of splitting into two other proteids, but is rather one of intro-molecular rearrangement.

Besides being obtainable from the serum and fibrin of shed blood, thrombase can be prepared from many of the tissues of the body. Halliburton has extracted it from leucocytes from lymphatic-glands, from the glands themselves, from the thymus, and from the stromata of the red corpuscles. He says that a proteid which can be obtained from muscle-juice, to which he has given the name *myosinogen*, has precisely the same action on diluted salted plasma as his cell-globulin or nucleo-proteid has, though he hesitates to identify it with fibrin-ferment. Buchanan found that many of the tissues of the body possessed the power of setting up clotting in serous fluids that were not spontaneously coagulable. He instances muscle, connective tissue, and certain parts of the central nervous system. Pekelharing has prepared a nucleo-proteid from muscles by macerating them in a ·25 per cent. solution of sodium carbonate, and adding acetic acid to the extract till a precipitate falls. He says this causes coagulation of the blood if it is injected into a vein.

The question of intravascular clotting and its relation to thrombase is still very obscure. Though most observers agree that the injection of a very active preparation of the ferment into the blood-vessels of an animal is not followed by coagulation of its blood, some exceptions have been noticed.

Edelberg has shown that intravenous injection of Schmidt's fibrin-ferment may produce coagulation of the blood in the vena cava, the right side of the heart, and the pulmonary arteries. General intravascular clotting may be caused by the injection into the vessels of certain proteids which were first prepared by Wooldridge and named by him " tissue fibrinogens." These now appear to be nucleo-proteid bodies, comparable with Pekelharing's product obtained from muscle, and perhaps also with the prothrombin which can be prepared from plasma and other sources as already described. If these bodies are injected rapidly in large quantity, almost instantaneous coagulation of the blood takes place. If on the other hand the injection is carried out slowly and not much of the nucleo-proteid is introduced, the blood becomes less coagulable. The nature of the changes which are thus set up and their relation to the formation or the action of thrombase need further investigation. Schäfer suggests that when these nucleo-proteids are either naturally formed in circulating blood, or are artificially injected into the vessels, they tend to interact at once with the calcium salts of the plasma and to form thrombase, intravascular coagulation being the result. It is difficult however to reconcile this with the observation that the injection of active thrombase into the vessels generally fails to produce such coagulation.

Lilienfeld has also associated nucleo-proteids with the phenomena of coagulation. He holds that a somewhat complex substance to which he gives the name of *nucleo-histon* is one of the products of the destruction of the leucocytes of the blood, and that this itself is capable of decomposition into two other substances one of which he terms *leuco-nuclein* and the other *histon*. The latter possesses anti-coagulant properties but the leuco-nuclein aids coagulation by splitting fibrinogen into two new proteids, one of which, called by him *thrombosin*, becomes fibrin by combining with a calcium salt. The action of the leuco-nuclein he attributes to the nucleic acid it contains. Other acids, particularly acetic, are said by him to have the same action.

If a solution of pure fibrinogen, prepared by precipitating it from blood plasma by semi-saturation with sodium chloride, is dissolved in a dilute saline solution and acetic acid is added, a precipitate is formed which is Lilienfeld's thrombosin. A solution of this in a dilute solution of sodium carbonate yields fibrin on the addition of some calcium chloride.

Lilienfeld's conclusions have not met with entire acceptance. They have been criticised particularly by Schäfer, who has repeated his experiments without entirely confirming his results. Schäfer is of opinion that the so-called thrombosin is only fibrinogen.

Crustacean Fibrin-ferment.

A coagulation which somewhat closely resembles that of the blood of Vertebrate animals can be observed to take place in the hæmolymph of many of the decapod Crustacea, the most convenient animals for examination being the lobster (*Homarus vulgaris*), the crab (*Carcinus mænas*), and the marine and fresh-water cray-fishes. The fluid in question, which is often spoken of as blood, is of a very pale red colour or nearly colourless. It is somewhat opalescent from the presence of numerous amœboid corpuscles.

When this blood or hæmolymph is shed from the animal and collected into a vessel it very quickly forms a clot in which the appearance of a network of fibres can be traced. The clot very soon contracts and squeezes out a clear fluid, which in a few more minutes sets into the form of a jelly. The process of coagulation appears to take place in two stages, the first clot only being fibrous. The translucency of the second jelly-like coagulum shows that the corpuscles of the blood are entangled in the first fibrous formation. After a few hours the second clot also shrinks and squeezes out a liquid which has no further power of coagulating, but which is the serum of the blood.

It was originally held that the two clots had different origins; that the first was made up of the corpuscles, which were thought to coalesce together after the manner of the plasmodium of a Myxomycetous fungus, while the second was due to the formation of a true fibrin, identical with that produced in the blood of the Vertebrata. This view was advanced by Fredericq and supported by Geddes and other writers.

Other observers suggested that the second process was only a continuation of the first and that both were due to a normal formation of fibrin, the chief difference between the clots being that the earliest-formed fibrin entangled the cells or leucocytes in its fibrils as they were produced. The accuracy of this view was established in 1885 by Halliburton, who made a very complete examination of the blood or hæmolymph and the phenomena of its coagulation.

The blood contains two principal proteids, one of which is a form of fibrinogen which differs but little from the fibrinogen of the blood of the Vertebrata; it possesses a somewhat higher heat-coagulation point, remaining unaltered up to a temperature of 65° C. It is not precipitated by sodium chloride until the liquid is saturated by the salt. Except for this peculiarity it can be precipitated from its solutions in exactly the same way as the fibrinogen of the higher animals. When extracted from the blood and dissolved in a dilute salt-solution it will undergo coagulation on the addition of a little thrombase, prepared from the blood of a vertebrate animal.

The process of coagulation presents the same features and peculiarities which are found in connection with the same phenomena in the blood of the Vertebrata.

It can be delayed or prevented by appropriate addition of neutral salts such as magnesium sulphate to the blood as it is shed. Usually a considerable quantity of the salt must be employed. It is affected by cold in the same way as the coagulation of the blood of a vertebrate animal.

The correspondence in behaviour with ordinary salted plasma, and especially the reaction with thrombase, suggests that the coagulation of crustacean blood is due to an enzyme, and the first occurrence of the fibrin in direct relationship with the leucocytes points to the latter bodies as its probable source.

Halliburton made a series of experiments upon these points and found that the results bore out the conclusions suggested. He was able to prepare an enzyme from either the blood or the serum separated out from it after coagulation. He used Schmidt's method of preparation, and found that either the powdered precipitate or an extract of it was easily able to set up coagulation in a solution of crustacean fibrinogen. He ascertained moreover that the enzyme which he thus prepared was so much like thrombase as to be able to cause coagulation in a salted plasma prepared from the blood of a cat, and in a solution of fibrinogen prepared from hydrocele fluid.

Halliburton says that the amœboid corpuscles of the blood are the source of the enzyme, but he does not quote any experiments which definitely lead to that conclusion. Hardy has made a very careful examination of the different kinds of corpuscles which the blood of the Crustacea contains and among them he finds some pale ones, oval in shape, which contain a small number of fine granules. When these, which occur in considerable numbers, are watched under the microscope they show some very peculiar features. From each a number of fine pseudopodia are shot out with some suddenness and little portions of the protoplasm travel rapidly along them, and burst, after swelling up to form a kind of vesicle. The pseudopodia vary in number and in shape; indeed the protoplasm of the surface of the cell sometimes swells and bursts, without any formation of them at all. These changes are associated with remarkable alterations of the nucleus of the corpuscle. Hardy has called them "explosive corpuscles," and has associated the formation of the fibrin-ferment with them, on account of the fact that they are the only ones which are disintegrated at all rapidly after the shedding of the blood. He says that there is a marked correspondence in time between their solution and the solidification of the plasma, and that substances such as very dilute solutions of iodine, which keep the corpuscles from exploding, delay in about the same degree the coagulation of the blood. He calls attention to the fine granules described as appearing in these cells, and shows that if the cells are prevented from breaking up while the granules are allowed to dissolve out of them slowly, the onset of coagulation corresponds with the time of the discharge of these granules from the corpuscles. There is thus a certain amount of evidence that these explosive corpuscles secrete a form of fibrin-ferment, which is probably identical with thrombase prepared from the higher animals.

The coagulation as we have said is seen to take place in two stages, the first of which is almost instantaneous and results in all the corpuscles being enclosed in threads of fibrin. This must be due to the rupture of the explosive cells and the liberation of their ferment, which causes the very rapid formation of the fibrin, the coagulation taking place before diffusion of the liberated ferment can go on to any great extent. The contraction of the fibrin threads of the clot squeezes out what plasma has not coagulated and at the same time the ferment spreads more fully through the liquid, hence the second slower phase of the coagulation is seen. Although the two coagula seem to the eye to be quite distinct from each other it is very doubtful whether the second process is not a continuation of the first one without any break or interruption, for the most careful efforts to extract the first clot from the liquid before the formation of the second have been entirely unsuccessful.

Myosin-ferment.

The formation of myosin, which is characteristic of the onset of rigor mortis, is a phenomenon which presents many features in common with the formation of fibrin in the blood. The similarity of the two processes has been pointed out by Kühne, who worked with the muscles of frogs, and by Halliburton, whose experiments were conducted on rabbits, cats and pigeons. Both observers were able to prepare from muscles by strong pressure a liquid or plasma which quickly became gelatinous in consistency and formed a clot. On standing the clot shrank and separated from a fluid comparable to serum. It was necessary in making the experiments to carry out all the operations at a very low temperature, scarcely above 0° C.

The two clots, obtained from blood and muscle plasma

respectively, differ however considerably in appearance and in properties. While that from blood is stringy and its fibrin portion almost insoluble after its formation, that from muscle is gelatinous and soft, and readily dissolves in salt solutions of moderate concentration. The substance which is formed and which is comparable to fibrin is known as *myosin.*

Halliburton showed a further difference between the two in that the salt solution of myosin can be made to clot a second time with comparative ease. In some cases he found that the process of redissolution and recoagulation could be repeated three or four times. A salt solution extract of rigid muscles, in which the myosin had been formed before their removal from the body, behaved similarly.

When a muscle plasma is prepared by squeezing it from a fresh muscle carefully freed from blood and manipulated at 0° C., it can be kept fluid for a long period in the same way as blood plasma by admixture with neutral salts, magnesium or sodium sulphate and sodium chloride being most advantageous. If such a salted plasma is diluted, coagulation slowly takes place.

The coagulation is due to the action of an enzyme which is present in the muscle, and which according to Halliburton can be extracted from it by the following method:—

Muscle is allowed to undergo rigor and is then minced finely and kept under absolute alcohol for a long time. The small pieces are dried over sulphuric acid and powdered. An aqueous extract of the powder contains the ferment.

If such an extract is added to a diluted salted plasma which normally clots in 12 hours, the coagulation takes place in less than one quarter of the time. Addition of the powdered muscle substance produces a similar effect.

The myosin-ferment is associated with a proteid which has the properties of a deutero-proteose, and which can be extracted from the muscle plasma or from coagulated muscle kept for a long time under alcohol. Halliburton is rather inclined to the view that this substance is itself the enzyme, much as he contends the nucleo-proteid or cell-globulin already described is the fibrin-ferment.

This enzyme is quite distinct from the fibrin-ferment, not

being able to cause coagulation of salted blood plasmas; nor can
fibrin-ferment set up clotting in muscle plasma. They differ
moreover in the temperature at which they are destroyed; while
thrombase is decomposed at 70°—80° C., the myosin-ferment is
not destroyed till a temperature of nearly 100° C. is reached.

The change which the ferment induces differs from that
which is involved in the formation of fibrin. Muscle plasma
contains four proteids which can be converted into coagulated proteid by heating. These have been investigated by
Halliburton, whose description of their reactions may be briefly
summarised as follows:

1. Paramyosinogen—a globulin coagulating on heating to
47° C.; precipitated by magnesium sulphate of 50 per cent.
concentration or by 26 per cent. of sodium chloride.

2. Myosinogen—a globulin coagulating at 56° C.; precipitated by 94 per cent. of magnesium sulphate or 36 per cent. of
sodium chloride.

3. Myoglobulin—a globulin coagulating at 63° C.; precipitated by saturation of its solution with either magnesium
sulphate or sodium chloride.

4. Myoalbumin—an albumin coagulating at 73° C. and
not separated from the plasma by saturation with either salt.

On removal of these a deutero-proteose remains in the
solution.

When the serum from the clot is examined it is found to
contain no proteid coagulating below 63° C. When the clot is
dissolved in dilute salt solution and the latter gradually heated.
coagula occur at 47° C. and at 56° C.

The clot so appears at first sight to consist of the two
myosinogens. But on separating them by fractional precipitation with magnesium sulphate and dissolving them separately,
clotting only occurs in the solution of the second of them.
With care this myosinogen can be prepared quite free from
ferment, when its solution remains fluid indefinitely. Addition
of ferment causes coagulation to set in. The myosin formed
in the clot of the normal plasma appears to entangle the
paramyosinogen in its substance. Paramyosinogen itself will
not clot either with or without addition of the enzyme.

The action of the latter is therefore, according to Halliburton, to convert myosinogen into myosin. It is not a splitting like the conversion of fibrinogen into fibrin, no second proteid being formed coincidently with the clotting. There is, however, a simultaneous formation of a form of lactic acid which appears to arise from the same or an accompanying decomposition of the proteid. The formation of the myosin is also attended by the liberation of a considerable quantity of carbon dioxide.

Vesiculase.

Camus and Gley have stated that they have discovered another enzyme of this class in the secretion of the prostate glands of the guinea-pig, rat and mouse. The liquid coming from this gland is clear and limpid and has a neutral reaction. When a drop of it is added to the semi-liquid contents of the vesiculae seminales a coagulation sets in instantaneously. The secretion of the prostate loses the coagulating property if it is heated for 15 minutes to 70° C. Camus and Gley give the enzyme the name of *vesiculase*. The proteids of the vesicular fluid have not been ascertained, but they do not appear to include fibrinogen.

CHAPTER XVII.

THE CLOTTING ENZYMES (*continued*). PECTASE.

THE formation of the vegetable jellies which can be prepared from so many ripe fruits is due to the action of an enzyme upon a certain substance which can be extracted from many vegetable tissues, in some cases from cell-walls, in others from the sap contained in the cells. This substance, which is known as *pectine*, is a member of a series of compounds which are very widely represented in plants, being usually associated with cellulose in the young cell-wall. The enzyme which is concerned in its transformations was originally investigated by Fremy in 1840. He says in his account of his researches that he finds a substance in cell-walls, which differs in many important respects from cellulose, and he gives it the name of *pectose*: it is easily converted into a closely related body, *pectin*, by the action of an acid. By the action of an enzyme, which he terms *pectase*, and which he extracted from certain cells of various plants, pectose or pectin can be converted into two gelatinous bodies, pectosic and pectic acids. The transformation takes place in two stages, the two acids being formed successively. They differ from pectin chiefly in the amount of water which they contain. Pectase according to his observations exists in two conditions in the vegetable organism, sometimes in solution in the neutral sap of such roots as the carrot or beet, sometimes in an insoluble state, as in the juice of acid fruits. If the juices or the pulp of these plants is added to a solution of pectin, a rapid gelatinisation of the latter takes place and, as stated above, pectosic and later on pectic acids are formed. Fremy says further that the enzyme can be prepared from the juice of young carrots by precipitation with alcohol. The solid precipitate possesses the ferment power, but will not give it up to water. Its optimum working temperature is 30° C., and it is destroyed by boiling. It can work in the absence of oxygen.

Since Fremy's work was published the occurrence of pectic compounds in plants has been the subject of careful investigation by many observers. In particular Mangin has contributed substantially to our knowledge of the several members of the group and their transformations under different reagents. The peculiarities of the action of pectase have also been studied and its distribution more completely determined.

The general properties of the pectic compounds show that they differ considerably from the celluloses, though they have often been confounded with them. They are uncrystallisable bodies, which can be precipitated from their solutions by various reagents, and then appear very often in a colloidal or gelatinous condition. They are however very rapidly altered by the solutions used for their extraction. Fremy, Scheibler and Reichardt considered them to be carbohydrates, allied to mucilages and gums. Mangin holds that their reactions separate them from the carbohydrate group; when oxidised by dilute nitric acid they give rise to mucic acid, while the carbohydrates are converted into oxalic acid. Pectic bodies are insoluble in ammonio-cupric-oxide, which dissolves the celluloses readily. The latter are coloured violet or blue by iodine in the presence of sulphuric or phosphoric acids; the pectic bodies yield no such coloration with any combination of iodine.

De Haas and Tollens from their analyses of pectine prepared from various sources support the view of Fremy and others that they are related to carbohydrates if not actual members of the group. They show that they contain hydrogen and oxygen approximately if not definitely in the proportion of two to one, while their carbon amounts to about 43 per cent. Hydrolysis of these pectines by dilute mineral acids yields sugars containing either five or six carbon atoms. Tollens suggests that they are carbohydrates chemically combined with acids.

A great number of these pectic bodies have been identified and described by various writers from 1825 onwards[1], and a certain confusion has been unavoidable in comparing those of one investigator with those described by another. Mangin's memoirs have put the whole group upon a surer foundation.

The pectic compounds can now be arranged in two series, one of the latter comprising bodies of a neutral reaction, while those of the other are feeble acids. In each there are probably several members which show among them every stage of physical condition between absolute insolubility and complete solubility in water, the intermediate bodies exhibiting gelatinous stages, characterised by the power of absorbing water in a greater or less degree.

Two of these bodies are of particular interest in connection with the study of the enzyme. One, *pectine*, belongs to the neutral, the other, *pectic acid*, to the acid series. According to Mangin their reactions are the following :—

Pectine. This substance swells up and dissolves in water, yielding a viscid liquid which is very difficult to filter and which tends to gelatinise when its solutions become at all concentrated. It is soluble also in dilute acids, from which it can be precipitated by alcohol. It gives no precipitate with neutral acetate of lead but is thrown down by the basic acetate in the form of white flocculi. If boiled for several hours in water it is converted into an isomer, *parapectine*, which is precipitated by the neutral plumbic acetate. Further boiling with dilute acids converts it into *metapectine*, which is precipitated by barium chloride.

Pectic acid. This body is insoluble in water, alcohol, and acids; it forms soluble pectates with alkalis, and insoluble ones with the metals of the alkaline earths, of which calcic pectate is most widely distributed. It dissolves in solutions of alkaline salts, such as the carbonates of sodium and potassium, stannates, alkaline phosphates, and most organic ammoniacal salts, forming

---

[1] For a *résumé* of these investigations see a paper by the writer on "The Cell Membrane," *Science Progress*, Vol. VI. p. 344.

with them double salts, which gelatinise more or less freely
with water. Its solutions in alkaline carbonates are mucilaginous
and difficult to filter, while when oxalate of ammonia is the
solvent, the liquid remains perfectly fluid and filters readily.

Fremy's discovery of the enzyme has been confirmed by
more recent investigators, and the nature of the changes set
up by its action are now more fully known.

The most complete information upon the peculiarities of
pectase has been furnished by Bertrand and Mallevre, who
published several memoirs upon the subject in 1895 and the
following years.

Their first preparations of the enzyme were obtained from
the juice of carrots, which were cultivated for the purpose and
gathered at the period of most vigorous growth. As already
mentioned, Fremy ascertained that the roots of these plants
contain pectase.

In its preparation Bertrand and Mallevre separated the
central cylinders of the roots from the cortex, and carefully
reduced them to pulp, afterwards extracting the sap by
pressure. About 70—80 per cent. of the bulk of the pulp
was thus squeezed out in the form of a turbid liquid, which
was then saturated with chloroform and filtered through
Bezelius filter-paper.

This juice when added to solutions of pectine quickly caused
the production of a jelly, just as Fremy had previously observed.

The change thus brought about was considered by Fremy
and his immediate successors to be the conversion of pectine
into pectic acid. The latter gelatinises more readily than
pectine and in solutions of less concentration. A perfectly
limpid preparation of pectine which shows no tendency to
gelatinise can be made to undergo the change readily, and
to set into a soft clot on the addition of a solution of
pectase.

But though confirming Fremy's work to the extent already
mentioned, Bertrand and Mallevre came to different conclusions
as to the nature of the product formed. They found that when
they dissolved the jelly in dilute hydrochloric acid, the solution
contained a certain quantity of calcium, which was not

precipitated on saturation with ammonia. Continued investigation satisfied them that the jelly was not pectic acid but a compound of this body with calcium ;—in fact calcic pectate.

Further researches were directed towards ascertaining whether the clotting was altogether dependant on the presence of calcium.

A purified solution of pectase was prepared from the crude one already described by adding to it a certain amount of the oxalate of one of the alkali metals to precipitate any calcium that the juice of the carrots might contain. After standing for a time till all sediment had settled down it was again filtered, when it yielded a clear liquid.

Pectine was then prepared from the residue left after the expression of the juice from the pulpy tissue of the root. This material was washed with alcohol and boiled for 15 minutes; and the mixture filtered before allowing it to cool.

The débris on the filter was thus freed from any adherent or soluble pectase, and was next extracted with a 2 per cent. solution of hydrochloric acid, in which pectine is soluble. Twenty-four hours' maceration was considered sufficient to extract the latter substance; at the end of that time the liquid was filtered from the residue and an equal volume of alcohol was added to it. This precipitated the pectine in the form of gelatinous flakes, which were collected and dried on a porous surface. To ensure their freedom from calcium salts they were then steeped in 50 per cent. alcohol containing 2 per cent. of hydrochloric acid. This treatment was continued till the pectine gave on incineration only a trace of ash which was free from calcium salts. It was then freed from acid by dissolving it in water, and then precipitating it with alcohol. Repeated treatment in this way eliminated most of the acid, and it was freed from the remainder by the addition of a few drops of dilute caustic potash.

Bertrand and Mallevre found that when prepared in that way a solution containing 2 per cent. of the pectine remained limpid and was in the best condition for further experiments. When they added to it the decalcified juice of carrots prepared as described, it remained indefinitely liquid, though such decalcified solution contained large quantities of pectase. The further addition of only a trace of a soluble calcic salt caused the formation of a jelly, the time taken depending upon the amount of calcium so added. The addition of a calcic salt alone, or of a calcic salt and some boiled carrot juice, always failed to induce the gelatinisation.

Barium and strontium were found to play the same part as calcium, in cases where excess of either was added to the pectine solution. Magnesium on the other hand had no action, or at most a very feeble one.

The clot formed by the action of pectase on pectine is therefore composed of pectate of calcium and not free pectic acid, as the earlier observers thought.

It will be remembered that, in the case of rennet, a similar combination of calcium with caseinogen takes place, the casein formed being according to Halliburton a caseate of lime.

Bertrand and Mallevre found that if a large quantity of a soluble calcic salt was added to the solution of pectine a gelatinisation took place without the presence of pectase. This was not however due to the formation of pectic acid or calcic pectate but to the production of another compound which they called a *pectinate*. The latter was also produced simultaneously with the calcic pectate if pectase was present and the calcium salt was in excess. The two gelatinous bodies can be distinguished from each other by the action of a dilute acid, such as a 2 per cent. solution of hydrochloric acid. This dissolves the pectinate, so that the jelly due to the presence of the latter disappears. It decomposes the calcic pectate by robbing it of the calcium, and leaves free pectic acid, which is insoluble in acids and consequently persists in the gelatinous condition.

Pectase is materially hindered in its working by the presence of free acid, in fact a neutral medium is almost essential for the production of the jelly. When a mixture of equal volumes of the expressed juice of carrots and a 2 per cent. solution of pectine is taken for experiment, the clot is usually formed in about an hour. If hydrochloric acid is added to such a mixture there is a considerable retardation, the longer as the amount of acid is greater; ·09 per cent. delays the gelatinisation for nearly two days, and 1 per cent. inhibits it entirely. The same results have been observed with sulphuric, nitric, tartaric and citric acids.

This observation is important, for it supplies a reason for the non-production of jelly in the juice of many fruits, as they contain besides the enzyme a larger percentage of acid than ·1 per cent.

The retarding influence of acid is lessened by the presence of a larger proportion of calcium salts, or of pectase. This explains why the juices of cherries and raspberries coagulate pectine though they are fairly acid in reaction. The gelatinisation depends ultimately on the relative proportion of pectase, calcium salts, and free acid present in the solution.

It has already been mentioned that Fremy denied the existence of soluble pectase in acid fruits. He said that in these the enzyme was present in an insoluble form and that it was not separable from the solid pulp. He explained in this way the fact that the latter would clot solutions of pectine, while he failed to get the coagulation with the expressed juice from such pulp. Bertrand and Mallevre repeated his experiments with the expressed sap of such fruits and carried them further. Though the juice as squeezed from the fruit will not clot the pectine as Fremy said, they found it would do so readily if a little alkali was added at the same time to neutralise the acid present. The failure was therefore not due to the absence of pectase, but to the presence of free acid which inhibited its action.

These observers contradict Fremy on another point; they have found that by prolonged maceration pectase can be extracted from the insoluble precipitate produced by alcohol from the expressed juice of the carrot. It is not therefore rendered absolutely insoluble by this treatment, as Fremy supposed.

Bertrand and Mallevre have discovered that instead of pectase being confined only to such pulpy tissues as have been mentioned, it has a very widespread distribution in the vegetable kingdom. Indeed they go so far as to say that it is universally present in green plants, being especially abundant in foliage leaves, which they think are the seats of its formation and from which they believe it spreads into the rest of the organs of the plant. Leaves which show rapid growth are usually the richest in the enzyme.

A very active preparation was obtained from the leaves of Lucerne (*Medicago sativa*) and Trefoil (*Trifolium pratense*). The plants were gathered when in the condition of most vigorous growth and bruised in a mortar to extract the sap. This was then saturated with chloroform and a flask filled with it, which was allowed to stand for 12—24 hours in the dark. A considerable sediment settled down, which was removed by filtration. A clear liquid was thus obtained which was poured into twice its volume of alcohol of 90 per cent. strength, a copious precipitate resulting. This was removed from the supernatant spirit and macerated for 12 hours in a small quantity of water. The pectase was thus dissolved out from the miscellaneous constituents of the sap, and on filtering the solution a nearly colourless liquid was obtained. The enzyme was again precipitated by the addition of a large excess of alcohol, and was collected on a filter and dried in vacuo. Using this method of preparation a litre of the filtered sap yielded 5—8 grammes of a white powder which was not hygroscopic, but dissolved readily in water. Its solution possessed great power of inducing pectic fermentation. A 1 per cent. solution of pectine was coagulated in 48 hours by the addition of $\frac{1}{1000}$ of its weight of Lucerne pectase or of $\frac{1}{1600}$ of its weight of Trefoil pectase.

By similar methods Bertrand and Mallevre ascertained that pectase is present in the following plants :—

Spirogyra

Chara fragilis

Marchantia polymorpha

Lolium perenne (leaf)

Zea mais (leaf)

Iris florentina (leaf)

Ginkgo biloba (leaf)

Thuja occidentalis (leafy shoot)

Pinus Laricio (needles)

Cydonia vulgaris (fruit)

Pyrus communis (fruit)

Malus communis (fruit)

Rubus idæus (fruit)

Armeniaca vulgaris (fruit)

Cucurbita Pepo (stem, leaf, flower, fruit)

Rheum rhapontiacum (leaf)

Beta vulgaris (leaf and root)

Plantago media (leaf)

Mentha Pulegium (flower heads)

Solanum Lycopersicum (ripe fruit)

Solanum tuberosum (leaf)
Ailanthus glandulosa (leaf)
Ampelopsis quinquefolia (leaf)
Brassica napus (leaf and root)
Brassica oleracea (leaf)
Syringa vulgaris (leaf)
Helianthus tuberosus (leaf)
Sambucus nigra (leaf)
Daucus carota (root)

Ribes rubrum (fruit)
Ceratophyllum submersum (whole plant)
Robinia pseudacacia (leaf)
Medicago sativa (shoot)
Trifolium pratense (shoot)
Vitis vinifera (leaf, fruit)
Acer pseudo-platanus (leaf)
Delphinium Staphisagria (leaf)

Its determination in *Thuja* was made a little doubtful by the viscous character of the cell-sap.

The amount of pectase in these different plants was ascertained by adding to the extract from a definite weight of tissue an equal volume of a watery solution of pectine of 2 per cent. concentration and noting the time that elapsed before gelatinisation took place. The results given by some of the tissues are subjoined :—

Solanum Lycopersicum (ripe fruit)   48 hours
Vitis vinifera (nearly ripe fruit)   24 „
Ribes rubrum (white currant fruits)   15 „
Rheum rhaponticum (leaf)   12 „
Marchantia polymorpha (thallus) $2\frac{1}{2}$ „
Daucus carota (adult cultivated root)   2 „
Delphinium Staphisagria (leaf) $1\frac{1}{2}$ „
Ginkgo biloba (leaf)   35 minutes
Syringa vulgaris (leaf)   20 „
Ailanthus glandulosa (leaf)   20 „
Daucus carota (young root)   15 „
Zea mais (leaf)   8 „
Iris florentina (leaf)   3 „
Trifolium pratense (shoot)   less than 1 minute
Medicago sativa (shoot)   „
Solanum tuberosum (leaf)   „
Brassica napus (leaf)   „
Plantago media (leaf)   „
Lolium perenne (leaf)   „

The activity so denoted probably gives a fairly accurate approximation to the quantity of pectase present. It varies

consequently within very wide limits. There are cases when
the coagulation is almost instantaneous. In others owing
probably to the smallness of the amount of pectase, the clotting
is slow and uncertain. In such cases the best results are
obtained when the mixture of sap and pectine is exactly neutralised and a little calcic salt added. It must not be at once
inferred that a slow process undoubtedly indicates a small
quantity of the enzyme, for the latter will not always act
uniformly, two determining factors being the reaction of the
medium and the amount of calcium salt which it contains.

The distribution of the enzyme varies also in the same
plant. In Cucurbita Pepo the time taken to clot the pectine
solution by extracts of equal weights of different parts was as
follows :—

Corolla	45 minutes	Stem apex	12 minutes
Young fruit	30 ,,	Leaf petiole	8 ,,
Stem-base	20 ,,	Leaf blade	1 ,,

It was mentioned above that the action of pectase was
largely influenced by the reaction of the solution in which it
was working, as well as by the presence of salts of calcium.
Bertrand and Mallevre have found that these determining
factors do not act independently of one another. We have
seen that calcic salts are essential to the coagulation,—in the
first place because they enter into the composition of the clot.
But this does not seem to be the only part they play, for the
clotting is accelerated by the addition of more calcium salt even
when sufficient to combine with the pectic acid is already
present. Conversely the presence of free acid has a retarding
influence on the action. Bertrand and Mallevre found in one
of their experiments that a mixture of equal volumes of the
sap of carrots and of 2 per cent. solution of pectine set into a
jelly after about one hour. On adding hydrochloric acid to a
similar mixture in increasing quantities there was a retardation
in the time of the occurrence of gelatinisation, which increased
pari passu with the amount of acid added. ·088 per cent. of
the acid delayed it 40 hours and ·1 per cent. stopped it
altogether. The same effect was noticed with sulphuric, nitric,
malic, tartaric and citric acids.

This retarding effect again was lessened by the presence of a larger proportion of calcic salts, or of pectase, so that the clotting depends upon the relative proportions of all three, pectase, acid and calcic salts.

The exact nature of the action remains still unknown, but the facts suggest as a tempting hypothesis that the pectase converts the pectine into pectic acid and that this then reacts with the calcium salt, forming calcic pectate.

The wide distribution of pectase, which indeed has been found in almost every region where it has been sought for, considered in connection with the universal presence of pectic bodies in conjunction with cellulose in cell-walls, suggests that its function in the living plant is connected with the changes which the cell-membrane undergoes during the life of the cell. The work of various investigators, from Braconnot in 1825 to Mangin during the last few years, has conclusively proved that at no time is the cell-wall a homogeneous membrane consisting of pure cellulose. While the latter enters very prominently into its composition, there are present in it a number of other substances, varying in nature and in relative proportions, which have been somewhat loosely described under the names of pectose, pectine, pectic acid, metapectic acid and their compounds. These have been lately investigated by Mangin, who has isolated several of them and described their reactions. They fall into two series, each comprising several members, which show among them every stage of physical condition between absolute insolubility and complete solubility in water. The intermediate bodies exhibit gelatinous stages, characterised by the power of absorbing water in a greater or less degree.

The first series, one of whose members is the pectine so often alluded to, is composed of bodies possessing a neutral reaction; the members of the other are feeble acids. Of these pectic acid is one of the most prominent. The two series are closely related to each other, for by the action of heat, acids or alkalis the various members of both can be prepared from pectose, the most insoluble of the neutral bodies.

These bodies exist together with cellulose in all the membranes which have not undergone change into lignin or suberin, and by treatment of the cell-walls with various reagents they can be separated from it. Pectic acid does not usually exist in the free state, but as pointed out long ago by Payen is usually present in combination with calcium, as calcic pectate.

Payen pointed out further that the so-called "middle lamella" between contiguous cells is almost or entirely composed of calcic pectate, and his opinion has been endorsed by Mangin in the course of his recent work. It is easy to demonstrate that this layer, whatever be its origin, has not the same composition as the rest of the cell-wall as it can easily be dissolved by reagents which leave the cells apparently intact, though isolated from each other. The most active of these reagents is the so-called "maceration fluid" of Schultze, which consists of a solution of potassic chlorate in nitric acid.

This difference of composition between the middle lamella and the rest of the wall has long been known. Before the name "middle lamella" was given to it by the writers of the time of Naegeli and Sachs it was called "intercellular substance" and was thought to be a kind of cement, binding contiguous cells together.

Besides the middle lamella other modifications of the original cell-wall have from time to time attracted attention. Chief among these we have the so-called "intercellular protoplasm" of Russow, which he described as forming in certain cases a delicate membrane, or lining layer, coating the intercellular passages. Russow's opinion that this substance is protoplasmic has been controverted by several subsequent observers, who have shown that it is much more probably a derivative of the cell-wall. Schenck held it to be of the same general nature as the middle lamella, and Mangin has suggested that it is composed of a mixture of pectic bodies, including calcic pectate.

The researches of Mangin have thrown a good deal of light upon the proportion of the neutral pectic bodies and the compounds of pectic acid which are present in various cell-walls. In the young unchanged membranes there is little pectic acid while pectose is present in larger amounts. In older cell-walls, especially in tissues in which intercellular spaces or passages have appeared the proportion of calcic pectate is more prominent. As said above the middle lamella is almost if not entirely composed of it, and it often collects over the surfaces of the intercellular spaces, being in such cases a continuation of the middle lamella of the wall which has split during the formation of the passage.

Even the youngest cells can be separated from each other by the reagents which dissolve calcic pectate, so that there is some reason to suppose that the cell-membrane is at no time absolutely homogeneous, but consists of a middle layer of calcic pectate, covered on both faces by a layer of a combination or mixture of cellulose and pectose. As it grows, this layer of calcic pectate becomes more pronounced and prominent till it can be made visible under the microscope. This change must be due to transformations which modify the composition of the layers abutting on the cell cavities and which result in the formation of the increased amount of calcic pectate.

The mode of deposit of the calcic pectate over the surface of the intercellular spaces may perhaps aid us to form a true conception of what takes place.

These pectates gradually tend towards the outside of the membrane, possibly passing as soluble pectic acid in its substance and being combined with the metallic base at the external surface, or in the intercellular space.

In the young growing cell, just behind the zone of cell-division at the growing point, there is the maximum of turgidity or osmotic pressure. It is quite conceivable that in a cell abutting on an intercellular space this is sufficient to cause a stream of soluble bodies to pass across the substance of the cell-membrane from within outwards. This then would lead to the extrusion of such soluble pectates or pectic acid as may be present in the wall. Bodies of this series are not unlikely to be formed from the pectine or pectose in the membrane by the action of the dilute acid of the cell sap, or more probably by the pectase which Bertrand and Mallevre have shown to be present in the apical region of the stem in many cases. Such a course of transformation is even more probable in the cells and cell-walls of the young leaves, in which pectase is often present in very considerable amount.

The same consideration may be applied with even greater probability to the formation of the middle lamella. Where the cells do not abut on an intercellular space, but have their neighbours pressing upon them, as they have in the young part of the growing zone, whether of stem, leaf or root, any of their membranes will be subject to a pressure from each side owing to the turgidity of the contiguous cells. In this case the stream of pectic acid or pectates would not pass out of the cell, but would tend to accumulate in the middle line between the two pressures, in the region, that is, where the middle lamella speedily becomes recognisable.

Though the transformation of pectine into pectic acid under the action of pectase has not been shown definitely to be the cause of the occurrence of the middle lamella, which belongs at present rather to the realm of hypothesis than of fact, it is certainly supported by the distribution of pectase already established by Bertrand and Mallevre and quoted in the present chapter. They have found it most abundant where cell growth is most vigorous and the more rapid the growth the more plentiful is the enzyme. They have found eight times as much in the leaf-blade as in the petiole, and more than twice as much in the apex of the stem as in the base. In the corolla of Cucurbita, in which scarcely any thickening of the cell-wall can occur, the amount compared with that in the foliage leaf was as 1 : 45.

The failure of a plant to thrive or even to grow beyond a slight extent in the absence of calcium may be partly connected with the same series of phenomena.

CHAPTER XVIII.

AMMONIACAL FERMENTATION. *UREASE.*

WHEN urine is first excreted by an animal its reaction is faintly acid; after standing for some time exposed to the air it becomes alkaline; the strength of the reaction gradually increases and after a somewhat longer interval a distinct odour of ammonia is evolved from it. If after being shed it is sterilised and kept from contact with air this change does not take place.

Chemical examination of the exposed urine shows that the alteration is due to the transformation of the urea into carbonate of ammonia, which takes place by a simple process of hydrolysis according to the equation

$$CO(NH_2)_2 + 2H_2O = (NH_4)_2CO_3.$$

This decomposition of urea has been regarded as a fermentation almost ever since the time when it was first observed. Before chemical investigation had been made into the composition of the changing urine, Fourcroy and Vauquelin suggested that the alteration of its reaction was due to a transformation of the urea, and that it was caused by a ferment action set up by the albuminous matter in the urine. Dumas took a similar view, but supposed that the ferment originated in the mucus which urine generally contains. Jacquemart associated the ferment action with the white deposit which gradually forms in the vessels when urine is allowed to stand until it becomes stale. He held the same view as Fourcroy that the ferment principle was amorphous.

About 1860 more accurate views began to be formulated.

A suggestion was made by a German chemist, Müller, that the action was probably due to a living organism, corresponding to the yeast of beer. Two years later Pasteur discovered such an organism in certain specimens of putrid urine. He described it as consisting of small spherical cells joined together in chains. The cells were in appearance a good deal like the cells of yeast but were much smaller. On account of this resemblance the organism was originally named *Torula ureæ*. It was studied with much care in 1864 by Van Tieghem, whose results were confirmatory of those of Pasteur. The organism was subsequently renamed *Micrococcus ureæ* by Cohn. It is composed of spherical or globular cells whose mean diameter is 1.5μ; they are united together into long curved chains which are dispersed throughout the liquid as long as the fermentation proceeds. When it is over they sink to the bottom of the vessel and the chains break up, so that a sediment is composed of free globules or short chains. The cells show no granulation, their cell-wall is hardly to be distinguished from their contents, and they multiply for the most part by budding.

Micrococcus ureæ can be cultivated most easily in urine, but it will grow in any nitrogenous fluid in which urea is dissolved, or in a solution of urea which also contains phosphates. According to Jaksch it is capable of thriving without urea if other amides or peptones are present instead.

The organism is aerobic.

A peculiar feature of its life which was noticed by Van Tieghem is its power of thriving in strongly alkaline solutions. He observed a fermentation to continue until the liquid contained 13 per cent. of carbonate of ammonia, a concentration which would be fatal to almost all other forms of vegetable life.

Van Tieghem has stated that this organism is capable also of decomposing hippuric acid, which is so prominent in the urine of herbivorous animals, the products of the decomposition being benzoic acid and glycin. This process also is one of hydrolysis and may be expressed by the following equation

$$C_9H_9NO_3 + H_2O = \begin{cases} C_6H_5 \\ COOH \end{cases} + \begin{cases} CH_2NH_2 \\ COOH. \end{cases}$$

Hippuric acid   Benzoic acid   Glycin

When hippurate of ammonia is dissolved in either yeast-water, or a solution of sugar which also contains phosphates, and is exposed to the air, a growth of the Micrococcus soon appears in the liquid, which quickly gives evidence of the decomposition just alluded to.

Other organisms have been ascertained by Sestini to effect the hydrolysis of uric acid, the resulting products being carbonate of ammonia and carbon dioxide.

This Micrococcus is not by any means the only microbe which has the power of decomposing urea. We owe to the researches of Miguel and other writers a knowledge of many other organisms which can effect its hydrolysis. Some of these are ordinary fungi, but most belong to the group of the *Schizomycetes* or fission-fungi. Miguel has described seven species of *Bacillus,* nine *Micrococci,* and one *Sarcina.* The Bacilli appear to act most energetically.

The organisms are very widely distributed in nature, being found in the air, in spring and river-water, and in the soil. According to Miguel 1—2 per cent. of the bacteria present in the soil and 15 per cent. of those present in cow-house manure, are capable of hydrolysing urea.

The urea does not appear to be normally a nutritive substance for the organisms. If other nitrogenous compounds, particularly proteids, are present in the solution in which they are growing, these are the sources from which they gain their nitrogen. In the absence of such compounds, however, they can use the nitrogen of the urea.

The activity of the Torula or Micrococcus ureæ has been ascertained to be due to a soluble enzyme which under certain conditions can be extracted from the cells. The enzyme, to which the name *urease* has been given, was first described by Musculus in 1874. When urine which was undergoing active ammoniacal fermentation was filtered through fine filter-paper and the paper subsequently well washed and dried, he found that he could excite a similar fermentation by immersing pieces of it in a neutral solution of urea. Musculus tested the progress of the fermentation by staining his paper with turmeric after treating it as already described. After a short

stay in the solution of urea the turmeric became brown owing
to the alkalinity of the liquid. The same result was obtained
when the paper was washed with strong alcohol before staining
with turmeric. Under this treatment the induced fermentation
could not be due to living cells left on the filter-paper as these
would not survive contact with the alcohol. In a subsequent
paper, published in 1876, Musculus described the preparation
of the enzyme from some highly alkaline urine which he
obtained from a pathological secretion. He added alcohol in
excess to such urine and obtained a viscous precipitate consisting largely of mucin, derived from the walls of the bladder.
When this precipitate was separated by filtration and dried, it
readily yielded to water a solution of the enzyme which was
extremely active. This solution when added to alcohol deposited an amorphous precipitate which possessed the power of
setting up the decomposition of urea. The source of the
enzyme in this case was apparently the alkaline urine, as
Musculus states his mucous urine did not contain any of the
cells of the microbe. This does not however prove that the
enzyme had any other origin, as they might have been present
in the bladder under the pathological conditions existing.

Musculus found that on acidifying the solution, the enzyme
was rapidly destroyed.

The behaviour of the Micrococcus was the subject of an
exhaustive series of experiments by Sheridan Lea in 1885.
He cultivated the organism in urine until he had obtained a
large quantity which was exciting a very vigorous fermentation,
when he poured the whole into an excess of alcohol. A copious
precipitate was thrown down, consisting partly of the organism
and partly of mineral and other matter contained in the urine,
together with a sediment which was already existing in the
fluid. The precipitate was thrown into a filter, well washed
with more alcohol and dried. A small quantity of the precipitate when mixed with a 2 per cent. neutral solution of urea
and kept at 38° C. developed a strong alkaline reaction in a few
minutes and this was followed a little later by the evolution of
a powerful odour of ammonia.

Some of the precipitate was next extracted with water, and

filtered. The reaction of the filtrate was very slightly alkaline but the liquid was clear and limpid. On adding some of this to a quantity of the same solution of urea as was used in the first experiment, the same sequence of phenomena occurred. The alcohol precipitate was in this way shown to contain an enzyme capable of hydrolysing urea and it was found possible to dissolve it out by treating the precipitate with water.

Lea claims to have been able to obtain the enzyme in an approximately pure condition by repeated solution of the precipitate and reprecipitation by alcohol. This treatment gradually removes the salts which are present as well as the enzyme. So prepared he found it to be a white powder, amorphous in character, giving a clear solution in distilled water, which however always showed the presence of a trace of proteid matter when tested with nitric acid and ammonia.

Lea's results confirm and extend those of Musculus, in so far as he shows that the enzyme can be prepared from the cells of the Micrococcus. In Musculus' experiments he says the cells were not present and the mucous urine itself was the source of the enzyme he obtained.

Lea made a further series of experiments to see if the enzyme is excreted by the Micrococcus into the urine or whether the normal action is intracellular. Musculus' results appear to indicate an excretion of the enzyme from the organism in the bladder, as it is hardly likely that the tissue of the latter produced it.

In these experiments Lea took a quantity of actively fermenting urine and filtered it till it was free from sediment. The organisms were left behind on the filter, very fine paper being used and each filter being composed of 12—15 thicknesses. The filtration was continued till no micrococci were visible under the microscope. The perfectly clear filtrate was then neutralised by very dilute acetic acid, and after the resulting effervescence had subsided, the remaining carbonic dioxide was extracted by exposure to a vacuum. Two per cent. of urea was then added to a measured quantity of the filtrate and the whole placed in a water-bath at 38° C. A control was prepared by taking another equal quantity of the filtrate without adding any urea. This was placed in a similar vessel in the water-bath side by side with the first one.

Both quantities remained neutral even after an exposure of six hours.

In further experiments Lea separated the organisms from the urine by filtering it through a porous earthenware cell, and found the same results as to the absence of the enzyme from the urine so prepared.

A quantity of similar urine was next filtered till free from cells and precipitated by the addition of an excess of alcohol. The resulting precipitate was soluble in water, but the solution failed to exert any decomposing action on a solution of urea.

Lea thus failed to obtain any evidence of the excretion of the enzyme from the cells of the organism and came to the conclusion that its action was altogether intracellular. This result was antagonistic to that obtained by Musculus. Only two modes of reconciliation of the two seem possible. Either Musculus' urine must have contained some cells which escaped his observation; or the excretion from the organisms in Lea's experiments was only small and what ferment there was was destroyed by the alcohol used in the precipitation. Neither explanation seems very satisfactory.

Lea came to the conclusion that the enzyme was incapable of passing out of the cell during life, on account of the cellulose membrane surrounding it.

When the organism was killed by the alcohol and its protoplasm to some extent decomposed by the action of the reagent, the enzyme could be extracted by a solvent such as water.

This explanation, however, leaves us in some uncertainty, as the treatment with the alcohol does not destroy the whole of the enzyme, nor does it modify the cell-wall, and it is hard to understand why if the ferment cannot pass the cell-wall during the life of the protoplasm it should be able to do so after its death, unless the latter modified the physical characters of the cellulose membrane. It seems more likely that the enzyme is retained very strongly by the living protoplasm, as Buchner has shown to be the case with the enzyme producing alcohol.   After death of the protoplasm it would be much more easily extracted.

Miguel has extracted the enzyme from fourteen different species of micro-organisms which present distinct morphological characters, and which are all capable of setting up ammoniacal fermentation.   He cultivated each of them in peptone solutions containing 2—3 grammes of ammonic carbonate per litre. Before inoculation with the microbe, the solutions were sterilised by being filtered through porcelain.   The cultures were continued for some days until the whole liquid had become turbid.   The peptone solution was then found to contain a quantity of the enzyme which had been excreted by the microbes.   In his experiments Miguel obtained sufficient enzyme per litre of peptone solution to convert 60—80 grammes of urea into ammonic carbonate in less than an hour.

The optimum temperature for the working of urease Miguel found to be 50—55° C., but even at this temperature the enzyme was gradually destroyed.   At a temperature near 0° C. the solution of urease retained its activity for several weeks ; at 75° C. it was destroyed in a few minutes and at 80° C. almost instantaneously.   The organisms themselves were not easily killed, surviving an exposure for 2—3 hours to a moist temperature of 95° C.   Lea found his preparation of urease was destroyed on heating to 80—85° C.

Bufalini has stated that besides decomposing urea and hippuric acid urease is capable of converting asparagin into succinic acid.

Schmiedeburg has found an enzyme in the kidney of the pig which he says is concerned in the splitting up of hippuric acid.   He has given it the name of *histozyme*.   Whether it is the same as urease is not at present determined, though it does not seem unlikely that they may be identical when we remember that Van Tieghem showed that Micrococcus ureæ can hydrolyse hippuric acid as already mentioned.

CHAPTER XIX.

OXIDASES, OR OXIDISING ENZYMES.

THE general course of action of the enzymes which we have discussed so far we have seen to be one of hydrolysis, or decomposition of the bodies attacked by them after a preliminary taking up of water into their molecule. This is fairly satisfactorily established with regard to most of the ferments which play a leading part in digestive changes in both animal and vegetable organisms. The changes effected by the proteolytic enzymes are not so clearly shown to be hydrolytic as are those brought about by ferments which act upon carbohydrates and fats, but we have seen reason to believe that the course of action is the same. The reaction is evidently more complicated in the case of the clotting enzymes, and for the present we must leave the matter doubtful so far as they are concerned.

A few of the decompositions we have examined do not appear to be concerned with hydration, particularly the action of myrosin, and we shall see later that the alcohol-producing enzyme of yeast does not initiate such a process.

During the last few years the existence of another class of enzymes has been indicated, all of which act by promoting direct oxidation of various substances, including various aromatic compounds and sugar. These have been called *oxidases*: they are somewhat widely distributed, occurring in both the animal and the vegetable body. They have recently been termed *respiratory* enzymes as most of those already discussed have been called *digestive*.

Laccase.

Of these *oxidases* the earliest to be recognised was *laccase*, the body which is concerned in the production of lacquer varnish from the crude sap of the lac tree of South-east Asia.

The existence of this oxidase was first pointed out in 1883 by a Japanese chemist, Yoshida, who investigated the latex of that plant and first ascertained the nature of the changes occurring in the production of the varnish.

The crude juice is obtained by making incisions into the trunk of several species of *Rhus*, and collecting the viscous matter which exudes. It has the appearance of a nearly white creamy liquid, possessing a faint odour resembling that of butyric acid. On exposure to air it rapidly changes colour, becoming brown, and ultimately black. Spread on a flat surface it dries with a brilliant black lustre. The juice is very difficult to experiment with, as it possesses a very irritating property which affects the skin, causing painful eruptions and sores.

Yoshida states that the juice, known by the name of *urushi*, consists in great part of a peculiar acid, which he has called *urushic acid*, and to which he ascribes the formula $C_{14}H_{18}O_2$. Separated by appropriate methods from the crude latex and dried at 110° C. it forms a dark, pasty substance, smelling of the original juice; it is then soluble in benzol, ether, alcohol, and carbon-disulphide, but is insoluble in water; it has a specific gravity of ·9851 at 23° C. When exposed to the air it does not dry nor show signs of change such as the original latex does.

Besides urushic acid, the crude sap contains a certain proportion of gum, and a variable quantity, usually about 3—8 per cent., of a peculiar nitrogenous body, which coagulates on heating to 63° C. If the latex is treated with excess of alcohol the gum and the nitrogenous constituent are precipitated. After filtration the latter can be separated from the former by the action of cold water, in which it dissolves, while the gum only becomes swollen.

The enzyme is associated with this nitrogenous constituent of the juice. If a solution of it is mixed with a small quantity of free urushic acid, the latter is under certain conditions converted into the varnish. The change does not take place if the solution is heated to 63° C.

The nature of this nitrogenous constituent has not been clearly established, but it seems to differ considerably from the proteids, containing a much smaller proportion of nitrogen, and more carbon than they do. Yoshida's analysis of it gives C 63·44, H 7·41, N 4·01, O 22·04, Ash 1·2 in 100 parts.

From his experiments Yoshida has come to the conclusion that urushi juice consists essentially of four substances, viz. urushic acid, gum, water, and a peculiar enzyme. The phenomenon of its drying is due to the oxidation of urushic acid, $C_{14}H_{18}O_2$, into oxy-urushic acid, $C_{14}H_{18}O_3$, which takes place by the aid of the enzyme in the presence of oxygen and moisture.

He supports this conclusion by two series of experiments, which may be quoted here.

A small quantity of the original juice was put into a covered beaker and subjected to the regulated heat of a water-bath, the water lost by evaporation being subsequently restored. The heating was carried to different temperatures, and subsequently the heated juice was spread thinly over a glass plate and left to dry in a box the air in which was kept moist. In each experiment the juice was heated for 3½—4 hours, and the drying was allowed to take place at a temperature of 20° C.

The results were as follows:—

Temp. of exposure.	Time of drying.
20° C.	2 hours
30° C.	4 „
40° C.	4½ „
55—59° C.	24 „
60—63° C.	Did not dry.

In the second series of experiments he found that unless moisture was present the latex did not dry; that in moist air it dried in 4 hours, in moist oxygen in 2 hours, in moist hydrogen or nitrogen it took 36 hours, and in moist CO_2 it was dry only after 2 days' exposure.

It follows from these experiments that the enzyme works

most energetically at a temperature of 20° C., but only when
oxygen and moisture are both present; a rise of temperature
above 20° C. is slowly prejudicial to it, and at 60°—63° C. it is
destroyed. It may be noted that it is at this temperature that
the nitrogenous matter coagulates.

Yoshida prepared oxy-urushic acid from urushic acid by
the action of strong chromic acid. He says that so prepared it
exhibits all the properties of the lacquer varnish.

The name *laccase* was given to the enzyme more than ten
years later by Bertrand, who made further investigations into the
peculiar behaviour of the latex and who ascertained several
additional facts about the enzyme.

In the main he confirms the earlier work of Yoshida as to
the constituents of the latex. The body described as urushic
acid he prefers to term *laccol*, but he has not examined it
minutely on account of its deleterious properties.

He prepared the enzyme by treating the latex with a large
excess of alcohol; this precipitated a gummy substance, which
he purified by redissolving it after filtration, and again throwing
it down by the addition of 10 volumes of alcohol. It separated
out in white opaque flakes which yielded on hydrolysis a
mixture of galactose and arabinose.

The enzyme was extracted from the gum by treatment with
cold water.

In the natural juice the laccol exists in the form of
an emulsion, which is probably due to the presence of the
gum.

The laccol remains unchanged if it is separated from the
latex by solution in alcohol and kept from the air. If a little
water is added to the solution in alcohol a white emulsion
results, which keeps for a considerable time unaltered; but if a
solution of laccase is substituted for the water, the resulting
emulsion turns brown at once and rapidly becomes black,
especially if air is admitted. With a boiled solution of laccase
no such change can be observed.

So far as Bertrand has investigated the properties of laccol,
it appears to be allied to certain polyatomic phenols. On this
account he has examined the action of laccase on several of

the latter, especially hydroquinone and pyrogallol. When the former is submitted to its influence, the colour of the solution quickly becomes rose-red, and after a short interval crystalline scales with a green metallic lustre appear, the quantity rapidly increasing. When this operation is carried out in a sealed tube the oxygen present is almost completely absorbed. The liquid gives off a strong characteristic odour, and quinone can be extracted from it by shaking it with ether after removal of the solid matter. The precipitate is quinhydrone.

In the absence of the laccase, the hydroquinone does not absorb oxygen, nor undergo alteration. The hydroquinone is therefore oxidised by the free oxygen under the influence of the laccase, according to the equation

$$2C_6H_4(OH)_2 + O_2 = 2H_2O + 2C_6H_4O_2.$$
$$\text{Hydroquinone} \qquad\qquad \text{Quinone}$$

The colour given to the liquid is due to the formation of the quinone. Some of the latter, combining with the excess of hydroquinone not oxidised, produces the less soluble crystals of quinhydrone.

When pyrogallol is used instead of hydroquinone similar results are obtained, a precipitate of purpurogalline being thrown down in the form of a powder which sublimes on heating, forming orange-red needles which are soluble in alcohol and acetic acid.

Laccase attacks many other polyphenols, but chiefly those whose hydroxyl groups are in the ortho- and para-positions. The corresponding meta-compounds are affected only with difficulty. The oxidisability of these bodies by laccase seems to depend on the facility with which they can be transformed into quinones. The monophenols are not oxidised by the enzyme, but it attacks gallic acid and tannin.

Bertrand's observations on the behaviour of laccase at different temperatures do not agree with those of Yoshida, as he finds it still active after heating it to 70° C.

Bertrand has sought for laccase with some success in other plants and has indicated a rather wide distribution for it. In his researches he has employed the guaiacum test and appears to a certain extent to rely upon this method of recognition. This is unfortunate, as most investigators do not find it give entirely satisfactory results. He says that an alcoholic tincture of gum guaiacum becomes blue in the presence of air and a little laccase; if much of the latter is present, it turns from blue to green and subsequently to yellow. In most cases however he has confirmed his results by isolating the enzyme and proving its presence by its action. This is really the only satisfactory method of demonstrating its existence. By the two methods conjointly he has found laccase in the roots of the beet, carrot and turnip; in the tubers of the potato and the Jerusalem artichoke; in the tuberous roots of *Dahlia*; in certain rhizomes; in the fruits of the apple, pear, quince and chestnut; in the vegetative parts of lucerne, clover, rye-grass, and asparagus; and in the flowers of *Gardenia.* It may be prepared from these sources by extraction with water and precipitation of the extract with alcohol. If the tissue is green, the extract may be saturated with chloroform and allowed to stand for 24 hours to free it from the colouring matter, after which the precipitation by alcohol may be carried out.

Rey-Pailharde has found laccase in germinating seeds, especially of plants of the *Leguminosæ.*

The activity of laccase appears to be associated in some way with the presence of manganese. Its ash always contains traces of an oxide of this metal, sometimes as much as 2 per cent. Bertrand states that the activity of a preparation of the enzyme is proportional to the amount of manganese which is present.

When prepared from Lucerne it is poor in this constituent, and the effect of the addition of a salt of the metal can be easily studied. Bertrand describes a typical experiment on this point. He gathered several kilograms of lucerne at the time of flowering and bruised them in a mortar, pressing out the sap, which was then saturated with chloroform and allowed to stand in the dark for 24 hours. The juice was next filtered and $2\frac{1}{2}$ volumes of alcohol added to precipitate the laccase. The precipitate was taken up with a little water, the solution filtered and the laccase again thrown down by large excess of alcohol. The final precipitate was collected and dried in vacuo. It contained a mere trace of manganese.

To 50 c.c. of a solution of hydroquinone ·1 gm. of this precipitate was added, and the whole was agitated for 24 hours in contact with air. There was then only a red coloration produced. To a further quantity of 50 c.c. of the hydroquinone solution ·1 gm. of the precipitated laccase and 1 mgr. of manganese in the form of the sulphate were added together, and in less than 2 hours crystals of quinhydrone were formed. In the latter case there was an evident oxidation, much more extensive than after 24 hours' agitation in the absence of the manganese.

In an experiment so arranged that the absorption of oxygen could be measured it was found after 6 hours' agitation with air at 15° C. that with laccase alone ·2 c.c. oxygen were taken up; with a salt of manganese alone ·3 c.c. were absorbed; but with both present together 6·3 c.c. of oxygen were fixed.

The manganese is thus seen to play a very active part in the ordinary action of the enzyme. No other metal was found to be capable of replacing it.

Manganese combined with various acid radicals was found in a further series of experiments to have a certain power of causing the oxidation of hydroquinone, the protoxide appearing to act as a carrier of the oxygen. Comparing the action of these salts of manganese with the conjoint action of manganese and laccase, Bertrand advances the theory that the oxidases may be conceived to be special combinations of manganese with certain proteid bodies containing acid radicals, the latter varying with the particular enzyme. In such combinations the acid radical has just the necessary affinity to keep the metal in solution. The work of conveying the oxygen is in Bertrand's opinion discharged by the manganese, while the proteid matter gives to the oxidase its other characters, such as are made evident by the action of heat, and the various reagents used to identify it.

Whether this hypothesis be accepted or not, the experiments show that laccase is at any rate much assisted in its working by the presence of manganese if its activity is not entirely dependant upon its association with that metal in some form.

Besides the plants already mentioned laccase appears to exist in a considerable number of Fungi. In these plants the phenomena of oxidation are very prominent, and in consequence of this fact Bourquelot and Bertrand instituted in 1896 an investigation of them with a view to ascertaining whether laccase or some similar enzyme plays a part in their metabolism. As in other cases, at the outset these observers laid considerable stress on the guaiacum reaction, and they found that the liquid that can be expressed from many fungi very rapidly oxidises the tincture with the formation of a blue colour, but that it does not bring about this change if it is first boiled.

The reactions of the expressed juice with other bodies than tincture of guaiacum leave no doubt that it contains the same principle as the sap of the lacquer tree. It causes the browning of the laccol prepared from the latex of Rhus; it yields crystals of purpurogalline when allowed to act upon pyrogallol, produces quinone and quinhydrone from hydroquinone, and gives a very distinct brown colour with gallic acid.

The fungus which yields laccase most readily is *Russula fœtens* Pers., one of the Basidiomycetes, which is fairly common in woods during the summer. 125 grams of this fungus extracted with its own weight of chloroform water yielded 60 c.c. of a liquid which was at first pale yellow in colour, but which gradually reddened on exposure to air. When made to act on gallic acid in a closed flask which was constantly shaken it was found that the oxygen was gradually absorbed, 15 c.c. disappearing during the first hour of action. It gave also the reactions just described with laccol, pyrogallol, &c.

When the extract so prepared was boiled, it gradually lost its enzymic powers. Bourquelot and Bertrand say however that it is more resistant to heat than most enzymes, and that to ensure complete destruction the boiling should be maintained for a short time.

When the extract of Russula is poured into an excess of alcohol it yields only a small amount of precipitate, but this when separated off gives up the enzyme to cold distilled water. The precipitation of the laccase is not complete however when the extract is so treated.

A very large number of species of Fungi have been examined, chiefly belonging to the Basidiomycetes, more than half of which have been found to contain laccase, capable of acting on the aromatic bodies mentioned. Of these the genera *Russula*, *Lactarius*, *Boletus*, and *Psalliota* are the most noteworthy.

The Gasteromycetes as a rule contain little, if any, and the Ascomycetes and Myxomycetes so far as they have been examined are free or nearly free from the enzyme.

Besides working at the effect of laccase on the aromatic bodies as described above, Bourquelot and Bertrand investigated the nature of the changes of colour which supervene when many of the fleshy fungi are cut and the damaged surfaces exposed to the air. The tissue of *Boletus* changes almost instantaneously under such conditions, assuming a blue colour, the depth of tint and rapidity of appearance varying somewhat in different species. *Lacterius* becomes violet when wounded, while *Russula* turns first red and finally black.

There have been several theories as to the cause of this change of colour. Schœnbein noticed the phenomenon as long ago as 1856 and he attributed it to the action of ozone upon a particular chromogen in the fungus, saying that the latter also contains a substance capable of transforming the oxygen of the air into ozone. In 1872 Ludwig made some investigations into the subject and confirmed Schœnbein as to the existence of a special chromogen in the tissue.

In the light of the recent work on the oxidases Bourquelot and Bertrand were led to the view that one of the latter probably is concerned in the alteration of the chromogens. According to Schœnbein there was evidently something concerned besides the chromogen, and in his opinion the work effected by the particular constituent in question was the transformation of oxygen into ozone. Whatever it may have been it cooperated with the oxygen of the air in causing the oxidation of the chromogen. As this is apparently the part played by laccase in the formation of the lacquer varnish, it seems probable that Schœnbein's hypothetical oxygen transformer was really an oxidising enzyme.

Working on this hypothesis Bourquelot and Bertrand carried out the following experiment. A definite weight of *Boletus cyanescens* Bull. was extracted with boiling alcohol of 95 per cent. concentration, the fungus being cut up as far as possible out of contact with air. The extraction was continued for a quarter of an hour, after which the liquid was cooled and filtered. The alcoholic extract so prepared was faintly yellow in colour, and it contained the substance which normally turns blue on exposure. So prepared it retained its colour for a considerable time, even when diluted with water and allowed to stand in contact with air.

The investigators added to such an extract, diluted with its own volume of water, a small quantity of the extract of Russula prepared as described above. In half-a-minute a purple coloration appeared which passed rapidly into blue. The same effect followed on the addition of a little laccase prepared from the latex of the lacquer tree. If the enzyme was added slowly without agitation the tint was seen to be assumed gradually, the upper layers of the liquid in contact with the air being coloured first and the tint spreading thence throughout the whole.

Hence Bourquelot and Bertrand infer that the oxidase which can effect these changes is identical with the laccase of Rhus and other plants, and that in addition to acting on aromatic bodies such as hydroquinone and pyrogallol it also assists to oxidise the chromogens of certain fungi, especially those which yield a blue or a red colouring matter. The laccase exists in the juice of the fungi side by side with the chromogen, but when the juice is boiled before exposure to the air has taken place the laccase is destroyed and the chromogen in consequence remains unchanged.

An enzyme similar in many respects to laccase has been described by Piéri and Portier as existing in the gills, labial palps, and blood of certain molluscs.

Tyrosinase.

In other Fungi there are different chromogens which do not turn blue on exposure to air but become red and finally black. Of these *Russula nigricans* Bull. is perhaps the most conspicuous example. The substance which gives rise to the black colour is almost insoluble in alcohol, but after the fungus has been boiled with this reagent it can be extracted from the residue by subsequent maceration with boiling water. When such an extract is treated with a little fresh cold water extract of the fungus, or a piece of the tissue is added to it the liquid turns red and after a time black. If the chromogen is extracted from the fungus by boiling water and rapidly pressed and the exuded liquor filtered and concentrated to a small bulk, it deposits colourless needle-shaped crystals, usually collected together into spheres. They are not soluble in alcohol nor readily in cold water but they dissolve freely in hot water. They have been identified by Bertrand with *tyrosin*.

Bertrand has observed the same general course of behaviour with the expressed juice of the roots of the beet, the tuberous roots of the dahlia, and the tubers of the potato. In these cases also he has identified tyrosin in the tissues.

The similarity of behaviour to that observed in the cases of Boletus, Lactarius, &c. points to a similar cause of the change of colour. Laccase however has no power to set up the blackening. Nor will simple oxidising agents bring it about. Bertrand asserts that it is an oxidation process due to the presence of a special oxidase, and he has named the enzyme in question *tyrosinase*.

If a little of the cold water extract of Russula nigricans is added to a solution of tyrosin, the mixture becomes at first red, and subsequently assumes an inky blackness, while finally a black amorphous precipitate settles out. If this is carried out in a glass vessel without agitation the colour first appears at the surface of the liquid. If it is conducted in a closed vessel from which air is excluded the change of colour does not take place. Nor is the change induced if the extract of the fungus is boiled before being added to the solution of tyrosin. In a closed vessel in the presence of air, the absorption of oxygen can be measured coincidently with the blackening of the liquid.

Tyrosinase can be extracted not only from Russula, but from the dahlia and the beetroot. It is immaterial which of the three serves as the source of the oxidase as the effect upon the tyrosin is the same in all cases. Russula appears to contain it in greatest quantity. The same mode of extraction can be employed with either material.

In some species of Russula the two oxidases so far discussed exist side by side. Bertrand has separated them by the following treatment. One and a half kilogrammes of freshly gathered *Russula delica* Fries. was reduced to pulp and macerated for half-an-hour with its own weight of chloroform-water at the ordinary temperature. On pressing it, about 2 litres of a mucilaginous fluid was obtained, to which 3 litres of 95 per cent. alcohol were added. A precipitate fell which was filtered off. The filtrate was concentrated to half a litre by distillation at 50° C. in vacuo, and when so obtained was found to be capable of acting with considerable energy on pyrogallol and hydroquinone, but to have no effect on tyrosin; it contained therefore only laccase.

The precipitate was washed with 200 c.c. of chloroform water and when it was well swollen up, forming a semi-solution, it was precipitated by addition of 400 c.c. of alcohol and pressed dry. It was further purified by a repetition of this treatment. Dried at 35° C. it weighed about 7 grms. This precipitate yielded to cold water after some hours' maceration, a principle which oxidised tyrosin rapidly, but had hardly any perceptible action on either hydroquinone or pyrogallol.

Tyrosinase is destroyed at a much lower temperature than laccase; it is injured at about 50° C. and perishes rapidly at higher points. It is possible to prepare laccase alone from a mixture of the two, by heating the liquid containing them to 70° C. It then oxidises hydroquinone, but is without action on tyrosin.

Bourquelot has recognised tyrosinase in many genera of Fungi, among which may be mentioned *Boletus, Russula, Lactarius, Paxillus, Coprinus, Psalliota, Hebeloma, Pholiota, Collybia, Clitocybe, Tricholoma* and *Amanita*; in all these it is associated with laccase, but in the case of Amanita, the latter enzyme is present only in small quantities.

Besides oxidising tyrosin, Bourquelot has found tyrosinase to act on all the cresols, resorcinol, guaiacol, metatoluidine, xylidine, ortho-, meta-, and para-xylenol, thymol, carvacrol, and *a* and *β* naphthol.

He has noted a further peculiarity in its behaviour in that it is effective when dissolved in a mixture of water and either ethyl or methyl alcohol, provided that not more than 50 per cent. of the spirit is present. The alcohols themselves are not affected by it.

Œnoxydase.

Another of these oxidising enzymes has been discovered to play a prominent part in causing a particular disorder in certain wines to which the name "casse" or " cassure " has been applied. According to Bouffard a wine affected in this way loses its characteristic colour, and after 3 or 4 hours it contains a red-brown precipitate. If the wine is at rest the decoloration begins at the surface, where a thin pellicle of colouring matter forms, and the disturbance gradually spreads to layers deeper and deeper in the liquid, until at last the walls of the vessel are covered by adherent matter, and the liquid is almost de-colorised, assuming a moderately characteristic yellow tint. The deposits are formed of the colouring matter of the wine, and are insoluble in solutions of tartaric acid, even if concentrated. The changes are not attended by any evolution of gas. Bouffard says that such wines can be preserved from the disorder by heating them to 60° C. or by the addition of traces of sulphurous acid. The change is not due to bacterial action, for it is not hindered by filtration through porcelain, nor by the addition of reagents which are fatal to microbes, such as salicylic acid or bichloride of mercury.

Gouirand has shown that this change is due to some principle which exists in the wine itself. He took some samples of affected wine and after filtering a quantity through porcelain, a large addition of alcohol threw down a precipitate of a flocculent character. When this was collected and washed, a small quantity of it added to sterilised sound wines very speedily produced the disorder.

This substance is decomposed by heating. In some of Gouirand's experiments he treated samples of sound wines with a small quantity of it, and dividing them into two parts, he heated half to 80° C. In periods varying from 12 to 72 hours the disorder was pronounced in the unheated samples, while the controls remained clear and limpid indefinitely. Warming the controls only to 60° C. gave variable results; in some it inhibited the action, in others it only retarded its progress. The substance was not affected by heating to 50° C.

When healthy wines were precipitated by alcohol in the same way as the unsound ones, the precipitate had no power of setting up the disorder when added to other samples.

Martinand has ascertained that this substance is present in ripe grapes. An extract of these gives all the reactions of laccase, oxidising hydroquinone, pyrogallol, &c., but it loses the power of producing these changes if heated to 100° C. If however there is added to the extract, after cooling, a little of the precipitate yielded when the juice of fresh grapes is treated with a large excess of alcohol, it regains the power. There is thus present in the grapes themselves as in the wine prepared from them, a certain amount of this oxidising substance, which from its behaviour must be classed with laccase and tyrosinase, as an oxidising enzyme.

The name *œnoxydase* has been given to this body. It appears to resemble laccase very closely but it is not certain that it is identical with it.

Martinand has proved it to be present in other fruits than grapes; plums, pears and apples especially may be mentioned. It appears to develop with the ripening of the fruit, unripe grapes containing very little. A good deal seems to be lost in the preliminary processes of wine-making, wine itself containing relatively little, when compared with the freshly expressed grape-juice.

Martinand finds that the œnoxydase can be removed from wine by shaking it with ether, which takes from it a body having some of the properties of tannin; this becomes olive-green or yellowish-brown on the addition of ferric chloride, is turned red by alkalis, and gives a white precipitate with albumin but not with gelatin. After the wine has undergone oxidation, most samples do not give up this body to ether, and many others yield only very small quantities of it.

Wine treated with ether in this way, and kept neutral, is not subject to self-oxidation. We may therefore infer that the enzyme is possibly associated in the wine with this body which is soluble in ether.

Martinand finds that the oxidase is destroyed when its solution is heated to 72° and kept at that temperature for four minutes. Exposure to 55° C. for $1\frac{1}{2}$ hours is also fatal to it. Intermediate temperatures bring about the same destruction after intermediate times of exposure.

Bouffard has observed that the temperature of destruction varies a good deal under different circumstances. He has found that wines beginning to be attacked with the disorder have been completely preserved by being heated to 60° C., and that warming them only to 55° C. materially helps them to resist it. He has further extracted normal wine by the alcohol method and side by side with it samples of the same wine after being heated to 60° C. The precipitate in the latter case had no oxidising power, while that in the former was very active. Further investigation showed him that the nature of the medium exercised a great influence on the destruction. When the enzyme was heated in an aqueous solution of neutral reaction it withstood all temperatures below 72·5° C. but when 10 per cent. of alcohol or ·5 per cent. of tartaric acid was present destruction was complete at 52·5° C. If double these percentages of alcohol or acid were present, the temperature necessary for destruction was reduced 5° C. He agrees with Martinand however in saying that it can be destroyed by prolonged heating in neutral media at 60° C.

Dealing with the action of various reagents upon œnoxydase, Bouffard has ascertained that it is destroyed by the action of very dilute sulphurous acid, the necessary amount being ·02 grm. per litre of the solution of the enzyme.

Cazeneuve has extracted the enzyme from unsound Beaujolais and examined many of its properties. He precipitated the wine by excess of strong alcohol and found the deposit was of a gummy consistency. He took up the gummy precipitate with water and reprecipitated it with alcohol, collected the deposit rapidly and dried it in vacuo.

He found the precipitate chiefly gum, impregnated with œnoxydase.

In most respects Cazeneuve's results agree with those already quoted, but he finds further that it acts slightly on alcohols and ethers and on the essences which give wines their peculiar *bouquet*. In its action on the wine he observes that it causes a disengagement of carbonic dioxide, and that after its action there is a diminution of the quantity of alcohol and acid.

He attributes the noticeable effects produced to the action of the enzyme on the tannins. As stated above Martinand has shown that if these are removed by ether the disorder of the wine does not occur. Whether this is due to the removal of the œnoxydase with the tannin or to the abstraction of the latter only seems uncertain.

Cazeneuve further establishes a fact which indicates clearly that the disorder is due to the enzyme. He has submitted sound wine to the influence of a current of oxygen for some time and also to the action of ozone and he finds that neither process causes "la casse."

The enzyme can be preserved unchanged for some considerable time if dissolved in weak alcohol or in wine which does not contain more than 9 per cent. of spirit. It is however rapidly altered by strong alcohol.

Laborde has suggested a different origin for œnoxydase. He finds the fungus *Botrytis cinerea* grows freely on grapes and on sterilised wine "must," and an investigation of its life-history has shown him that it normally secretes the enzyme. In his experiments he employed a culture fluid in which this Botrytis had been growing freely and he compared the action of this liquid before and after boiling it. In the first case he mixed a certain volume of the culture fluid with an equal quantity of a perfectly sound wine and kept it in contact with the air for 4 hours at the ordinary temperature. At the expiration of that time all the colouring matter had been precipitated. When he boiled the culture medium before adding it to the wine no such change took place.

Laborde found that the oxidising power of the enzyme as it existed in the mixture of the wine and the culture medium was

21—2

destroyed by heating the latter to 70° C. If no wine was present it would resist even a higher temperature than this. He has given the following table to show the effect of gradually heating the culture-medium alone.

60° C. destroys about half of the oxydasic power.
65° C.   „   „   two-thirds   „   „
70° C.   „   .,   four-fifths   „   „
85° C.   „   the whole   „   „

The oxidase is slowly destroyed by absorbing oxygen, losing about half its power in two days and nearly all in twelve days. The destruction is greater in proportion during early than late periods of oxidation.

The fungus contains most oxidase when it is in full fructification.

It was mentioned above that Martinand had found œnoxydase in the juice of apples, pears and plums. Either the same enzyme or a similar one has been described by Lindet as causing oxidation of the tannin in the cider-apple. If slices of apple, or a mass of the pulp, or sterilised sponges soaked in the expressed juice are placed under a bell-jar over mercury, the material rapidly reddens, and there is a simultaneous absorption of oxygen and an evolution of carbonic dioxide. The phenomena are the same if the juice in which the sterilised sponges are soaked has been filtered through porcelain, or if antiseptics are added, so that it is evident that the changes are not due to the presence of micro-organisms. If boiled juice is used, it remains uncoloured and there is no exchange of the gases mentioned.

The juice may be precipitated by alcohol and the precipitate collected and washed in the usual way and it is then found to be capable of setting up similar changes in boiled juice.

It is of course a common experience that the behaviour of the pulp of a raw apple on exposure to air is very different from that of a cooked one. The latter remains uncoloured, while the surface of the raw pulp soon turns a reddish-brown, particularly if it is unripe.

Lindet holds that the enzyme attaches itself to the tannin, and explains the change of colour seen on wounding the fruit

by the suggestion that in the intact pulp, the tannin and the
enzyme are situated in different cells, being brought into contact in consequence of the wound. This suggestion seems
however unnecessary, as the oxidases have been shown to work
upon the aromatic bodies they attack only when they are in the
presence of oxygen. The access of the latter is only possible
when the surface of the pulp is exposed.

Other vegetable oxidases.

The various flavours and the odours which constitute the
"bouquet" of different wines have been generally associated
with peculiarities of the fermentations induced by the different
yeasts employed. Recently Tolomei has brought forward
reasons for thinking that some of them may be due to definite
oxidases which can be extracted from the yeasts.

He first demonstrated the existence of oxidases in yeast by
cultivating in sterilised wine must, a crop of *Saccharomyces
ellipsoideus* that had originally developed in muscatel grape
juice. After a few days there was a considerable growth of the
organism and he separated it from the must and exposed it to
the air. After a time he extracted it with chloroform-water,
and found that the liquid then contained an oxidase which gave
Bertrand's reactions for laccase. He obtained similar results
with *S. cerevisiae*, and *S. apiculatus*. In the case of the former,
a young beer-yeast was suspended in a solution of glucose, and
a little alcohol was added; the whole was then kept at 0° C. for
3 days, and was finally filtered through porcelain to free it from
yeast-cells. It was found to absorb oxygen from the air and to
give off carbon dioxide, and to form sulphuretted hydrogen
when in contact with sulphur. If heated to 72° C. it lost all
these properties. When alcohol was added to the liquor a
precipitate fell which contained the oxidase, and when a little
of this precipitate was added to some of the original liquor that
had been sterilised by heat, it restored to it the properties
which had been lost during the sterilisation.

Tolomei showed further that the bouquet of muscatel wine
was caused by the oxidase of *S. ellipsoideus*. He extracted some

of the enzyme from this muscatel yeast, and added it to an ordinary white wine, at the same time exposing it to the air. The wine acquired a muscatel bouquet which it did not previously possess.

Jacquemin attributes the development of the special bouquets of wines to the presence of certain glucosides in the grapes. He says that these are present also in the leaves of the vines, and that if these organs or extracts of them containing the glucosides are added to the must, the fermentation being subsequently carried out with pure yeasts, the various flavours and bouquets are acquired by the fermenting liquids. He attributes the transmission of these features therefore to the secretion of glucoside-splitting enzymes and not oxidases by the yeasts.

Buchner also says that beer-yeast contains an oxidase, to the action of which he attributes the fact that an extract of this fungus turns brown after a prolonged exposure to the air.

Effront has observed a considerable absorption of oxygen by yeast when it was finely fragmented and subsequently exposed to the air. The absorption was accompanied by a sensible increase of temperature. Effront suggests that the phenomenon is due to the action of an oxidase.

Bréaudat has recently found an oxidase in the leaves of *Isatis alpina* and other indigo-yielding plants. As already mentioned in a preceding chapter the formation of indigo involves the decomposition of a glucoside, *indican*, which is effected by a ferment resembling emulsin. The products of this decomposition are indigo-white or *leucindigo*, and *indiglucin*, which is a sugar. The indigo-white is converted into indigo-blue by an oxidase which is present in the leaves. Bréaudat says that it acts most advantageously in the presence of a weak alkali such as lime-water. The alkali alone does not produce the effect.

Tolomei has found an oxidase in ripe olives, which appears to oxidise the oil, giving rise to oleic, acetic, and sebacic acids. Bouffard and Semichon state that purple grapes contain an oxidase which in the presence of a current of air oxidises and precipitates the colouring matter, so that white wine can be prepared from the juice. Lepinois has found a similar body in

the expressed juice of aconite and belladonna which destroys
the green colouring matter. Boutroux says that the colour of
brown bread is due to another oxidase which is in the bran.

Within the last year oxidases have been discovered in malt
extract (Grüss) in the leaves of *Corchorus* (Khouri), *Helleborus*
(Vadam), *Digitalis* (Brissemonet and Joanne), *Vitis* (Cornu), in
the latex of *Schinus molle* (Sarthou), in the root of *Valeriana
officinalis* (Carles), and in the Coli bacillus (Roux). Woods
attributes the changes in the chlorophyll of leaves to the
same cause.

Animal oxidases.

Oxidative processes have long been known to take place in
blood when shed and exposed to the air. Claud Bernard first
pointed out that under these conditions sugar disappeared, and
his results have been confirmed by many subsequent observers.
Only within recent years, however, has it been suggested that
this disappearance of sugar is due to the action of an enzyme,
but this view is now put forward by several observers.

Lepine and Barral in 1890, in the course of an investigation
into the changes taking place in sugar in the blood, found that
glycolysis could be detected after blood had been shed, and that
it was more rapid in proportion as the temperature was raised
till it reached 54° C., at which point it suddenly stopped. The
physical condition of the fluid was not appreciably different
at this point from what it was at 52° C. when glycolysis was
very evident. Lepine and Barral found further that the blood
drawn from the portal vein lost more sugar than that taken
from the splenic vein under identical conditions.

Arthus showed a little later that the presence of actually
living elements in the blood were not essential to the process.
He collected blood aseptically into sterilised vessels and defibrinated it. Maintaining it then at 10° C. for several days,
with antiseptic precautions, it continued to lose sugar. In
other experiments, he showed that glycolysis took place in
serum, in oxalated plasma free from corpuscles, and in blood
diluted with several volumes of water. He found that freshly-

drawn blood showed comparatively little glycolytic power, but that it was developed on standing for some time.

These experiments are strikingly suggestive of the presence of an oxidase; the process shows a minimum point of activity at 0° C., an optimum at 40°—50° C., and a maximum one at 54° C. Its gradual development is remarkably like that of the formation of the pancreatic digestive ferments.

The action was also examined with some completeness in 1892 by Seegen, who found that the disappearance of the sugar was not influenced by the presence of chloroform. This reagent prevents the action of living cells and micro-organisms, but does not inhibit the work of enzymes. Seegen found that the exclusion of bacteria by other means did not prevent the glycolysis, and argued consequently in favour of the presence of a sugar-destroying enzyme. He demonstrated the same influence of temperature as had been found by Lepine and Barral and by Arthus.

Both Seegen and Arthus suggested that the enzyme was formed in consequence of post mortem changes taking place in the blood. Arthus held the source to be the white corpuscles or leucocytes. Lepine and Barral found they could extract it from the corpuscles in greater quantity than from the serum.

Further experiments made by many observers have shown that the power of destroying sugar is not confined to the blood. Lepine found that the removal of the pancreas of the dog was followed by an intense diabetes, the ratio of sugar to urea in the urine increasing enormously. He at once associated the pancreas with the secretion of an oxidase. In its absence, sugar, which would in the general course of events be destroyed in the blood, passed out of the system in the urine. This view was supported by the observation already quoted, that the destruction of sugar in the portal vein is much greater than in the splenic. Lepine found further that the chyle, which must receive some of the pancreatic secretion through the intestinal wall, also contained the glycolytic property. The injection of chyle into the jugular vein of an animal caused a marked reduction in the sugar in the living blood. The addition of chyle to a 1 per cent. solution of glucose kept *in vitro* at 38° C. caused a decided reduction of the amount of the sugar.

Lepine and Barral in a subsequent paper stated that, in the cases of both blood and chyle, the glycolytic power was increased by a rise of temperature and impeded by the presence of carbon dioxide.

Lepine subsequently carried out a series of researches on the pancreas of the dog, which confirmed his view that the secretion of a glycolytic oxidase is one of the functions of that organ. He ground up the pancreas, with aseptic precautions, immediately after its removal from the body, and macerated it for two to three hours at 38° C. in water containing ·2 per cent. of a mineral acid, and then neutralised the extract with sodium hydrate. To 100 c.c. of the resulting liquid he added half a gramme of glucose and digested it for an hour at 38° C. In a series of such experiments he found there was a disappearance of sugar, ranging from 10 to 50 per cent. A fresh pancreas similarly extracted with water instead of dilute acid, yielded an extract with very little glycolytic power. Lepine inferred that a glycolytic enzyme could be prepared from the tissue he used, just as similar treatment yields trypsin from the same gland.

He supported his hypothesis still further by an experiment in which he compared the glycolytic power of the blood leaving the pancreas during active secretion with that possessed by it when the gland was at rest. He found that during the secretion caused by stimulation of the vagus, blood drawn from the pancreatic vein possessed little glycolytic power, but that the latter became considerable in the blood from the same vein during the hours immediately following the cessation of the secretion.

The work of Abelous and Biarnés, carried out in 1894, has advanced our knowledge of this enzyme still further. These investigators examined the action of the blood on salicyl-aldehyde. This body was not oxidised to salicylic acid by the air, nor by distilled water, nor by normal saline solution (a solution of sodium chloride containing ·6 per cent. of the salt). But when defibrinated blood, or blood-serum was added to the aldehyde and the mixture kept at a temperature of 37° C. salicylic acid was formed. The oxidation was found to vary in amount with the blood of different animals.

Abelous and Biarnés, following up Lepine's work on the pancreas, looked for the presence of glycolytic power in other tissues. They found that it was manifested also by the testes, the thyroid glands, the liver, kidney, lungs, and spleen.

Spitzer has still more recently confirmed the statements of previous observers, but he originally opposed the view that the oxidation is due to the action of an enzyme, comparing it preferably with the oxidation produced by hydrogen-peroxide and other oxidising agents.

In a paper which appeared in 1897 he published the result of some researches on the oxidative powers of various tissues, estimating their capabilities by measuring the quantity of oxygen they could liberate from hydrogen-peroxide. In some cases he also ascertained the quantity of salicyl-aldehyde which they could convert into salicylic acid. He found the tissues could be arranged in this respect in an order differing but little from that published by Abelous and Biarnés. Spitzer has stated more recently that he has found them capable of converting arsenious into arsenic acid.

Spitzer made a very important advance by ascertaining that these different tissues owed their oxidising powers to nucleoproteid substances which they contained. He was able to prepare nucleo-proteids from every tissue which he found capable of inducing glycolysis, and he ascertained that the oxidative energy of a tissue was proportional to the amount of nucleoproteid which it gave up to an appropriate solvent. He found further that these compounds showed the same peculiar relation to temperature as the original organs exhibited, and that they were affected in precisely the same way by various antiseptics and poisons.

Seegen attributes the glycolytic power of the blood to a similar substance.

Jaquet, and Salkowski and Katsusaburo Yamagiwa have confirmed the results of the investigators quoted and have concluded from their own researches that the action is due to a soluble enzyme which is destroyed by boiling and by prolonged contact with alcohol.

Hammarsten has stated that the gastric mucous membrane contains another enzyme belonging to this group. It has the property of converting lactose (milk-sugar) into lactic acid.

Another oxidase has recently been described by Piéri and Portier as existing in the gills, labial palps, and blood of several of the acephalous molluscs, particularly *Artemis exoleta*, *Mya arenaria*, *Tubes pullastra*, *Ostrea edulis*, and *Pecten jacobœus*, besides one or two other species. It has many points of resemblance to laccase, especially with regard to its action on hydroquinone. The authors say that if one of the gills or a piece of one of the labial palps of any of these molluscs is warmed to 50°—60° C. in a solution of hydroquinone of 1 per cent. concentration, the odour of quinone is soon evolved, and if the liquid is subsequently concentrated by evaporation, crystals of quinhydrone are deposited over the surface of the tissue.

The oxidase can be extracted from the gills or the labial palps of the animals by chloroform-water, or by solutions of sodium fluoride or of salicylic acid, and can be precipitated from its solvents by excess of alcohol. It is active in either neutral or faintly acid media. Like laccase it does not decompose tyrosin.

Some observations of Abelous and Gérard call for notice here, as indicating possibly the existence of enzymes possessing the power of obtaining oxygen from somewhat stable compounds instead of utilising the free oxygen of the air. Many observers, among whom may be mentioned Gautier, Bokorny, Ehrlich, and Binz have shown that various animal tissues have reducing properties, being able to convert alkaline nitrates into nitrites. Abelous and Gérard have examined various organs of the horse from this point of view, and find many of them possess this power but to different extents. They arrange the tissues in the following order; liver, kidney, supra-renal capsules, lung, testicles, intestines, ovary, sub-maxillary glands, pancreas, spleen, striated muscle, brain. An extract of these organs, prepared with either water or glycerin, in the presence of chloroform, and filtered till free from all traces of cells, when mixed with a solution of nitrates and allowed to stand for some

time reduces them to nitrites. The extract loses the power on
being boiled. The action is not exerted at 0° C. but above that
point rises gradually and reaches a maximum at about 40° C.;
from this point it diminishes being very little at 60° C. and
ceasing a little below 70° C. Hydrogen increases and carbon
dioxide diminishes its intensity. They attribute the action
consequently to an enzyme which they say works most advantageously in a faintly alkaline medium. The enzyme can
be precipitated from the extracts by alcohol, but contact
with the spirit soon weakens it. The nitrates most easily
reduced are those of potassium and ammonium. Besides
attacking nitrates it reduces butyric acid to butyric aldehyde
and bleaches methylene blue.

CHAPTER XX.

ALCOHOLIC FERMENTATION.

ALCOHOLIC fermentation is the one which has been longest known and most closely studied. In an earlier chapter we have briefly summarised the steps by which our present knowledge of it has been obtained, and we have seen how closely it has been associated during the present century with the various problems presented by the biology of yeast.

The first observations upon its true nature were those made by Becher in 1682, in the course of which he ascertained the fundamental facts that it was only possible in saccharine fluids, and that the alcohol proceeded from a decomposition of some constituent present in them.

More than a hundred years passed before any very important addition was made to our knowledge of the chemical changes involved in the process.

The researches of Lavoisier mark a definite epoch in the gradual advance of our information, for they first showed the relation existing between the sugar and the products found after the fermentation of the latter had taken place. Lavoisier was the first observer who studied the process quantitatively, and applied new methods of analysis in his research. By determining the percentages of carbon, hydrogen, and oxygen in the sugar, and in the products resulting from the fermentation, he arrived at the conclusion that the operation consisted of the separation of the sugar into two parts, one of which became oxygenated at the expense of the other to form carbon dioxide, while the second became converted into alcohol after parting with a portion of its oxygen. He said further that if it were possible to recombine these two substances, alcohol and carbon dioxide, sugar would again be formed.

Though the methods of analysis used by Lavoisier were imperfect and his figures were inaccurate in consequence, we know that his general conclusions were sound. About the year 1815 analyses by Gay-Lussac, Thénard and de Saussure fixed definitely the composition of sugar and alcohol. These more accurate analyses confirmed Lavoisier's position, but revealed a discrepancy which for a long time remained unexplained. Computation of the composition of cane-sugar based upon the weights of carbon dioxide and alcohol formed during its fermentation, pointed to its having the formula $C_6H_{12}O_6$ (taking the modern values of the atomic weights) the decomposition being capable of expression by the equation

$$C_6H_{12}O_6 = 2C_2H_6O + 2CO_2.$$

The analyses made by Gay-Lussac and Thénard of cane-sugar itself, demanded the formula $C_{12}H_{22}O_{11}$. These authors were unable to account for the discrepancy except on the assumption that the analyses of the sugar were not quite accurate. Dumas and Boullay suggested as a more probable explanation, that the fermentation was accompanied by the absorption of water before or at the same time as the splitting of the sugar. Thus the equation representing the reaction would become

$$C_{12}H_{22}O_{11} + H_2O = 4C_2H_6O + 4CO_2.$$

Very shortly afterwards the discovery was made by Dubrunfaut that before cane-sugar could ferment it became transformed into another sugar that was not crystallisable, and Biot showed that under the action of weak acids it became decomposed into two other sugars, now known as glucose and fructose.

Berthelot was the first to show that besides inorganic reagents an enzyme possesses the power of bringing about this transformation and that the living yeast-cell forms this enzyme, already described in a preceding chapter under the name of invertase.

The fermentation of cane-sugar has thus been proved to take place in two stages; in the first it is split up with hydrolysis into the two hexoses glucose and fructose, according to the equation $C_{12}H_{22}O_{11} + H_2O = C_6H_{12}O_6 + C_6H_{12}O_6$; in the second both of these are decomposed, forming alcohol and carbon dioxide as suggested by Gay-Lussac.

Another fact came to light sometime after the publication of the results of Dumas, Dubrunfaut, and Biot, the significance of which for some time met with but slight recognition. In 1847 Schmidt of Dorpat found the presence of succinic acid in small amount in fermenting liquids. It remained however for Pasteur to complete the discovery of which this fact was part. To his careful researches we are indebted for our knowledge that from 4 to 5 per cent. of the cane-sugar used in a fermentation does not undergo the ordinary decomposition, but splits up into glycerine, succinic acid and carbon dioxide.

Pasteur has suggested the following equation as representing the change.

$$98C_6H_{12}O_6 + 60H_2O = 24C_4H_6O_4 + 144C_3H_8O_3 + 60CO_2$$
$$\text{succinic acid} \qquad \text{glycerine.}$$

Monoyer has represented the reaction rather differently thus :—

$$8C_6H_{12}O_6 + 6H_2O = 2C_4H_6O_4 + 12C_3H_8O_3 + 4CO_2 + O_2$$

According to this hypothesis the change is associated with the liberation of a small quantity of oxygen, a fact the importance of which will appear later.

Bechamp and later Duclaux have found that acetic acid also is formed in small quantities during alcoholic fermentation. Duclaux states that its presence is constant, but that the amount never exceeds ·05 per cent. of the weight of sugar, if the fermentation is stopped as soon as all the latter has been decomposed. Bechamp says that under appropriate conditions this quantity can be largely exceeded.

Other observers have noticed variable but always minute quantities of aldehyde as accompaniments of fermentation.

The ordinary alcohol produced is often found to be mixed with higher alcohols, propylic, isobutylic, amylic, caproic, œnanthylic, and caprylic all having been identified, amylic alcohol sometimes occurring in some quantrty.

The amounts of all these less prominent constituents seem to vary a good deal under different conditions of fermentation. More succinic acid is produced when the operation is slow than when it is rapid. The different kinds of yeast show considerable differences among themselves with regard to these various products. Indeed the latter seem to be more closely related to the biological work of the cells than the ethylic alcohol, the appearance of which is the most conspicuous feature of the process. The relation of the alcohol to the ordinary metabolism still remains a subject of discussion.

We have seen that for a long time alcoholic fermentation was associated only with cane-sugar. Much light has been thrown upon the composition of various sugars in recent years, especially by the researches of Emil Fischer. Without entering into details, which would be beyond the scope of the present work, we may divide the sugars that are of general occurrence into two main groups, the simplest of which have the empirical formula $C_nH_{2n}O_n$, the value of n in those known up to the present ranging from 2 to 9. The members of the other group are more complex, being theoretically composed by the combination of either 2 or 3 molecules of a hexose with elimination of one or two molecules of water. Thus we have cane-sugar $C_{12}H_{22}O_{11}$ corresponding to $2(C_6H_{12}O_6) - H_2O$, and raffinose, $C_{18}H_{32}O_{16}$ represented by $3(C_6H_{12}O_6) - 2H_2O$. This group is generally known as the *polysaccharides*.

The sugars which are fermentable appear to belong entirely to the first group and to be represented especially by the hexoses $C_6H_{12}O_6$. Some of the others are however capable of fermentation, but only those in which the value of n is divisible by 3.

Certain of the polysaccharides appear at first sight to be fermentable, particularly cane-sugar and maltose. It has been found however by Fischer that they are not directly attacked by the alcohol-producing principle, but undergo a preliminary hydrolysis with the consequent formation of hexoses. We have seen that this fact was established for cane-sugar by Dubrunfaut and later by Berthelot. This sugar by the action of invertase yields glucose and fructose. Similarly each molecule of maltose is converted by hydrolysis into two molecules of glucose; lactose or milk-sugar is resolved into glucose and galactose. The other polysaccharides behave similarly.

The different hexoses also show greater or less facility of conversion into alcohol and carbon dioxide. Glucose is most readily attacked, fructose less so, and galactose undergoes fermentation only with great difficulty.

The progress of a fermentation has been found to depend also, and in very large measure, on the species of yeast which is employed. A great development of our knowledge in this direction has been made during the last twenty years, which has been due very largely to the labours of Hansen.

It is difficult to overestimate the importance of his researches into the biology of yeast. Indeed it may be said that since Pasteur no one has done so much to elucidate the problems of alcoholic fermentation. Hansen was one of the pioneers who elaborated exact methods for the cultivation of micro-organisms in a state of purity or isolation. By the methods which he adopted he made it clear that the Saccharomycetes really form a separate and distinct group of alcohol-producing fungi, and he brought out new points of view with regard to species and races. It is not in the province of this work to discuss his results in detail, as they turn more on the biological peculiarities of the yeast-plant than on the wider question of the chemical and physical peculiarities of the changes which it induces. As an illustration of the scope of his researches we may however mention the investigations he made upon the conditions of the formation of spores and films, and upon the morphology of these developmental structures and their course of life, taking especially the yeast known as *Saccharomyces apiculatus* as a representative of the group. His experiments upon the alterations induced by differences of conditions of life, such as exposure to the cold of winter, and upon variations following such primary alteration, have thrown a flood of light upon this section of the subject.

Hansen's researches have had even greater effect on the industrial methods of brewing and of the manufacture of wine. It is hardly too much to say that he and his pupils have almost revolutionised the methods in use.

Hansen's work has not however been all that has been carried out in this field of research. Other investigators have described other species, while other fungi than yeasts, especially species of *Mucor*, have been found capable of setting up fermentation. A copious literature has sprung up on the subject, which cannot however be considered here in any detail.

The course of action on mixtures of fermentable sugars in the presence of different yeasts is often very different. Most of the latter contain hydrolytic enzymes which effect the hydrolysis of the polysaccharides cane-sugar and maltose. If either of these inverting enzymes is not present, however, the yeast cannot ferment the corresponding sugar. The yeasts for example which do not form invertase have no action on cane-sugar. Hansen's six true species of Saccharomyces all hydrolyse both cane-sugar and maltose, and subsequently cause their products to undergo alcoholic fermentation. *S. Marxianus* differs in not attacking maltose, while *S. membranœfaciens* does not ferment either of them. Other organisms have similar idiosyncracies.

As the fermentation of sugars by yeast is thus seen to be a function of the living cell, everything that affects the life of the latter has a certain influence on its fermentative power. Substances which serve as food for the yeast promote fermentation; acids and bases are deleterious to the cell, their influence being the more marked as their concentration is increased. The proportion of acids which are effective in this direction have been given differently by various observers. Hayduck found that ·1 °/₀ of sulphuric acid reduced the energy of fermentation, while ·7 °/₀ inhibited it altogether. Hydrochloric acid was found to be less deleterious, but fermentation was unfavourably influenced by ·2 °/₀. The development of the yeast-plant was promoted by ·02 °/₀ of sulphuric acid or by ·1—·5 °/₀ of lactic acid, but unfavourably influenced by ·07 °/₀ of the former or 1·5 °/₀ of the latter. In the presence of ·2 °/₀ of sulphuric acid or 4 °/₀ of lactic acid the yeast-cells did not undergo multiplication. Bokorny found the influence of sulphuric acid was deleterious, and fermentation was impeded when as little as one part in 20000 was present, while four times that amount stopped it entirely. The writer found that the development of the yeast-plant was markedly impeded by the presence of ·5 % of lactic acid.

Among deleterious bodies Bokorny mentions corrosive sublimate, potassic permanganate, free chlorine, and iodine. Phosphorus also impedes fermentation. Bokorny prepared a culture medium containing one part of it in 20000, besides cane-sugar and mineral salts. A feeble fermentation was excited in this liquid 6 hours after the addition of yeast.

Bromine is less harmful than chloride or iodine. One part in 10000 parts of a fermenting solution does not cause inhibition; solutions of potassic chlorate or potassic iodate of 1 per cent. concentration are also without deterrent effect. Very dilute solutions of potassic cyanide or strychnine nitrate (1 : 50000) do not cause complete inhibition, nor does quinine acetate in a concentration of ·1 per cent.

Other carbohydrates than sugars have been stated by various writers to be capable of undergoing alcoholic fermentation. Lévy has produced alcohol by fermenting artichoke tubers with yeast; Buchner has shown that the glycogen of the yeast can give rise to alcohol.

Pasteur found that when a considerable quantity of yeast was allowed to act upon a very small amount of sugar the fermentation proceeded for some time after the sugar had all disappeared. The later stages of the process were marked consequently by the fermentation of other carbohydrates which the yeast contained. He showed this to be the case by examining the liquid for sugar as the fermentation proceeded. Eventually three times as much carbon dioxide was evolved as could have resulted from the sugar employed. The alcohol formed corresponded in amount to the volume of the carbon dioxide.

Pasteur showed further that when fresh yeast was mixed with water and kept for a time at 25° C., carbon dioxide was very soon evolved and on distillation of the liquid after a not very prolonged exposure, alcohol was found to be present. Under the conditions of the experiment this must have proceeded from the fermentation of the other carbohydrates of the yeast employed.

These experiments are found upon examination however to be strictly comparable to those in which the polysaccharides take part. The material is not directly fermentable, but gives rise first to sugars of the hexose type. The carbohydrate which the artichoke tuber contains in greatest amount is inulin, and we have seen in a preceding chapter that the zymogen of inulase is also present in its cells. Inulase is produced from the latter and converts the inulin into fructose, which is easily fermentable by the yeast employed. The method adopted in practice is to treat the washed and sliced tubers with 4 times their weight of water containing ·2 per cent. of potassium bitartrate. The liquid is allowed to stand for four or five hours at a temperature of about 16° C. and is then decanted off, and the extraction repeated with a further quantity of the same solution. The resulting extract, containing the fructose resulting from the hydrolysis of the inulin, has a specific gravity of about 1030. It is sterilised by heating it to the boiling point three times on successive days and the yeast is then added, after which fermentation proceeds rapidly. The latter process is conducted at a temperature of 20—25° C. and is continued for 3 days, the alcohol being subsequently separated by distillation. The inulin is converted into fructose during the preliminary extraction with the dilute solution of potassium bitartrate. Similarly in Pasteur's experiments there can be no doubt the fermentation supervened upon a preliminary hydrolysis.

The fermentation of milk-sugar or lactose underlies the production of the beverages known as *Koumiss* and *Kephir*, which are manufactured and used largely in Russia and in central Asia. Lactose is a polysaccharide having the same empirical formula as cane-sugar and like it is decomposed by an enzyme of the same type as invertase, as has already been mentioned. The products of its hydrolysis are glucose and galactose, both of which can give rise to alcohol though the last named is only fermentable with difficulty.

Koumiss is chiefly prepared in the steppes of South Western Siberia and the countries adjoining. The milk used in its manufacture is yielded by mares, and there are variations in its mode of preparation in different parts. It is apparently due to the action of several organisms, one of which is a yeast, while others are lactic-acid-producing bacteria. The organisms are not isolated for the purpose of the fermentation, but a certain quantity of old *koumiss* is mixed with the fresh milk, usually one part of the former to ten parts of the latter. The fermentation is conducted in small casks or vats which are fitted with a stirring apparatus. It is found advisable to employ the latter for about 5 minutes at the commencement of the operation so that the whole mass of liquid may be well aerated. When the fermentation is nearly finished the liquid is transferred to strong bottles which are closely corked and the corks wired tightly down. The continuation of the fermentation makes the koumiss an effervescent beverage. The change which takes place in the milk is the production of alcohol, carbon dioxide, and lactic acid at the expense of the milk-sugar. The latter exists in the milk to the extent of 5·5 per cent. In the koumiss there is usually about 1·3 per cent. of sugar, 1·6 per cent. of alcohol, and nearly 1 per cent. of both lactic acid and carbon-dioxide.

There appears to be a slight diminution in the proteids of the milk, amounting to about ·1 per cent., but the nature of the changes they undergo has not been ascertained. There is reason to think the casein is converted into acid-albumin and peptone, as it does not curdle under the influence of the lactic acid as normal milk would do. The agent effecting this is probably a proteolytic ferment, such as we have seen to be secreted by both yeast and bacteria.

Kephir is another effervescent beverage, prepared from cow's milk, and very much used in the mountains of the Caucasus. The nature of the organism employed in the fermentation is not well known, but it appears to consist of a yeast which is associated symbiotically with one or more bacteria. It exists in the form of gelatinous lumps which are mainly composed of a modi-fication of cellulose, probably forming the sheaths or zoogloea of the bacteria, and which entangle the yeast mechanically. The method of preparation of the kephir is much the same as that of koumiss. The milk is kept at a temperature of 18—19° C. and a quantity of the organism is added to it in the fermenting vat, and the liquid kept in agitation. The fermentation is usually well advanced in about 24 hours, when the liquid is bottled and the fermentation completed in the bottles.

Mix has described a similar organism which is in use in North America, which can ferment lactose and glucose.

The products of the kephir fermentation differ somewhat from those of koumiss. The sugar of the milk falls from 4 to about 2 per cent.; the fatty matters are diminished in about the same proportion, and the proteids are reduced to about three-fourths of their original quantity. The alcohol generally reaches a concentration of nearly one per cent., and the lactic acid is formed in slightly larger amount, the ratio between them being 8 : 9. Koumiss we have seen contains sensibly less lactic acid than alcohol. The proteids do not appear to be peptonised by the kephir organism, though some change takes place in the casein in consequence of which it is very incompletely precipitated by the lactic acid.

In both these processes we have almost certainly to recognize two different fermentations proceeding side by side, the one due to the yeast, and resulting in the formation of the alcohol; the other due to the bacteria, and giving rise to the lactic acid.

A third instance of an alcoholic fermentation combined with other chemical changes is afforded by the curious organism which is used in many rural districts of England for the manufacture of an effervescent beverage known as *ginger-beer*. In several respects it presents many resemblances to the kephir of the Caucasus. Though it has been in use for many years it was only in 1887 that attention was called to its peculiar composition. A complete investigation into the nature of the organisms composing it was carried out by Marshall Ward in 1891.

The ginger-beer plant resembles the kephir organism in its outward appearance, being found in white semi-translucent masses of irregular shape and almost gelatinous consistency. In size they vary from the dimensions of a pin's head to a diameter of nearly an inch. In a fermenting liquid they grow very freely, and become much softer and more slimy than they are in the dry condition.

The appearance of a fermentation is thus described by Marshall Ward :—" The most striking characteristics of the above-described lumps of ginger-beer plant however only become evident when they are placed in saccharine solutions and they are perhaps best shown roughly as follows. A soda-water bottle is filled three parts full of Pasteur's fluid, or any other similar solution of sugar in water, and a lump of ginger added. Into this mixture are placed a few lumps of the ginger-beer plant; the bottle is then well corked and laid in a warm place, and observed from time to time. In from 24 to 28 hours depending on the season, temperature, &c., the liquid is observed to become more and more turbid, and bubbles of gas begin to ascend, the fermentation soon goes on rapidly, and unless the cork is well secured by string or wire, it will be blown out. This primary turbidity is found to be due almost entirely to innumerable yeast-cells, and further examination proves that these yeast-cells are shed from the lumps of ginger-beer plant (which rise and fall with varying buoyancy in the liquid) and then multiply in the medium, and soon form a greyish deposit at the bottom. The buoyant dancing of the lumps is seen to be determined by the copious evolution of gas bubbles from their surfaces....The liquid...is not only surcharged with carbon dioxide, but is evidently more or less viscous, with a viscosity which is different from any property directly imparted to it by the sugar and other materials added.... As time goes on this viscosity increases and it sometimes happens that the liquid becomes so thick that the gas-bubbles rise comparatively slowly. The viscosity is not due to the mere presence of yeast-cells, because they fall to the bottom; it is due to the presence of innumerable swollen or slimy vermi-form bodies distributed through the mass of the liquor. Myriads of rod-shaped bodies (Bacteria) are also observable....The ginger beer is distinctly acid as well as viscous; the colour of the liquid is paler than that of the original solution."

A very detailed and elaborate investigation into the composition of the lumps of the " Plant" showed it to be composed of several organisms, of which two specific forms constitute the ginger-beer plant proper, while the rest are merely due to admixture of other forms from outside. The two in question are a Yeast, *Saccharomyces pyriformis*, and a Schizomycete which has been named *Bacterium vermiforme*. Marshall Ward found it possible indeed under proper conditions to reconstruct the "Plant" from pure cultures of these separate organisms originally isolated from a fermenting liquid.

The bacterium is a peculiarly vermiform organism consisting of filamentous or rod-like, often homogeneous, bodies much coiled and twisted together, which are sometimes found to be broken up into short rodlets or even cocci, the cells being thus arranged in chains. The separate filaments are invested with a pellucid swollen gelatinous sheath, which gives to the complete organism its peculiar consistency. In the midst of this gelatinous matrix of coiled filaments the yeast-cells appear to be mechanically entangled. The whole forms a symbiotic plant, which has a superficial resemblance to the thallus of a lichen, in which a yeast has replaced the alga.

The ginger-beer plant is capable of causing fermentation in solutions of grape-sugar and of cane-sugar, but it acts most readily on the latter. It differs from kephir in not being able to ferment sugar-of-milk (lactose).

In addition to the carbon dioxide evolved, the solution at the end of a fermentation contains alcohol and acetic acid, with relatively large quantities of another acid which has not been completely examined, but which appears to resemble lactic acid. The alcohol and acetic acid are formed during the earlier stages of the fermentation, the other acid appearing somewhat later.

Another organism of a similar nature has recently been investigated by the writer in collaboration with Marshall Ward. It occurs as a growth parasitic upon the sugar-cane, and was originally procured from Madagascar. It consists of gelatinous masses closely resembling the ginger-beer plant, and like it, is composed of a yeast and a bacterium, the latter again being very much like the bacterium last described. It ferments cane-sugar, maltose, glucose, and fructose, but not lactose. The products of the fermentation are alcohol, carbon dioxide, and certain acids, of which the chief are acetic and succinic. When cultivated in solution of cane-sugar it produces a quantity of viscous material, which apparently consists in

great part of the extremely diffluent sheaths of the bacterium.
This substance can be obtained in large quantities from pure
cultures of the latter, and it has been found to consist of two
modifications of cellulose, resembling the *dextrans*, or *hemicelluloses* of various writers. The yeast produces little besides
alcohol and carbon dioxide, and small quantities of succinic
acid. The bacterium gives rise to the volatile acetic acid, but
forms the latter directly from the sugar and not indirectly through
the medium of the alcohol, which, unlike the ordinary acetic
microbe, it is unable to decompose. The bacterium is peculiar
also in being one of the very few which secrete invertase.

Besides the yeasts and such symbiotic fungi as have been
described, other fungi have the power of exciting alcoholic fermentation. In the preparation of a spirituous beverage much
in use in the far East a material is employed by the Chinese
to which they give the name of *Koji*. This is prepared by
inoculating freshly steamed rice with the spores of *Eurotium*
(*Aspergillus*) *oryzeœ*. The grain, husked and bleached, is
steamed for 12 hours till it is soft; it is then cooled on straw
mats. As soon as the temperature has fallen to about 28—35° C.
a little is mixed with the spores of the fungus and stirred into
the rest of the mass. The mats with their coating of rice are
then allowed to stand in a sort of cellar, frequently dug out of
the ground, for 18—24 hours, at the end of which time a good
mycelium has developed from the spores. The grains are next
worked up by the hands and exposed on trays in the warmest
part of the cellar. This process is repeated about 12 hours
later, and the mass is cooled and moistened with a little water.
A third working up of the mass of grains and fungus is carried
out after a further similar interval, and after 14—16 hours more
incubation the Koji is ready for use. The active principle in
the fermenting mass is derived from the fungus, which as we
have already seen contains diastasic, inverting, and other enzymes. It has also the power of forming alcohol at the expense
of the sugars which are produced by the action of these enzymes
on the amylaceous and saccharine materials with which it is
mixed in the processes of manufacture. It does not affect
lactose.

Another fungus possessing similar properties has been de-scribed by Wehmer, who has named it *Aspergillus wentii*. It closely resembles *A. oryzeæ*, but can only be obtained by covering beans, which have been boiled and superficially sun-dried, with the leaves of *Hibiscus tileacus*. The fungus invari-ably makes its appearance on the leaves. It is characterised by possessing a light chocolate colour. The method of collecting the mould exactly resembles that employed by the Chinese in the manufacture of arrack.

Arrack is prepared in Java from rice starch by the action of a substance which the natives call " raggi." This is a material in which many microbes and fungi abound, but the one to which the alcoholic fermentation is due is not known with certainty. " Raggi " has been examined lately by Went and Prinsen Geerligs, who have made cultures of several of the organisms it contains. One of these has been called *Chlamy-domucor oryzeæ*; it is a fungus which developes a much-branched unseptated mycelium. It is aerobic, and secretes rennet and diastase, but does not invert cane-sugar nor set up the alcoholic fermentation of glucose. Another fungus present has been named *Monilia javanica*. It has the power of fermenting glucose, levulose, raffinose, maltose and cane-sugar, the latter of which it inverts. A true *Saccharomyces*, which the authors call *S. vordemannii*, appears to be the principal agent in the production of the alcohol of the arrack.

When *Mucor racemosus* is cultivated in a solution of sugar, instead of forming an ordinary mycelium, it is found in the form of round or ovoid cells which resemble very closely those of brewers' yeast, and which like the latter multiply rapidly by budding. Other species of Mucor behave similarly. These fungi have the power of fermenting glucose but apparently cannot attack fructose. When cultivated in invert-sugar they conse-quently can only ferment one moiety of it. Generally the fermentation is much slower than that excited by the true yeasts and is more easily stopped by the alcohol produced. *M. erectus* can produce as much as 8 per cent. of spirit in a saccharine solution; *M. racemosus* only about 3 per cent.; *M. stolonifer* only 1·3 per cent.

Mucor racemosus does not produce alcohol and carbon dioxide in the same proportion as yeast, viz. 100 : 96·3, but in relatively larger amount, the ratio being stated by Fitz as 123·1 : 100. The spirit contains a little aldehyde, and succinic acid is an accompanying product. Whether or no glycerine is formed is doubtful.

The different species of Mucor are not all able to ferment cane-sugar, as some of them do not secrete invertase and cannot therefore effect the preliminary hydrolysis.

Penicilium glaucum and *Rhizopus nigricans* also are said to be capable of exciting alcoholic fermentation.

A curious bacillus was described in 1891 by Perdrix, who isolated it from the water of the Seine. It has among other properties the power of producing alcohol when cultivated in a starchy medium. The starch is first hydrolysed to sugar and the latter is partially fermented. A solution of 4·5 grms. of starch in 200 c.c. of water fermented by it in the absence of air was found to yield the following constituents at the end of the fermentation :—

Sugar	3·52 grms.	Acetic acid	·08 grm.
Ethyl alcohol	·347 „	Butyric acid	·175 „
Amyl alcohol	·082 „		

During the process the liquid gave off ·407 grm. of carbon dioxide and ·022 grm. of hydrogen.

Perdrix named the microbe *Bacillus amylozyme*.

It will ferment the starches of wheat, maize, rye and barley. When yeast is added to a mixture in which it has excited a fermentation the two organisms will work symbiotically and under these conditions 90 per cent. of the theoretically possible quantity of alcohol may be formed.

Fitz has stated that glycerine as well as sugar is capable of giving rise to alcohol, the organism producing it being a microbe to which he has given the name *Bacillus aethylicus*, and which can be cultivated from an infusion of hay. He cultivated it in ·1 per cent. solution of Leibig's extract to which 3 per cent. of glycerine had been added. 200 grammes of glycerine yielded 25·8 grammes of alcohol.

Fitz's organism has been cultivated also by Emmerling, who found it capable of forming alcohol from mannitol as well as from glycerine. Emmerling has shown that *Granulobacter butylicus* produces butylic alcohol from glucose.

Alcohol also appears as a bye-product in several other fermentations which will be considered in a subsequent chapter.

The formation of alcohol by the vegetable cell is not, however, confined to the lowly organisms which we have so far described. There are many facts which lead us to the view that it is a property of living protoplasm, which is, however, only exercised when the latter is exposed to abnormal conditions, the chief of which is the absence of oxygen. The earliest of the observations recorded on this point were those of Bérard in 1821. These are very meagre and incomplete, but we learn from them that he found that when certain fruits were shut up in a closed vessel containing carbon dioxide or some inert gas, there was a continuous production of carbon dioxide, which he attributed to a kind of fermentation.

Lechartier and Bellamy were the first observers to record carefully the course of events under these conditions. Their observations were published in 1869 and 1872, and they throw a good deal of light upon alcoholic fermentation in general. They examined principally succulent fruits such as apples, pears, cherries, and gooseberries: also potatoes, and grains of wheat. Placing these in closed vessels from which tubes were led off to receivers containing mercury, they found that they gradually absorbed all the free oxygen around them, and at the same time evolved carbon dioxide. The disengagement of the latter gas continued after all the oxygen had disappeared, but became gradually less as time went on. During the experiments, the fruits were kept from contact with each other, and from the walls of the vessels in which they were enclosed. After several months the fruits were removed, mashed to a pulp, and the mass distilled, when a considerable quantity of alcohol was separated from them. Very careful microscopic examination showed the absence of all micro-organisms.

One of their experiments may be quoted in detail. On Nov. 12, two pears, weighing 157 grammes and 125 grammes respectively, were suspended in a glass vessel so as not to be in contact with its sides, and a quantity of calcic chloride was at the same time introduced, so that only dry air should surround the fruits. The vessel was then closely stoppered, and from the stopper a leading tube was taken to a receiver containing mercury. The apparatus was kept closed till the 19th of the following July. During that time, 1762 c.c. of carbon dioxide were evolved, and the fruits at the moment of opening were found to contain 2·62 grammes of alcohol. The pears had kept their natural colour; the skin was wrinkled but not moist; their condition was that of normally withered fruits. On weighing them it was found they had given off 134 grammes of water, but still contained a quantity equal to nearly 70 per cent. of their weight. Towards the end of the experiment the evolution of carbon dioxide gradually diminished. From March 3 to April 8, only 28 c.c. were disengaged and the formation ceased entirely at the latter date. The authors point out that this fact was incompatible with the view that the fermentation was due to an alcohol-producing micro-organism; moreover, microscopic investigation of the pulp taken at different distances from the centre of the fruit showed that no such bodies were present. In some of their experiments spores or cells of yeast made their way into the fruits, and in these cases the progress of the fermentation was altogether different from the one described, the evolution of carbon dioxide continuously increasing after the fermentation was once initiated.

During the time when the spontaneous fermentation was taking place, the fruit became sensibly modified, the cellular tissue being wholly or in part dissociated, and becoming syrupy, while the seeds lost the power of germination.

The experiments of Lechartier and Bellamy were subsequently repeated by Pasteur and their results confirmed.

When fruits were gathered and allowed to ripen without being deprived of oxygen, there was no formation of alcohol accompanying the process. The alcoholic fermentation in the cases described was thus undoubtedly set up by the peculiar conditions of the experiments.

Investigations published by Brefeld in 1876, and by de Luca

in 1878, have shown that similar results are met with in
experiments with seeds, leaves, and branches. About the same
time Müntz found that when he cultivated plants in pots in an
atmosphere of nitrogen instead of carbon dioxide, alcohol was
produced in their living cells. He carried out his investigations
on many different plants, including maize, beets, cabbages, and
chicory, all of which remained healthy so long as they were
under observation. He detected alcohol in their tissues by
the iodoform reaction.

The occurrence of alcohol in the more purely vegetative
parts of plants, observed more than 20 years ago by Brefeld
and by de Luca as already mentioned, has met with abundant
confirmation in recent years.

Berthelot determined its presence in 1893 in the leaves of
wheat. He took leaves freshly gathered from the plants and
transferred them at once to a flask, and replaced the air by
hydrogen. He then exposed the flask to a temperature of
110° C. and collected the distillate, which he found to contain
alcohol. In an experiment conducted on a larger scale he
obtained sufficient alcohol to prepare ethylene from it.

Mazé in 1899 claimed to show that the leaves of the vine
normally contain alcohol. He obtained between 50 and 100 milli-
grams of spirit from 35 grams of the fresh tissue. He found
also that alcohol is present in measurable quantities in the
germinating seeds of the pea when access of oxygen to them
is prevented.

Devaux has shown that alcohol can be formed in woody
stems and that its presence is probably attendant upon partial
asphyxiation. He found the air in the stems of such trees to
be relatively poor in oxygen, containing often not more than
half the normal amount. By raising the temperature of such
stems so as to quicken their respiration, the oxygen content of
their interior rapidly sank, becoming after 3 days at 35° C. as
little as 1 per cent. in several cases. The simultaneous output
of carbon dioxide was greatly increased. Investigation showed
a coincident formation of alcohol.

Devaux has shown that this internal asphyxia exists in
some trees at the normal temperature (12°—20° C.) and that

the quantity of alcohol in the fresh wood of such trees amounts sometimes to 1 c.c. of absolute spirit per kilogram of wood.

Some experiments carried out in 1896 by Gerber, have also shown the power of vegetable protoplasm in the higher plants to set up alcoholic fermentation. This observer carried out a long series of researches on the changes undergone by various fruits in the course of their maturation, and paid special attention to the respiratory changes which can be observed during this period. Attention may be called especially to the ripening of fruits which contain considerable quantities of tannin. Among these may be noted especially the fruit of the Chinese date-plum (*Diospyros kaki*), which when ripe contains a very agreeable volatile aromatic perfume. When this fruit was allowed to ripen in a confined atmosphere, and the pulp examined, it was found to yield ten per cent. of ethylic alcohol, mixed with other alcohols among which amylic was conspicuous. The pulp also contained acetic acid. The aromatic principle was found to be a mixture of amyl and ethyl acetates, with traces of œnanthylates and pelargonates. The tannin disappeared during the ripening, but the alcohols formed did not appear to be developed from it, but rather from the sugars of the fruit. Gerber associates its formation with the absence of oxygen during the later stages of the ripening. He obtained similar results with *Diospyros costata*. When the fruits of this plant were gathered and allowed to ripen at a temperature of 30° C., the ratio of oxygen absorbed to carbon dioxide evolved was $1 : 3.12$; when the temperature was maintained at 15° C. it was $1 : 1.26$. Alcohol was produced in the former case but not in the latter. The fermentation appeared to be due to the supply of oxygen not being sufficient for the needs of the plant when its metabolism was exalted by the higher temperature, though it was so with the less active metabolism at 15° C. The order of events in the ripening of these fruits was the following. In the early stages, the respiration showed that they exhaled less carbon dioxide than they absorbed oxygen, the tannin gradually disappearing during this period, and being in Gerber's opinion completely oxidised to carbon dioxide and water. After the disappearance of the tannin, the pectose which was present in the walls of the cells was converted into pectine, which caused the softened character of the pulp. The swelling characteristic of pectine partially obstructed the intercellular passages in the pulp, thus interfering very seriously with the transport of air to the cells of the interior, and causing the supply of oxygen to be considerably lessened. At the temperature of 15° C. sufficient oxygen was able to pass along the narrowed intercellular passages to minister to the needs of the cells, but at the higher point of 30° C. this was not so, and a sort of asphyxiation set in. This set up the fermentation noticed at that temperature; at once the output of carbon dioxide increased, and the ratio of the two gases concerned rose from 1 : 1·26 to 1 : 3·12. The increased amount of carbon dioxide was made up of that which resulted from the fermentation of the sugar, added to the quantity proceeding from the respiration of the cells, the latter being the only factor when the lower ratio was observed. This view of the cause of the increase was supported by the observation that when air was admitted more freely to the cells by subdividing the tissue into small pieces, the ratio at once fell slightly, although as such subdivision did not completely remedy the obstruction of the intercellular passages, the difference was not very marked.

Gerber obtained similar results with bananas and melons, observed under corresponding conditions.

He came to the conclusion reached by Lechartier and Bellamy, and by Pasteur, that the cause of alcoholic fermentation in these fruits is to be looked for in the diminished supply of oxygen to the cells, that it is indeed a struggle against incipient asphyxiation. The alcohols which he obtained were not free as in the experiments of Lechartier and Bellamy, but were combined with other volatile bodies to form the various etherial substances which constitute the perfumes of the fruits.

Returning to the consideration of the part played by the yeast-cells in ordinary fermentation, we find that it has been a subject of considerable speculation, and that two hypotheses have been put forward to explain its action. Pasteur came to the conclusion that the fermentative power was connected with nutri-

tion in the absence of free oxygen, and that the decomposition
of the sugar under these conditions was the expression of the
effort of the organism to obtain oxygen for respiratory purposes.
We have seen that according to Monoyer, the formation of
succinic acid and glycerin, which Pasteur first discovered to
accompany the alcoholic fermentation, is accompanied by the
liberation of a small quantity of free oxygen. Pasteur strengthened himself in this opinion by experiments on the cultivation
of yeast in the absence of free oxygen, carried on side by
side with others in which the gas was freely supplied to the
organism. In these experiments he found that the relative
weights of yeast formed and sugar decomposed were very
different under the two conditions. When no oxygen was
supplied, fermentation was very slow, and not much yeast was
produced, but for each part of yeast formed, 60 to 80 parts of
sugar disappeared. When oxygen was admitted, the fermentation was very rapid, the yeast grew luxuriantly, but for one
part formed, not more than 4 to 10 parts of sugar were
decomposed. The difference was not due to any weakening of
the energy of the yeast in the second case, for when some of
it was removed, and made to act on sugar in the absence of
oxygen, it behaved just as did that which was used in the first
experiment. Pasteur was thus led to hold that fermentation
is a phenomenon of life in the absence of oxygen, and devoted
especially to the production of the latter from the carbohydrate
material of the cell itself, or that existing in its environment.

This view was not however allowed to pass without chal-
lenge. Schützenberger argues against it with some force,
pointing out that Pasteur's inference does not take into account
all the observed facts. In the presence of free oxygen we have
a very active fermentation set up, while the yeast is said to
possess, bulk for bulk, less fermentative power. Schützenberger
claims that the power to decompose sugar is not the same
thing as the power of respiration, and that the two should be
kept distinct in forming a hypothesis of the nature of the
action. The yeast-cell possesses both these powers, and so
long as it exercises both freely the fermentative power is not
directly related to the respiratory process. If the decomposition

of the sugar were only the result of the effort of the yeast-cells
to respire at the expense of part of the oxygen contained in
the sugar molecule, it would appear that fermentation should
either not have taken place at all in the presence of free
oxygen, or that it should have been much less than in the
other case, whereas the reverse is what is found.

Schützenberger supports his objection by experiments
carried out to ascertain how respiration is affected under
changed conditions. The results he obtained were briefly the
following :—

(1) In a watery liquid without sugar, but containing oxygen
in solution, the quantity of oxygen absorbed in unit time by a
gramme of yeast is constant, whatever proportion of oxygen is
present.

(2) In a saccharine liquid, containing also albuminous
matter, and with oxygen in solution, the same result is obtained, except that the quantity of the gas absorbed in unit
time is greater.

(3) In two fermentations, carried on side by side for some
time, one only being continuously supplied with oxygen, the
latter produced most alcohol. Hence Schützenberger came to
the conclusion that the sugar is alimentary.

On a consideration of all the phenomena described by
these several observers together with the facts ascertained by
Lechartier and Bellamy and by Gerber it seems possible to
reconcile these two hypotheses. We may agree with Schützenberger that when the yeast-cell is cultivated under normal
conditions its peculiar metabolism demands a certain vigorous
decomposition of sugar to minister to its nutrition. Under
these conditions it is freely supplied with oxygen and with food
material. Its life is consequently very vigorous; the cells
grow and multiply and the coincident fermentation is extremely
active. But it does not seem at all a strained view to suppose
that its protoplasm is like that of the cells of the fruits and
other organs examined by the other observers. We have
seen that when these are deprived of access to free oxygen
the ratio of the gaseous interchange is rapidly altered, the
carbon dioxide being increased two or three-fold. The stimulus

of incipient asphyxiation leads in them to the setting up
of fermentation, evidence of which is afforded by the increase
of the carbon dioxide evolved; supplementary evidence is
given by the coincident appearance of alcohol. Similar difficulty is placed in the way of the yeast when it is deprived
of oxygen. The commencing asphyxiation checks the multiplication of the yeast-cells; instead of the weight of newly
formed yeast being one-fourth to one-tenth that of the decomposed sugar it falls to one-eightieth. At the same time
an intensification of the fermentation takes place, a so-called
respiratory process being developed in addition to the already
existing alimentary one. The difference between the yeast and
the fruits will thus be seen to be that the nutritive processes of
the former involve production of alcohol from sugar while those
of the latter do not. The increased fermentative activity of the
yeast when deprived of oxygen corresponds to the only fermentative power of the cells of the fruit, which does not appear so
long as its environment is normal. That the fermentation by
yeast under incipient asphyxiation is intensified may appear
to be contrary to experiment, as the ordinary fermentation in
free oxygen seems most vigorous. But the quantity of the
yeast engaged is much greater under these conditions, as the
cells multiply very rapidly, and this must be taken into account
in making the comparison. Taking the ratio of yeast to sugar,
we have seen that in incipient asphyxiation the yeast can
decompose 80 times its weight of sugar, whereas in the presence
of oxygen it can only ferment 4—10 times its weight.

We see thus that the action of the yeast-cell is not always
the same. In the normal condition, it is exciting fermentation in connection with its nutrition. This is a special case,
apparently confined to such organisms as the lower fungi.
A much more wide-spread action of protoplasm can also be
noted, when the production of alcohol is coincident with the
effort of the living cell to supply itself with the oxygen which
is not presented by its environment.

By many writers this latter process is spoken of as "intramolecular respiration." This is rather an unfortunate phrase,
for the process has nothing in common with the intimate

respiratory process. The latter consists of the absorption of oxygen by the protoplasm, and of a series of decompositions of the substance of the latter, whereby continually simpler bodies are produced, the ultimate terms reached being carbon dioxide and water. The carbon dioxide liberated in respiration therefore proceeds immediately from the protoplasm. In the process of fermentation this is not the case; the decomposition appears to be conducted outside the protoplasmic molecule and to be set up with the view of liberating from some external body the oxygen which the living substance needs for respiration. The process is essentially a fermentative and not a respiratory one.

There is however another question of considerable importance which must not be overlooked in this discussion. During its life the yeast plant not only requires nutritive material for its development and growth, but it must also be furnished with sufficient energy to carry out its vital processes. This is a primary necessity for all living organisms.

In the case of the yeast-plant, we find that a temperature somewhat higher than that of the surrounding air is eminently favourable for its growth and multiplication, though the particular temperature desired is not the same for all races of the plant. The maintenance of this slightly raised temperature demands the expenditure of a certain additional amount of energy in the form of heat, and this must be obtained by the plant in the course of its metabolic changes.

Normally the energy required by an aerobic plant which contains no chlorophyll is furnished largely by its respiratory processes. The same is true of such cells of green plants as do not possess chlorophyll corpuscles, or are not exposed to light of sufficient intensity to enable them to absorb radiant energy. In the respiratory process, we find a continual breaking up of complex bodies into simpler ones, in which decompositions the protoplasm is immediately concerned. The living substance is itself the seat of such decompositions and its own breaking down and reconstruction are largely involved in them. The decompositions appear to be provoked by the access of oxygen to the protoplasm. It is well known that the addition of oxygen to a compound increases its instability. The ultimate expression of these respiratory changes is the production of carbon dioxide and water, and the decompositions are undoubtedly accompanied by the liberation of energy.

When sufficient oxygen is supplied to yeast the gentle respiratory processes of the healthy cells are supplemented by the decomposition of the sugar, which liberates energy without a direct process of oxidation. If no oxygen is supplied or if the quantity is an inadequate one, the decomposition of the sugar alone must be the source to which the plant must look for the energy it needs. We can understand that increased fermentative activity will accordingly be manifested under the stimulus of commencing asphyxiation. The yeast may be regarded under these conditions as decomposing more sugar either for the purpose of obtaining oxygen for its respiratory processes, or perhaps for that of liberating energy directly without the intervention of respiration. The same explanations may be applied in the case of the ripening fruits and the other parts of plants already instanced.

In the case of anaerobic plants the decompositions which are provoked may in like manner have for their object a gradual supply of oxygen to the organism. They may on the other hand set up the decompositions for the purpose of supplying energy directly, without the intervention of any oxygen obtained from the decomposing body. In either case the main object sought seems to be the energy needed for the various vital processes.

There is reason, as we have seen, to believe that the fermentation of sugar with its coincident production of alcohol is at any rate at the outset connected with the alimentation of the yeast plant. How far this is the case with the higher plants cannot be regarded as definitely ascertained. The researches of Berthelot and of Mazé already quoted show that in the leaves of the vine at any rate alcohol is of frequent if not constant occurrence under normal conditions. Its most probable antecedent is sugar, but the exact nature of the decomposition which occurs, and the definite meaning of the alcohol, have still to be ascertained. Its occurrence however suggests a parallelism between the nutritive processes of the vine and of the yeast plant.

The method by which the yeast-cell effects the decomposition of the sugar has till quite recently remained undiscovered.
During the gradual development of our knowledge of enzymes,
the minds of many physiologists have been directed to the
question of the possible existence of one of these bodies, capable
of giving rise to alcoholic fermentation. The yeast-cell is known
to be the seat of formation of several enzymes. At different
times invertase, diastase, glucase, and trypsin have been discovered in it as we have already seen. Till 1896 all efforts
to prepare from yeast an alcohol-producing enzyme failed.

Claud Bernard was engaged during the last few months of
his life in investigating the formation of alcohol in the juice of
grapes of different degrees of ripeness and decay. During some
of his experiments he found the quantity of spirit increase,
though no yeast was present, and he attributed this to the action
of a soluble ferment. He did not however isolate it. His experiments so far as they went were published posthumously by
Berthelot, in July 1878. In 1896 however Buchner was successful in extracting the enzyme. His process was the following:—

One kilogramme of purified pressed yeast was mixed with
an equal weight of fine quartz sand and with 250 grms. of
kieselguhr, a fine infusorial earth. The mixture was carefully
ground in a mortar till a moist plastic mass was obtained. The
microscope showed that most of the yeast-cells were disintegrated or at least ruptured. 100 c.c. of distilled water was
then stirred into the pasty mass, and it was wrapped up in a
cloth and submitted to strong pressure in a hydraulic press.
The pressure was gradually raised till it reached 500 atmospheres. The liquid thus squeezed out measured 300 c.c. The
cake was removed from the press and ground up again with
another 100 c.c. of water and the pressure renewed, when a
further 150 c.c. of liquid was obtained. In all therefore 250 c.c.
of the contents of the yeast-cells were yielded by the treatment.
The liquid was shaken up with 4 grammes of kieselguhr and
filtered through fine filter-paper. The resulting extract was a
clear, slightly opalescent liquid of a yellow colour, possessing a
yeasty aroma. It had a specific gravity of 1·046 at 17° C.
When a small quantity of this solution was mixed with an

equal volume of a 37 per cent. solution of cane-sugar, a regular evolution of carbon dioxide commenced in about half-an-hour. When the operation was carried on at a low temperature the disengagement of the gas was very gentle, but at about 35° C. it was so energetic as to cause the liquid to froth.

The preparation of such an extract as this presents considerable difficulty, the operation of grinding the yeast being very tedious. Since Buchner's results were published it has however been successfully carried out by Delbruck in Berlin, and by the writer and by MacFadyen, Morris and Rowland in this country.

The fermentation is due to an enzyme in the yeast extract, to which Buchner has given the name *zymase*. The name is not perhaps well chosen, as the same term has been applied to the whole group of soluble ferments, though it has been replaced in this connection by the word "enzyme."

The decomposition of the sugar effected by zymase is its disruption into alcohol and carbon dioxide. If a quantity of the yeast extract is mixed with a solution of a fermentable sugar and placed in a flask fitted with a manometer, the mercury very soon rises in the distal limb of the latter, and the pressure of the gas keeps it there for several days, bubbles of gas from time to time escaping through the mercury column. At the end of some days a measurable amount of alcohol can be separated from the liquid by distillation. The action goes on until all the sugar has been decomposed.

The extract is perfectly active when saturated with chloroform, which inhibits the action of the yeast cell itself. Other antiseptics may be used instead of chloroform. Buchner recommends the use of potassium metarsenite in the proportion of 2 gms. per 100 c.c. of solution.

When chloroform is used there ensues a gradual precipitation of the proteids of the yeast solution, which mechanically carry down a large portion of the enzyme. If such a solution is filtered and the precipitate suspended in a quantity of water, equal in bulk to the original volume before filtration, the latter has much more fermentative power than the filtrate, when sugar is added to both.

If the solution of the enzyme is heated gradually, a copious coagulum of proteid matter is formed at 35°—40° C., and after separation of this the extract is found to have lost its fermentative power.

The extract is very difficult to preserve, as the enzyme decomposes very rapidly. It may be kept for 2 to 5 days if the temperature be maintained at 0° C., but at the temperature of the laboratory it is usually destroyed in half that time. If it is kept in contact with sugar Buchner says it can be preserved for a considerable time longer, retaining some ferment power after two weeks.

In some cases the yeast extract has been found to be active after filtration through a porcelain or kieselguhr filter. Different preparations however show differences in this respect.

The enzyme, like so many others, is capable of resisting desiccation to a certain extent. The yeast extract may be evaporated under diminished pressure till it consists of a thin paste, and the latter may then be spread out in a thin layer and further dried at 35° C. The solid residue generally represents about 10 per cent. of the weight of the liquid. It can be preserved unchanged for three weeks. If during that time it is dissolved in 5 times its weight of water and the solution filtered, the latter contains the ferment in an active condition.

In his earlier experiments Buchner was unable to precipitate the zymase from the extract of the yeast. More recently, working in conjunction with Albert, he has succeeded in doing so. He poured yeast extract into 12 times its volume of a mixture of 10 volumes of absolute alcohol and 2 volumes of ether, when a flocculent precipitate slowly settled out, which contained nearly all the zymase, which could be filtered off and dried without losing much of its activity. If less alcohol was used the precipitation was very incomplete.

The precipitate so obtained was found to be soluble in glycerin or in water. Filtration of the solution in glycerin did not impair its power of setting up fermentation, but the watery solution was not so potent after passing through a filter.

The sugars which are capable of fermentation by zymase are those which the yeast itself can decompose. Buchner has

fermented cane-sugar, maltose, glucose and fructose, but not
lactose nor mannitol. It appears at present that the fermentation is not accompanied by the formation of succinic acid and
glycerin which is characteristic of that brought about by yeast.

At the outset of his investigations Buchner found that the
enzyme could not be prepared from every sample of yeast with
which he worked. Many other observers, following his method
of preparation with scrupulous care, also failed to extract it.

The ordinary yeast of our breweries taken from the fermenting wort at the close of the fermentation usually fails to yield
any enzyme. If however it is taken while in very active condition, when the fermentation is at its height and the yeast
frothing vigorously, a considerable quantity of zymase may be
extracted from it. Buchner in one of his early papers suggested
that the amount of the enzyme may vary in different yeasts and
from time to time in the same yeast. The secretion is apparently intermittent, and only called forth by the conditions which
excite the cells to vigorous growth and multiplication. This
again calls to mind the behaviour of many gland cells, such as
those of the stomach, which do not begin to secrete their characteristic enzyme till stimulated by the absorption or supply of
nutritive material.

Since Buchner's discovery attention has been directed again
to the production of alcohol in fruits and other parts of plants
when they are deprived of oxygen. Effront has recently claimed
to have demonstrated the secretion of zymase by the cells
concerned in setting up alcoholic fermentation in these organs.
His first experiments were carried out with cherries, which he
sterilised by washing them with a dilute solution of formic
aldehyde. They were then carefully dried and plunged into
flasks containing olive oil. After three days a considerable
evolution of carbon dioxide was observed, which continued for
about three weeks. At the end of that time the cherries were
crushed and the juice extracted from them by pressure. The
juice was freed from oil by treatment with ether, and was dried
at a low temperature *in vacuo*. The resulting powder was
macerated with water for 12 hours and then subjected to strong
pressure. The expressed liquid after filtration was found to

excite alcoholic fermentation when added to cane-sugar; evolution of carbon dioxide, production of alcohol, and diminution of the specific gravity of the liquid being all observed. If the extract was kept at 40° C. for an hour before being mixed with the sugar it produced no change in the latter.

Effront has prepared zymase from peas and from barley by a similar method.

From these researches it appears certain that the production of alcohol whether in the presence or absence of oxygen is brought about by the activity of an enzyme. Its secretion by the cells of yeast attends the ordinary nutritive processes as well as the abnormal decompositions set up by incipient asphyxiation. The latter condition induces its formation in other parts of plants. The absence of oxygen stimulates the protoplasm of the cells to secrete it, the ultimate effect of its appearance being the liberation of energy as already stated.

CHAPTER XXI.

THE FERMENTATIVE POWER OF PROTOPLASM.

IN the introductory chapter of this volume we saw that the metabolism of the living cell consisted of two different series of events, the one constructive, embracing the building up of simple substances into complex ones and culminating in the formation or repair of the living substance itself; the other destructive, marked by the decomposition of complex substances, with the ultimate formation of others of much greater simplicity. We have seen further that many of the latter processes are carried out by means of enzymes, whose modes of action we have since examined. But if we turn our attention especially to intracellular changes we must recognize that at present all such changes cannot be traced to enzymic activity. Indeed a very important factor of the intracellular mechanism is the breaking down of the protoplasm itself, or its "self-decomposition" as Pfluger has expressed it.

The living substance must be regarded as the seat of a multitude of chemical changes, some constructive and others destructive in their nature. The incorporation into its substance of the ultimately assimilated proteids, fats, and carbohydrates, goes on continuously, and no sooner is it constructed than it undergoes a species of self-decomposition, giving rise to many residues of various kinds, some of which are re-incorporated with new material into its molecules, while others are broken down still farther and ultimately eliminated.

Not only does the substance of the protoplasm itself undergo these decompositions, but different materials in the cells are involved in them, and the residues of these are absorbed into, and incorporated with, the living substance in the same way as those which spring from its self-decomposition. We find that in many cells there is stored a reserve of nutritive material, not quite in a suitable state for immediate assimilation, but ready to form a nutritive pabulum when split up in the manner indicated. Nor are such materials of use only as supplying nutritive material; by their decomposition a supply of energy is afforded to the protoplasm, of which it avails itself in carrying out constructive processes. Under the condition of an insufficient supply of oxygen also, the protoplasm frequently effects decompositions of the same kind, as we shall see in the present chapter.

These changes are very largely of the same nature as the decompositions effected by means of enzymes, and we may consequently speak of them as exhibiting a *fermentative* power of protoplasm.

In discussing this power or property of the living substance, we may confine it to cases in which protoplasm provokes decomposition of various bodies for the purposes of preparing nourishment for itself, of liberating energy for its various processes, or of obtaining oxygen for respiratory purposes.

Such operations can be observed more or less easily in almost all organisms, from the highest to the lowest, and in both the animal and the vegetable body.

We have seen that the liver of mammals contains a quantity of glycogen stored in its cells. Glycogen is also present in the muscles of the trunk and limbs. During life there is a constant demand for carbohydrate material by all the living substance, and this is supplied partly by the consumption of the stored glycogen. In an earlier chapter we have seen that the glycogen of the liver is probably converted into sugar by an enzyme formed in the substance of the cell-protoplasm. Many physiologists hold the opinion however that this is not the only agent in the transformation, but that some at all events is directly hydrolysed by the protoplasm. The same appears to be true of the muscles, for up to the present these have not been shown to be capable of giving up diastase to an extracting liquid to more than a trifling extent.

The consumption of the fat stored in the various adipose tissues appears to be due to a similar cause. The fat ministers to the nutrition not only of the cells in which it is deposited but also of the rest of the body. It must for this purpose be removed from the seat of storage, and this removal almost certainly involves its decomposition with a view to facility of transport. The enzyme lipase however is by no means universally present in adipose tissue.

Other nutritive materials are also acted upon by protoplasm. In cases of starvation the substance of the body does not diminish uniformly; the fat disappears most readily, and the muscular tissue is drawn upon very seriously before there is any wasting of the brain. Of the muscles those of the heart do not suffer till some time later than those of the trunk and limbs. The latter are practically digested with the view of nourishing the other tissues, and the more important cardiac muscles receive aliment derived from the consumption of those which are more easily spared. The living substance itself must therefore be capable of breaking up the proteid materials of various parts, in the absence of similar substances normally supplied by the food.

The vegetable organism affords more examples of this power of protoplasm than the animal body. What is perhaps the exception in the latter case appears among the normal functions of various vegetable cells. The power of the protoplasm to effect the disruption of carbohydrates is seen in the transformation and reconstruction of the transitory starch which is constantly going on in various parts of the higher plants, though no doubt in many cases diastase here takes part. The same power may be noted in the changes found to take place among the various sugars that the higher plants contain; the various acids which are normal constituents of the cell-sap of succulent parenchyma are probably if not certainly produced from sugars in the cells. Kohl has suggested that they are formed much in the same way as lactic acid is formed in certain fermentations carried out by bacteria. These acids are not usually such simple ones as are formed by microbes; we find malic, citric, tartaric, &c. instead of acetic, lactic, butyric,

&c. This may however be due to the character of the metabolism of the two classes of cells respectively, for the action is
intracellular in both cases.

The intracellular digestion of proteids in the absence of an
enzyme can be seen in the case of the tuber of the potato.
The outer portion of the tuber contains crystals of proteid of a
cubical form, the proteid being probably one of the vegetable
globulins. At the onset of germination of the tubers these
crystals are found to disappear, being made use of in the nutrition of the young shoots.

The fermentative power of the protoplasm upon cellulose
can be observed in almost all growing cells which assume an
irregular shape. We have noticed in considering the development of branches upon the hyphæ of different species of Botrytis
that the accumulation of cytase at the points of branching leads
to a softening of the cell wall at those spots, and that then the
hydrostatic pressure in the hypha causes its gradual extension.
The shapes of such cells as form the stellate hairs upon the
leaves of *Deutzias* can be explained only by taking into account
the power of the protoplasm to soften the wall of the originally
nearly spherical cell at particular spots, so that in these areas
it becomes more easily extensible than over the remainder of
its surface. The hydrostatic pressure due to turgescence consequently causes protrusions at the softened spots, and hence
the appearance of the mature hair.

Less conspicuous cases are of frequent occurrence in the
interior of the vegetable tissues. This cytolytic action of
protoplasm is evidently the same thing as the softening brought
about by cytase.

Till very recently the fermentations induced in ripe fruits
as demonstrated by Lechartier and Bellamy, by Pasteur, and
by Gerber, were held to be illustrations of this property of
protoplasm. Effront's researches point however to their being
due to the presence of zymase.

If we pass to cases observable among more lowly forms we
find evidence of the same property of living substance. The
leucocytes of the blood of vertebrates may be regarded in some
sense as independent bodies, capable of carrying out their own

processes of nutrition. Metschnikoff has observed that certain
of these corpuscles, which he calls *phagocytes*, are able to
absorb and digest bacteria which have made their way into the
blood stream. The bacteria themselves are capable of bringing
about an almost infinite variety of similar decompositions, many
of them causing specific fermentations. In a few cases they
have been shown to effect these changes by means of enzymes,
but this is very far from being universally the case. The direct
action of their protoplasm must consequently be recognized as
the exciting cause of these decompositions. Various processes
of intracellular digestion occurring in the hyphæ of *Mucor* and
other filamentous Fungi demand a similar explanation.

The influence of deficient oxygenation upon the metabolic
processes of plants throws a further light upon this property.
Many phenomena have been recorded which show how pro-
foundly the ordinary metabolism is modified when circumstances
prevent their obtaining oxygen freely. Boehm, and later de Luca,
have shown that if any part of a living plant is insufficiently
supplied with oxygen, hydrogen and sometimes marsh-gas are
evolved from it. Boussingault and Schulz have observed
similar phenomena. From plants containing mannite also
hydrogen is given off, while according to de Luca, acetic acid
is formed from malic acid in the fruits, flowers, and leaves of
the Privet. In the decomposition of proteid too, Boehm found
ammonia exhaled.

As already mentioned, we find many specific fermentations
set up by micro-organisms without any secretion of enzymes,
several of which call for a brief notice. Attention may first be
given to those which in their normal metabolism cause the
production of quantities of various acids. The phenomena of
putrefaction also may be mentioned in this connection.

Lactic fermentation.

The frequent appearance of lactic acid in solutions of saccharine matters has been noticed from very early times. Its
occurrence explains the souring of milk which is allowed to
remain exposed to the air, and it was from this phenomenon that

the acid formed derived its name, lactic acid being for the first time prepared from whey by Scheele in 1780. The acid is not however formed exclusively from milk-sugar, but it can be produced from glucoses and from almost all substances convertible into them. Braconnot found it in a liquid obtained by steeping rice in water; also in the juice of the sugar-beet, and in infusions of peas, and seeds of other leguminous plants. The polysaccharides offer more resistance to its formation than the glucoses. Lactose, which is itself a member of this group, lends itself much more readily to lactic than to alcoholic fermentation. Bourquelot states that the formation of lactic acid is not preceded by hydrolysis as is the production of alcohol. The reactions which take place may be expressed by the equations $C_6H_{12}O_6 = 2C_3H_6O_3$, and in the case of polysaccharides $C_{12}H_{22}O_{11} + H_2O = 4C_3H_6O_3$, the last term being the formula for lactic acid.

Besides the sugars mentioned, lactic acid may be produced from sorbite, inosit, mannite, and dulcite; also from malic acid, carbon dioxide in this case being evolved, according to the equation $C_4H_6O_5 = C_3H_6O_3 + CO_2$. It occurs also in the butyric fermentation of many carbohydrates.

Lactic fermentation arises from the activity of many species of microbe, including the bacteria found in pus, *Micrococcus prodigiosus*, Erberth's bacillus, *Bacterium coli commune*, and others. The most energetic form, the one which is most frequently found in sour milk, is the organism observed and described by Pasteur, *Bacterium acidi lactici*. This microbe appears under the microscope as rounded globules, or short rods slightly constricted in the middle; they are generally isolated, but sometimes form chains two or three cells in length. They are capable of forming spores, which are found at their extremities. These offer a considerable resistance to the destructive action of heat.

The microbe is aerobic, and requires considerable quantities of oxygen. It is killed by exclusion of this gas from a liquid in which it is living. It flourishes best in a neutral medium, and consequently in cultivating it it is advisable to add a quantity of powdered chalk to the milk or saccharine solution. This neutralises the lactic acid as fast as it is formed, and the

XXI] BUTYRIC FERMENTATION. 369

fermentation can consequently proceed. An alkaline medium is also unfavourable to its activity. If the fermentation is allowed to go on without neutralisation of the acid, it is found that the action of the organism is usually stopped when about 8 per cent. of lactic acid has been formed.

For the cultivation of the lactic microbe it is necessary for the fluid to contain a certain amount of proteid material besides the sugar. It grows most readily at a temperature of about 35° C.

Lactic acid has often been detected in the contents of the mammalian stomach during digestion, and considerable discussion has taken place as to its cause. Many observers have argued in favour of its being a product of bacterial action there. Others, among whom may be mentioned Hammarsten, have attributed its presence to the secretion of a special enzyme by the gastric mucous membrane. It is said that certain gastric extracts have been treated with sodium hydrate solution of such concentration as to destroy both pepsin and rennet. After removal of the excess of soda, the liquid has been capable of bringing about the conversion of milk-sugar into lactic acid. The enzyme cannot however be said at present to have been satisfactorily isolated. Maly has stated that when glucose has been fermented in the presence of the mucous membrane of the stomach of the pig, a variety of lactic acid has appeared among the products. The conditions of his experiments do not seem however to have entirely excluded the possibility of the presence of micro-organisms. Other observers have attributed the same property to the gastric mucous membrane of herbivora.

Butyric fermentation.

It is generally found that in the souring of milk there is a formation of butyric acid, which appears to take place at the expense of some of the lactic acid which is first formed. The chief agent in its production is a special microbe, which has been named by different observers *Clostridium butyricum, Vibrio butyricus, Amylobacter clostridium,* and *Bacillus amylobacter.* It consists of slender cylindrical rods which are occasionally united into short rows. Sometimes they are found in active movement; at others they are gathered together into a zoogloea. They are often very irregular in shape; when they produce their spores, the latter are solitary in each cell, and are generally found at about the middle of its long diameter, causing the rod to have a somewhat fusiform appearance. This organism is very widespread, and may generally be found accompanying the lactic acid microbe. It differs in its habit of life from the latter, being anaerobic and easily killed by the presence of free oxygen.

Butyric acid is produced by this organism, not only from lactic acid and substances which can give rise to the latter, such as sugar and starchy bodies, but also from higher acids, such as tartaric, citric, malic and mucic acids. The decomposition which it provokes in these substances is of a somewhat complex character, being usually marked by the disengagement not only of carbon dioxide, but of free hydrogen as well. According to Van Tieghem, it has also the power of splitting up certain varieties of cellulose, or of pectic compounds associated with the latter, particularly the middle lamella of the cells of many tissues. It does not, however, attack starch.

The bacillus is constantly found in the alimentary canal of animals, particularly such as are herbivorous. It no doubt plays a very important part in the digestion of cellulose in the mammalian body. De Bary suggests that this digestion occurs in two stages; hydrolysis of the cellulose first takes place by virtue of a cytolytic enzyme, resulting in the formation of dextrin and glucose, and these subsequently undergo butyric acid fermentation. This, however, cannot at present be considered proved.

Many other microbes than *Bacillus amylobacter* are capable of giving rise to the butyric fermentation. Among them may be mentioned the *Bacillus butylicus* of Fitz, which can ferment glycerine, liberating carbon dioxide and hydrogen, and leaving in the solution butylic alcohol, butyric and lactic acids, propylenic glycol, and traces of two other alcohols. This bacillus is one of the few forms at present known to secrete invertase.

Hueppe's bacillus is another interesting form, from the fact that it not only forms butyric acid from lactates, but it liquefies gelatin, and excretes the soluble enzymes trypsin and rennet.

Bacillus ethylicus is capable of exciting butyric fermentation in a thin starch-paste, provided the reaction is kept neutral by the presence of powdered chalk. The fermentation is preceded by hydrolysis of the starch, brought about by diastase which is secreted by the microbe. The formation of the butyric acid in this case is direct, no intermediate occurrence of lactic acid being noticeable.

Among other bodies capable of undergoing butyric fermentation by various microbes, may be mentioned mannite, quercite, and glycerine.

The optimum temperature of butyric fermentation is a little higher than that of lactic, being 39—40° C.

The products of the butyric fermentation are generally complicated by secondary reactions taking place in the liquid. The nascent hydrogen frequently effects processes of reduction among the various bodies present. Succinic and valeric acids can often be detected among these.

Propionic fermentation.

Under the action of certain bacilli, lactic acid, when in combination with lime, undergoes a different decomposition, being converted into several fatty acids, among which are acetic and valeric, the most prominent of them however being propionic acid. This decomposition was first observed by Strecker in 1854, in a fermenting liquid kept for several months at a low temperature. Fitz has described a form of bacillus which can bring about this transformation; it is an elongated slender organism, forming curved chains consisting of several cells joined end to end. In the fermentation, the lactic acid splits up into two parts of propionic and one of acetic acid, carbon dioxide and water being simultaneously formed. It differs from the butyric fermentation of the same acid in not disengaging hydrogen.

Another bacillus, also investigated by Fitz, decomposes malic acid into almost the same products, but the relative quantity of propionic acid is greater, and traces of butyric acid and alcohol are simultaneously formed.

Citric fermentation.

During the year 1892, it was shown by Wehmer that under certain circumstances glucose can be made to give rise to citric acid. This change is brought about by two separate species of moulds or filamentous fungi, to which their discoverer gave the names of *Citromyces pfefferianus* and *C. glaber*, respectively. They form veils of dense mycelium of a green colour, about five millimetres thick, on the surface of suitable solutions. The most favourable sugar for their development is glucose, but they are not dependent upon that form alone. The fermentation is accompanied by the evolution of carbon dioxide, and the citric acid produced is identical in properties and composition with that obtained from lemon juice. Under proper conditions the quantity of acid formed corresponds to more than fifty per cent. of the glucose employed. Eleven kilogrammes of sugar yielded six kilogrammes of pure citric acid in one experiment, without the appearance of any secondary products. To secure this result, however, it is necessary to precipitate the citric acid when it has reached a certain percentage, as a concentration of 20 per cent. is deleterious to the fungi and checks their growth.

The fungus cannot excite the fermentation in the absence of air; it is deleteriously affected even by the accumulation of the carbon dioxide which is evolved during the process. It is independent of light, but the temperature needs to be carefully regulated.

The presence of inorganic acids in the fermenting liquid is very deleterious, while neutral chlorides favour the production of citric acid.

If the fermentation is allowed to proceed too far, the citric acid itself is attacked and decomposed.

The process of acidification is not uniform, but its rate describes a rather suddenly ascending and descending curve, which is in close relation to the amount of the decomposition of the sugar effected by the fungus.

In connection with this fermentation it is interesting to note that citric acid is occasionally present in the juice of the sugar-cane, indicating an action of the protoplasm of the cells of the latter similar to that of the fungi under discussion.

Oxalic fermentation.

Several fungi, particularly *Penicilium* and *Sclerotinia*, are known to be capable of giving rise to oxalic acid when cultivated in solutions of sugar. The same power resides in certain of the Saccharomycetes, one of which, *S. Hansenii*, has been described by Zopf. When cultivated in beer-wort, solutions of sugar, or mannite, it forms a pellicle on the surface, which soon sinks through the liquid. If chalk is added to the fluid in which this organism is growing so as to neutralise the acid as it is produced, large quantities of oxalate of calcium are formed. The yeast can form oxalic acid from galactose, glucose, cane-sugar, lactose and maltose; also from dulcite, mannite, and glycerine.

The oxalic acid fermentation contrasts somewhat sharply with the citric in certain respects. While a rise of temperature and the presence of neutral chlorides are advantageous to the latter, they promote the destruction of the oxalic acid and prevent its accumulation in the cultivations.

Besides these fermentations many others have been observed in which various microbes are concerned, but no very complete study has been made of the morphology of the latter, or the exact nature of the changes they effect. Many of these fermentations are probably very complex, several different microbes taking part in them either simultaneously or successively, and many interactions taking place between the various products.

A particular bacillus is said to decompose malic acid with the formation of succinic and acetic acids, carbon dioxide and water being simultaneously produced. Others which may be prepared from hay have the power of fermenting erythrite, giving rise to butyric acid with traces of formic and acetic acids. Others again decompose erythrite into butyric, acetic, caproic and succinic acids, traces of alcohol also being formed. Glycerine also can be split up by several of these organisms besides the butyric microbe already alluded to. A particular micrococcus is known, which slowly decomposes it with formation of ethyl and other alcohols, and butyric, formic and acetic acids, the last two being in small quantity. The microbe of pus, *B. pyocyaneus*, produces succinic acid in addition. Certain other micro-organisms can decompose the calcium salt of quinic acid; some anaerobic forms yield formic, acetic, and propionic acids at its expense; others which are aerobic oxidise it to pyrocatechic acid. Several other acid fermentations of a similar kind are known.

Acetic fermentation.

When alcoholic liquids are allowed to remain exposed to the air, they speedily become attacked by micro-organisms. Very soon, if the concentration of the spirit is not too great, a kind of pellicle or skin can be found upon the surface of the liquid, which sometimes remains thin and delicate and easily broken, and at other times attains a considerable thickness. In some cases the formation of the pellicle is accompanied by a growth of the organism in the body of the liquid, when a large amount of viscous matter may be produced. Coincidently with the growth of the microbes, the alcohol of the liquid becomes decomposed and is replaced partly or wholly by acetic acid.

The pellicle was named *Mycoderma* by Persoon in 1822; he noted its botanical nature, and rightly concluding it to be a fungoid growth, named it accordingly.

Kützing reinvestigated the subject in 1837, and described the organism as consisting of a number of small circular bodies arranged together in the form of chains. He was the first observer to note that the development of the acetic acid was connected with the presence of the microbe. He erroneously classed the latter among the Algæ, naming it *Ulvina aceti*.

About the year 1821 light was first thrown upon the process of acetification. Before that date it was known in a general way that acetic acid was formed from alcohol, and that

the access of air was necessary for the process, but no reasons
for the phenomena were known. Lavoisier stated that acetic
fermentation was effected by the absorption of oxygen from the
air. In 1821, however, Edmund Davy discovered that when
platinum black was moistened with alcohol, it became incandescent, and the consumption of the spirit thereby induced was
attended by the formation of acetic acid. His discovery was
confirmed two years afterwards by Dœbereiner, who carried it
further by showing that the alcohol absorbed oxygen, and
became decomposed into acetic acid and water.

By most of the writers who succeeded Dœbereiner the
membrane of the mycoderma was compared with platinum
black as an excitant of the decomposition, and was held to act
in a similar way. Berzelius in 1829 taught that the action of
the pellicle was due to acetic acid enclosed within its pores.
Liebig considered the mycoderma to be devoid of life, a structureless precipitate of albuminous matter, and held that it was
a porous substance like platinum black and that its action was
connected with its porosity and the oxygen thus mechanically
entangled in it. He believed that the formation of the acetic
acid was due to a movement or vibration set up by matter
undergoing decomposition, as he taught in the case of alcoholic
fermentation. The vibration in the present case provoked the
oxidation of the alcohol.

Pasteur himself failed to see the actual part played by the
organism. He recognised the accuracy of Kützing's work and
showed experimentally that the opinion of the latter as to the
function of the mycoderma was accurate. He agreed however
with Liebig that the fungus acted after the manner of platinum
black. He showed moreover that not only was alcohol oxidised
during its activity, but that if the action was prolonged the
acetic acid also was decomposed, carbon dioxide and water being
the only products.

It was not till 1873 that it was clearly shown that the two
modes of producing acetic acid from alcohol are not comparable.
Von Knieriem and Ad. Mayer pointed out several differences.
Platinum black oxidises both concentrated and dilute alcohol,
while the mycoderma cannot attack alcohol of greater concen-

tration than 14 per cent. Acetic fermentation proceeds most
actively at 35° C. and the organism cannot work at a point
higher than 40° C.; platinum black becomes more and more
efficacious as the temperature rises, and the action may become
so violent that the metal becomes incandescent.

The phenomenon gradually came to be considered a biological process, corresponding in all essential features with
the alcoholic fermentation induced by yeast.

The action was at first held to be always due to one particular fungus, but later work carried out by many observers
has shown this to be erroneous. The form which is most
commonly met with was named *Mycoderma aceti* by Pasteur,
but was referred later to the genus *Bacterium* by Zopf. Hansen
distinguished two species of this genus which are capable of
acetifying alcoholic solutions, and he named them *B. aceti* and
B. Pasteurianum. Two other species have been described by
A. J. Brown under the names of *B. aceti* and *B. xylinum.* The
former does not appear, in the light of later researches, to be
identical with Hansen's organism bearing the same name.
Lafar has isolated and cultivated an acetifying organism belonging to the *Yeasts*; he found it present in an acid beer.

Bertrand has described a fission fungus which has the power
of attacking sorbitol, an alcohol isomeric with mannitol; it
converts it into a ketose or ketone sugar known as sorbinose
or sorbose. This fungus has not however been accurately
identified.

A very careful investigation into the acetous fermentation
set up by *Bacterium aceti* was made in 1886 by Adrian J. Brown.
He obtained the organism in a pure condition by a combination
of the fractional method of Klebs and the dilution method of
v. Nägeli. So cultivated, it formed a somewhat greasy pellicle
covering the surface of the liquid and inclined in its early
stages of growth to creep up the moist sides of the vessel containing it. When grown on diluted claret the coating or veil
attained the thickness of stout paper. Slight agitation easily
broke the surface of the pellicle, and when the latter was completely wetted it sank to the bottom of the liquid. A fresh
growth however speedily appeared upon the surface. The

liquid below the veil became turbid from the presence of isolated cells of the organism, and after some weeks a considerable quantity of the latter had sunk to the bottom of the vessel. When the pellicle was examined under the microscope it appeared to be a mass of cells averaging 2μ in length, but showing a good deal of difference in this respect. The cells were slightly contracted in the middle, and presented roughly the appearance of a figure of 8. They were united together in chains of variable length. Sometimes the cells were divided in the middle, producing chains of micrococcus-like forms. In the liquid below the surface-film longer cells were found, sometimes reaching a dimension of $10—15\mu$; in some cases their form was that of leptothrix threads of equal thickness throughout; in others the long cells were swollen out in two or three places along their length, giving them a very irregular appearance. These cells were generally of a dark grey colour. At their ends short chains of small rods or micrococci were sometimes observed. Other forms frequently seen were short rods of about 3μ in length and micrococci of about 1μ in diameter, which floated freely in the culture liquid. The shorter rods and cells were motile.

A litre flask half full of a 5 per cent. sterile solution of pure ethyl alcohol in yeast water, possessing a neutral reaction, was inoculated with a trace of a pure culture of the organism, and kept at a temperature of 28° C. for 10 days. It was then found to contain rather more than 1 per cent. of acetic acid and a very slight trace of a non-volatile acid which appeared to be succinic.

When the microbe was cultivated in a solution containing ·75 per cent. of acetic acid but no alcohol, and was kept at the same temperature for six weeks, two-thirds of the acid disappeared, being oxidised to carbon dioxide and water.

When an alcoholic solution was allowed to ferment until all the alcohol was converted into acetic acid, and the experiment then continued, the acid so formed was decomposed even more quickly, the same products being formed.

The oxidation of the alcohol takes place in two stages; there is first a formation of aldehyde, which is subsequently

converted into acetic acid, the reactions being expressed by the
equations

$$2CH_3CH_2OH + O_2 = 2CH_3COH + 2H_2O \ ;$$
$$2CH_3COH + O_2 = 2CH_3COOH.$$

B. aceti was found by Brown to be capable of oxidising
propylic alcohol to propionic acid, but it was not able to decompose either methylic, butylic, or amylic alcohol.

Boutroux stated in 1880 that when B. aceti was cultivated in
a solution of glucose it converted the latter into gluconic acid.
During the progress of the action the solution was kept neutral
by the presence of calcic carbonate, which neutralised the acid
as it was formed.

Blondeau also noted that sugar could be converted into an
acid by the organism, without passing through the stage of
alcohol. Brown confirmed these observations and suggested
the following equation as expressing the reaction—

$$2CH_2OH(CHOH)_4COH + O_2 = 2CH_2OH(CHOH)_4COOH.$$

The oxidation is thus seen to affect the aldehyde group in the
sugar just as in the case of the second stage of the oxidation of
ethylic alcohol.

The organism was found to be incapable of attacking cane-
sugar.

The alcohol *mannitol*, which corresponds to the aldehyde
glucose, was also examined. Brown cultivated the organism in
a 2 per cent. solution of mannitol, Pasteur's mineral medium
and a little gelatin being added as food for the bacillus. The
culture was continued for six weeks, at the expiration of
which time the solution had acquired a very sweet taste and
reduced Fehling's fluid freely. It was then found that the
mannitol had completely disappeared, having been oxidised to
levulose. The sugar was identified by its reducing power
and its optical activity, and further by its power of forming a
crystalline compound with lime, according to Dubrunfaut's
process.

The organism cannot decompose lactose, starch, dulcite, or
levulose ; it can convert glycol into glycollic acid.

Another organism which possesses the power of forming

acetic acid at the expense of ethylic alcohol is the so-called *vinegar plant*, which also has been investigated by Brown, who has given it the name of *Bacterium xylinum*. This, like B. aceti, is an aerobic organism, found growing on the surface of the culture fluid, and forming there a jelly-like translucent mass, sometimes as much as 25 mm. in thickness. This membrane is slightly heavier than water and sinks when completely wetted, a further growth commencing at once above the old layer. Frequently 5 or 6 layers of growth are thus formed, so that the veil or pellicle appears laminated. If the organism is cultivated in a liquid unfavourable to its free growth, such as yeast-water, it appears as a jelly-like transparent mass at the bottom of the solution. This gradually increases in size till the liquid is almost entirely filled with it. The jelly when treated with sulphuric acid and iodine takes on a deep blue coloration, indicating that it is composed of cellulose. It seems to be made up of extremely diffluent membranes of the organism, forming a kind of zoogloea.

When a film of the vinegar plant is examined microscopically it is found to consist of bacteria arranged more or less in lines and lying embedded in a transparent structureless sheath. These bacteria are most commonly found as rods about 2μ in length, several often being united together. In old cultivations the rods are frequently to a large extent replaced by micrococci about $\cdot5\mu$ in diameter. When the organism has been cultivated in an unsuitable medium, such as yeast-water, it appears as long twisted threads, from $10—30\mu$ in length and of a leptothrix nature, differing however in appearance from the similar threads of B. aceti.

The organism grows best at a temperature of about 28° C. Above 36° C. it ceases to develop.

The fermentations set up by B. xylinum are similar to those caused by B. aceti. Ethylic alcohol is oxidised to acetic acid, and the acid subsequently entirely destroyed. Glucose is oxidised to gluconic acid and mannitol to levulose. The organism has no oxidising action on cane-sugar, starch, or levulose. The great difference between the two is the large amount of cellulose formed by B. xylinum.

Nitric fermentation.

The decay of organic matter in the soil, whether it is derived from the débris of vegetable or animal matter, is associated with the formation of ammonia or its compounds. Ammonium salts, especially the sulphate, are constantly in use among agriculturists for purposes of manure. The compounds of ammonia are however not so easily made use of by green plants as are the salts of nitric acid. It has long been known that the soil is the seat of a variety of chemical changes, among which the formation of such salts from compounds of ammonia is especially prominent.

Nitrification, as this process is called, was formerly regarded as a purely chemical process. Like the oxidation of alcohol, it can be effected through the agency of platinum black. If a mixture of air and ammonia gas is heated gently in a tube containing spongy platinum, the latter becomes white-hot, and forms ammonium nitrate at the expense of the gaseous mixture. The influence of porosity, and the free oxygen entangled in the interstices of the porous body were held to be the explanation of the oxidation in the soil as they were in the case of acetification.

That the phenomenon is to be regarded as biological and not purely chemical may be attributed to the researches of Schlösing and Müntz, which were published in 1877. They found that the conditions of nitrification strongly suggested bacterial agency. The process is hardly noticeable at a low temperature, such as 5° C.; as the temperature rises it becomes more marked, and reaches a maximum at 37° C. Beyond this point it gradually weakens and cannot be observed above 55° C. They found that soil which had been heated to 100° C. was not subsequently capable of exhibiting the process.

By introducing small quantities of soil into a fluid medium containing a salt of ammonium, preferably the chloride, nitrification of the latter can be made to take place.

Experiments made subsequently by many observers have shown that the compounds of nitrogen found in liquids and soils which are the seat of nitrification are not always the same. In some cases *nitrites* and in others *nitrates* are produced. A long series of investigations has shown that the action takes place in two stages; in the first the ammonium salt is oxidised to the condition of a nitrite and subsequently further oxidation produces nitrates from the latter.

Schlösing and Müntz originally attributed the action to a single organism, which they considered oxidised the ammonium salt to the condition of a nitrate, and they held that the nitrites found were the result of purely chemical processes of reduction. They described the microbe as consisting of corpuscles, round or slightly elongated, of varying dimensions, largest when grown in media rich in organic matter, but at all times very small; occurring either singly or in pairs, and multiplying by division.

The researches of Warrington in England and Winogradsky on the continent have satisfactorily shown that two distinct micro-organisms play a part in the process of nitrification, and that probably more than one species of each exists.

One of these has the power of oxidising salts of ammonium to the condition of compounds of nitrous acid. When in a pure culture this stage has been reached no further oxidation takes place. Winogradsky has established two genera of these organisms, *Nitrosomonas* and *Nitrosococcus*, the former being somewhat strangely only found in the soils of Europe, Asia, and Africa, the latter in those of America and Australia.

The form of nitrosomonas as cultivated by Winogradsky in an ammoniacal solution is generally that of an elongated ellipsoid, but the youngest cells are nearly spherical. The breadth does not exceed 1μ nor the length $1\cdot1—1\cdot8\mu$. The longest are those about to divide and they then exhibit a dumb-bell form. More rarely the organism is spindle-shaped with blunt ends. A chain of 3—4 individuals is very rare.

When the Nitrosomonas is cultivated in an aqueous solution of sulphate of ammonium to which some carbonate of magnesium has been added, three stages can be noted in its growth. After about 4 days isolated colonies appear in the liquid, each of which is included in a gelatinous mass of the nature of a zoogloea. Three days or so later the liquid becomes turbid, and is found to contain a motile form which is furnished with cilia.

The colonies at this stage have broken up. After a further 2 days the liquid again becomes clear and the microbes are found to have formed another zoogloea which lies as a gelatinous layer at the bottom of the culture vessel.

The Nitrosococcus forms behave differently. They do not form a zoogloea, nor do they possess cilia. They are slightly larger than the others, ranging from 1·5 to 2μ in diameter.

Besides these organisms, which are responsible for the first stage only of nitrification, there are others which have the power of converting the nitrites therein formed into nitrates, thus completing the work. These are among the smallest of all living organisms so far discovered. The cells are elongated, and oval or somewhat pear-shaped, being about ·5μ in length and from ·15 to ·25μ in breadth. The name *Nitrobacter* has been given to the genus.

Most of our knowledge of this microbe is due to the researches of Winogradsky. It can be cultivated in liquid media, when it forms a thin mucinous skin which adheres to the wall of the vessel.

The two kinds of organism are usually both present in the same soil, those of the second type immediately oxidising the nitrites which those of the first form from ammonium salts. The Nitrobacter forms not only cannot oxidise the latter bodies but they are very injuriously affected by the presence of free ammonia. Except in this respect the two classes show great similarity. Both flourish in inorganic solutions, and are aided in their work by the presence of acid carbonates. Neutral or alkaline carbonates hinder or prevent their development, and the chlorides of potassium and calcium are injurious to their action. Neither of them apparently will grow on gelatin.

The conditions of their action are remarkably similar; below 5° C. hardly any change can be initiated by either; at about 12° C. nitrification becomes sensible and increases rapidly as the temperature rises to 37° C. At this point the rate of change is ten times as great as at 14° C. Above 37° C. a rapid decline of the rate may be observed, and according to Winogradsky the fermentation stops at 55° C. Warrington failed to observe nitrification above 40° C.

A very interesting peculiarity attaching to the bacteria of nitrification is their marked distaste for organic nutriment. Neither of them will grow upon gelatin or similar material, apparently disliking all organic substances. They can be cultivated readily on masses of gelatinous silica, impregnated with the appropriate compounds of nitrogen. As a suitable culture fluid for the Nitroso-forms Winogradsky recommends a mixture of 2—2·5 grammes of ammonium sulphate, 2 grammes of sodium chloride and a sufficient quantity of magnesium carbonate per litre of well-water. The latter salt is needed to neutralise the acids as they are formed. For Nitrobacteria the ammonium salt must be replaced by sodium nitrite.

When either organism is cultivated in such a medium it can grow and multiply, and the development is greatest in the absence of light. The source of the carbon of the increased bulk of the plant is carbon dioxide, derived partly from the carbonate in the solution, and, according to Godlewski, partly from the air. We have in these plants therefore a power which appears special to them, viz. that of decomposing carbon dioxide and availing themselves of its carbon in the construction of new cells, in the absence of light. They show indeed that the power of decomposing carbon dioxide can go on in these plants in the absence of any chlorophyll apparatus. The steps by which this carbon dioxide is built up into a compound capable of being assimilated by the living substance are not known. The energy necessary for the process appears to be supplied by the oxidation of the molecules containing nitrogen, so that it is dependent upon such oxidation taking place. Winogradsky has investigated this point with much care, and he has come to the conclusion that about 35 milligrammes of nitrogen are oxidised for each milligramme of carbon absorbed and fixed.

Besides the oxidation of ammonium salts, another process is also constantly going on in the soil which is of exactly the opposite character. We meet continually with a process of denitrification, marked sometimes by the evolution of ammonia and sometimes by the disengagement of free nitrogen. The decomposition of proteid matter and its derivatives is effected by many species of Schizomycetes, and by many of the higher fungi. Of them all *Bacillus mycoides* especially may be mentioned, as it gives rise to a very considerable volatilisation of ammonia. This process has very much in common with the ordinary putrefactive processes, which will be discussed subsequently. The production of free nitrogen is different, as the sources from which it is derived are the nitrates and nitrites of the soil. The *Bacillus denitrificans a* and *β* of Guyon and Dupetit are especially noteworthy in this connection. Both species reduce nitrates, *a* reducing them to nitric oxide and nitrogen, *β* liberating free nitrogen only.

Several other species possessing similar powers have been described by various writers, among them being *Bacterium coli commune*, which reduces nitrates to nitrites when cultivated in a nutrient solution by itself and in the absence of air. When in the soil it frequently grows symbiotically with a bacillus, known as *B. denitrificans I.*, and the two can then decompose the nitrate completely, free nitrogen being evolved. Loew has shown that the process of reduction goes so far as the formation of ammonia.

Viscous fermentation.

A peculiar phenomenon occurs frequently in certain wines, in beer-worts, and in some natural vegetable juices. The liquid on exposure to the air becomes extremely viscous, in some cases indeed it appears to be transformed into a jelly.

This phenomenon, which must not be confused with the action of pectase, has been investigated by several observers, and has been found to be due to the presence of microbes, several of which have been isolated. Among the earlier workers in this field may be mentioned Braconnot, Guy-Lussac, Fremy, and Pasteur.

According to Pasteur this viscous fermentation is caused by a particular organism which acts upon either glucose or invert-sugar and transforms it into a kind of dextrin or gum, mannite and carbon dioxide being formed at the same time. The gum appears to resemble dextrin rather than gum-arabic, as nitric acid oxidises it to oxalic and not to mucic acid.

Héry found that the peculiar ropiness which is formed in certain inks is due to similar agency.

The same phenomenon is often met with in the manufacture of cane-sugar, masses of gelatinous consistency being formed in the crude juice. These were first accurately investigated in 1878 by Cienkowski, who found them to be composed of microbes with extraordinarily swollen and gelatinised cell-walls, which appeared as masses of jelly in which the organisms were embedded. Van Tieghem, who also studied them, and who followed out the life-history of the microbes, held that they resembled in most important respects the alga *Nostoc*, except that they did not contain any of the blue-green colouring matter so characteristic of that plant. He gave to the microbe consequently the name of *Leuconostoc*. Further investigations were carried out by Liesenberg and Zopf in 1891.

The microbe, according to the last-named observers, does not form the gelatinous material unless the nutrient material on which it is cultivated contains either cane-sugar or glucose. In the absence of these carbohydrates it is found in chains, sometimes of considerable length, sometimes of only two or three cells. When either sugar is present large sheets of zooglœa are formed, in which the cells are arranged in pairs. The cells have greatly swollen walls, forming mucinous capsules, which gradually coalesce, giving rise to the gelatinous masses.

The jelly does not consist of unchanged cellulose, but of a material which was originally called *dextran* by Scheibler, and which is probably a derivative of cellulose.

An organism which has recently been examined by the writer in collaboration with Marshall Ward, and to which reference has been made in the preceding chapter, was also found to have the power of forming a mucous or slimy material when cultivated in solutions of cane-sugar. After about 3 days' growth, the whole contents of a flask in which the culture was conducted became extremely viscid. The solution was found to contain two carbohydrates resembling in many respects Scheibler's dextran. They were thrown out of solution by the addition of alcohol, and were separated by subsequent treatment with water, in which one of them dissolved. The other was soluble only in dilute alkali, 1 per cent. of caustic soda taking it up slowly in the cold and more rapidly on boiling. It was freely soluble in the cold in a 10 per cent. solution of the alkali.

The first of these bodies had a specific rotatory power of $(a)_D = + 130$ in a 1 per cent. solution. It gave a pink coloration when treated with iodine, and did not reduce Fehling's solution.

The other, when in solution in 1 per cent. caustic soda, had no action on polarised light. It was coloured violet by iodine, and like the first had no cupric-reducing power.

Both bodies when boiled for some time with a dilute mineral acid were converted into substances which reduced Fehling's fluid.

From their reactions they appeared to belong to the group of the hemicelluloses, but they did not yield oxalic acid when oxidised by nitric acid.

Various organisms have been found to set up similar formations in milk, causing a peculiar "ropiness" in the liquid. Their influence appears to be chiefly confined to the milk-sugar.

It is probable that in most of these cases the viscous matter is nothing more than the extremely diffluent cell-walls of the organisms, and it is therefore doubtful whether these phenomena should be classed among such fermentations as we are considering. Whether or no enzymes having the properties of cytase or pectase are secreted by the microbes we have no evidence to show, nor whether the changes in the cell-walls are due to cytasic or pectasic powers of the protoplasm.

There are however cases known in which a similar production of viscous or mucous matter has been found to take place extracellularly. A microbe was described in 1889 by Kramer as possessing this property. Microscopic investigation shows it to be a bacillus, occurring in short rods which are joined together in chains. The cell-walls of the organism do not swell up as in the formation of a normal zoogloea, but when it is cultivated in the presence of cane-sugar a peculiar mucoid

material is formed in the culture medium which does not seem
to include the bacilli.

Another organism was ascertained by Glaser to possess the
same property. He named it *Bacterium gelatinosum betæ*, from
its occurrence in the juice of the Beet. The mucous material
which it forms shows considerable resemblance to dextran.

Putrefaction.

Putrefaction is a process which is associated with the changes
which various micro-organisms set up in proteids or albuminoid
bodies. The various decompositions usually begin with the
formation of peptones in a manner which resembles the gastric
and pancreatic digestions already described. The changes do
not stop at this stage, nor do they follow the same course as
those brought about subsequently by trypsin; leucin and
tyrosin appear to be formed, but at the same time many
other bodies appear, probably simultaneously. Among these
are *indol, skatol, phenol,* and a variety of substances belonging
to the aromatic series. Besides such comparatively simple
bodies whose composition is known, we find others of more
complex nature, many of which are eminently toxic. These
include the so-called animal alkaloids or *ptomaines,* a great
number of which have been isolated from various putrefying
materials. The proteid molecule is broken down yet further as
the fermentations proceed, several of the higher fatty acids,
ammonia and its compounds, various amines, and different
nitrates, making their appearance. The sulphur of the albuminoid matter is generally evolved in the form of sulphuretted
hydrogen, though mercaptan has been detected in some
cases.

The putrefactive process is accompanied by the development
of a peculiar and characteristic odour, partly due to the presence of certain of the bodies mentioned, and partly to the
sulphuretted hydrogen, ammonia, and other gases which are
disengaged. This feature is especially noticeable when the
microbes carry out their work in the absence of air, the process
being comparatively inodorous when a free access of oxygen is

25—2

permitted. Indeed putrefaction in the strict sense of the term
is largely an anaerobic process.

Many microscopic organisms take part in these putrefactive
changes and their action is far from completely ascertained at
present. It is not known whether a particular series of changes
is due entirely to one kind of microbe, or whether the process
is initiated by one and carried further by another; whether
the decompositions take place simultaneously or successively.

The earliest researches on the question which may be
quoted are those of Ehrenberg, who was the first to notice, in
1830, the microbe long known as *Bacterium termo.* This was
described 11 years later by Dujardin as possessing a cylindrical
form about 2—3μ in length and 1—1·2μ in thickness. The
cells were said to be frequently joined together in pairs, and
the short chain so formed exhibited a tremulous movement.
Cohn stated in 1892 that putrefaction was a special process set
up by this bacterium.

Bacterium termo was the subject of study by many
observers during the next decade. Improved methods of
cultivation showed that several separate forms were associated
under the one name, and in 1884 Rosenbach described three
distinct species, under the names of *Bacillus saprogenes* I,
II, and III. Hauser a year later described three species of
Schizomycetes taking part in putrefaction, which he referred to
a new genus, *Proteus.* These are eminently motile, the cells
being furnished with a variable number of cilia. *P. vulgaris*
is often met with in the form of elongated rods, ranging from
3—6μ in length, with a transverse diameter of ·9μ. All the
forms however are very polymorphic.

Escherich has described another microbe under the name of
Bacterium coli commune, which is always present in the mammalian intestine and which can be detected in large numbers
in the evacuated fæces. It is a short rod ranging in length
from ·5μ to 2—3μ and having an average breadth of ·4μ.

Several other species, not so clearly identified, have been
described by many writers. All agree however in being extremely polymorphic, the form they take varying according
to conditions of cultivation and nourishment. The difficulty of

ascertaining the limits of any species is consequently extremely great.

The complicated nature of the digestive changes in the mammalian alimentary canal must be associated with the presence there of many of these bacterial forms. As we have seen the final action of the pancreatic juice is to convert part of the proteid molecule into leucin and tyrosin. The occurrence of indol, skatol, and other malodorous products in the large and in part of the small intestines, can no doubt be explained by the action of these microbes, which work simultaneously with the trypsin of the pancreatic secretion and continue to act after the enzyme has ceased to be effective. The absence of the microbes from the region of the small intestine in which the bile remains is explained by the antiseptic powers of this secretion.

Though no doubt the direct intervention of the protoplasm of the microbe is in most cases the exciting cause of the fermentation, or putrefaction, it must not be forgotten that many of these organisms have been shown to excrete enzymes, to the action of which definite decompositions have been traced. The observations of Lauder Brunton and MacFadyen, and of Wood, have been alluded to in a former chapter, and it has been shown there that various microbes have been proved to secrete trypsin. Various toxic albumoses have been shown by Hankin, Martin, and others, to result from the activity of *Bacillus anthracis*, and Hankin was able to extract from this microbe an enzyme that had the property of forming albumoses from fibrin. Another enzyme has been prepared by Courmont and Doyon from the bacillus of tetanus which effects a different decomposition, producing a toxic substance which is comparable to strychnine. This enzyme, which has received no name, is destroyed on heating to 65° C. Its discoverers attribute the peculiar symptoms of tetanus to its presence, and state that immunity from this disease is the result of the action of causes that inhibit the activity of the enzyme. It is not itself toxic, but produces the alkaloidal substance in the body, and the deleterious material can be extracted from tetanised muscles and from the blood and the urine of an animal which has been thus poisoned.

The toxic products of putrefaction include not only albumoses, but a number of derivatives of various albuminoid
bodies to which the name of *ptomaines* was given by Selmi in
1878. There have been many workers in this field, chief among
whom may be named Panum, Nencki, and Brieger.

The majority of the ptomaines at present known were discovered by the last-mentioned investigator, to whose labours we
are indebted for new methods of separating these substances from
putrefying liquids. Brieger isolated *choline, putrescine, neuri-
dine,* and *cadaverine.* From the decomposition of the first of
these *muscarine* may be obtained, the poison of the red Agaric.
Some of these toxic bodies occur in stored cheese which has
been attacked by certain microbes.

Bodies allied to the ptomaines, though not identical with
them, which are produced in the body have been called *leuco-
maines.* They differ from the ptomaines in being products of the
metabolic activity of the tissues and not of microbes. They are
produced, that is, by animal, and not by vegetable protoplasm.

Many of the ptomaines and similar bodies appear to belong
to the pyridine group of aromatic compounds.

The fermentative power of protoplasm is thus seen to be
constantly manifested and to conduct a great variety of decompositions, some relatively simple, and others extremely
complex. The living substance may be that of unicellular or
of multicellular organisms, of microbes of almost immeasurable
smallness, or of plants of very large dimensions, the differentiation of the plant affecting the distribution but not the
character of the active protoplasm. Certain features connected
with the manifestation of this property by unicellular microbes
led to their being for a long time considered to be especially
"ferments" and to possess a property which was peculiar to
them as a class. Among these features was the phenomenon
that the changes they caused were on such a relatively large
scale that they seemed to be altogether out of proportion to
their biological needs, and hence to require an explanation of
a different kind. The prominent fact was the magnitude of

the disturbance they caused, and this obscured the metabolic work to which the induced changes ministered.

The same facts meet us even to-day and we are not yet prepared with more than a tentative explanation of them. But we can see from comparing these humble forms with the higher plants that their activity is paralleled by the activities of the cells of many of the tissues of the latter; that the decompositions set up by both are of the same character; and that in some cases the changes are brought about by both through the agency of secreted enzymes.

The decompositions of these so-called *organised ferments* are then only peculiar in two respects : (1) the great and apparently wasteful extent to which they are conducted, (2) the more complicated character of the products produced by some of them. The second of these statements must not however be regarded as proved; it may be that the complex products of putrefaction, for instance, are the result of the energy of several microbic forms working in succession. In this case the second statement merely indicates that the protoplasm of different organisms possesses very varied powers of exciting very different decompositions. And even this view is incomplete, for the number of enzymes known is continually increasing and may ultimately be found to correspond to the variety of decompositions now known to be set up by living substance.

Other specific peculiarities have been associated with microbes, vindicating their claim to be regarded as organised ferments. These have been brought forward with a view to demonstrating that "organised" and "unorganised" ferments are radically different. They have been especially emphasised by Naegeli and Sachs, and may be mentioned here to show how completely this old distinction has broken down, how indeed two things have been put in comparison that are so fundamentally different that comparison is impossible, whereby the true relation between them has been entirely obscured.

Sachs states the first of these peculiarities in the following terms :—"I find with Naegeli a further particularly striking point of difference in the fact that (apart from split-ferments such as emulsin and myrosin) the plastic matters are transferred by the action of unorganised ferments from the passive into the active condition. Fermentation by means of Fungi has, as Naegeli insists, just the opposite character; its products are, without exception, less nutritious compounds, and it destroys especially the most nutritious substances."

Naegeli states, " The contrast appears most striking in the case of carbohydrates and proteid substances. While the action of unorganised ferments produces from them glucoses and peptones, fermentation by means of Fungi breaks up these compounds into alcohol, mannite, lactic acid, and into leucin, tyrosin, &c.—and in some cases several fermentations follow one another; their products then became less nutritious step by step. We may say, generally, that the yeast fungi render the medium in which they occur chemically less suitable for nutrition by every process of fermentation which they effect."

In the light of the knowledge which has been obtained during recent years these statements can no longer be regarded as marking differences between fermentation by microbes and by enzymes respectively. Sachs's first criticism loses a good deal of force on account of the exceptions he mentions. There is no reason why emulsin and myrosin should be considered exceptional. We have seen that the conditions under which they are active are the same as those which are needed by diastase and by pepsin. The action is similar in kind as far as we can judge; diastase and emulsin both cause hydrolysis and subsequent decomposition of the bodies which they attack. The property of giving rise to less nutritious compounds does not belong to Fungi alone; enzymic activity often results in similar decompositions. Urease produces ammonium carbonate from urea; the lactic enzyme discovered by Hammarsten effects the same change as the lactic microbe; Buchner's zymase produces alcohol from sugar. On the other hand, again, certain microbes convert starch into sugar, while others peptonise albumin.

The contrast drawn by Naegeli also fails. We have just seen that the saccharification of starch and the formation of peptone from albumin can be brought about by members of both classes.

The conversion of proteids into leucin, tyrosin, asparagin, &c., is not a special property of Fungi; it is in fact the characteristic feature of both animal and vegetable trypsins.

Both the botanists quoted lay stress on the formation of nutritive substances from plastic ones by enzymes. The action of the latter is by no means confined to effecting transformations of this kind. We may mention here the recently discovered group of the oxidases, which are not concerned with nutritive processes at all. Nor so far as we know are the clotting enzymes thrombase and pectase to be considered as forming nutritive materials for the organism in which they occur.

Sachs says again, "Naegeli puts his view of fermentation due to organisms as opposed to that produced by unorganised ferments in the following statement:—'Fermentation is therefore the transference of the movements of the molecules, atomic groups, and atoms of various compounds composing living protoplasm (which remain chemically unaltered in the process) to the fermentable material, by which means the equilibrium between its molecules is destroyed and it is broken up.'" It has been shown by Fischer and other writers that this statement, which presents Naegeli's theory of fermentation, if true at all, applies equally well to enzymes. This point however will be more fully examined in a subsequent chapter.

If we turn again to the consideration of the fermentative processes set up by the protoplasm in the higher plants we find that the latter exhibit the same variety as the microbes in the nature of the products that are formed. Boehm and de Luca have shown that if any part of a living plant is insufficiently supplied with oxygen, hydrogen and sometimes marsh-gas are evolved from it. Boussingault and Schulz have observed similar phenomena. Hydrogen is given off also from plants containing mannite, while malic acid gives rise to acetic acid in the fruits, flowers, and leaves of the Privet. In the decomposition of proteids, Boehm found ammonia exhaled. The condition under which these results are obtained, viz. the lack of oxygen, is the normal condition of many microbes, they being anaerobic in their mode of life. When oxygen is present we find the same agreement. The inversion of cane-sugar is the same process whether that action is brought about by the action of the living cell of a leaf or by invertase extracted from either a higher plant or from yeast.

Sachs claims as a peculiarity of all fermentation set up by fungi that carbon dioxide appears as a bye-product. This, however, we have seen to be rather an effect brought about through an insufficient supply of oxygen, and easily made evident under the same conditions in the fermentative actions of the protoplasm of the higher plants also. We can see therefore that in both the lower and the higher plants we have to recognise essentially the same constitution, the differences between them depending only on differentiation and consequent division of labour. In the lowly forms the great prominence of their metabolic decompositions has overshadowed all their other functions, and they have therefore been regarded as endowed with special properties. In the higher plants investigation has shown us that precisely similar decompositions can be brought about, not now by the whole plant-body, but by special cells or parts of it. The agent in the decomposition is the same, viz. protoplasm; the conditions are similar and the resulting products are strictly comparable. The secretion of an enzyme, which is a power exerted by both lowly and more highly organised plants, is a mark of differentiation within the living substance, just as in the slow movements of amoeboid protoplasm we recognise something which in the higher and more differentiated organism appears as the contraction of muscular fibre.

A fact which at first appears to constitute the microbes into a special class of fermenting organisms has several times been incidentally alluded to. That is the power which many of them exhibit of exciting more than one kind of decomposition. Lauder Brunton, Wood, and other writers already quoted, have shown that the same bacillus may saccharise starch and peptonise albumin; many others can cause the clotting of milk in addition. *Bacillus mesentericus vulgatus* has been shown to be possessed of diastasic, inverting, cytasic, and peptonising power, and to clot milk.

In these cases it is important to recognise the biological

peculiarities of the organisms. They exercise these powers according to the medium in which they are cultivated; many of them carry out the changes by the excretion of enzymes, and the formation of these depends in some way upon the stimulus of the culture medium. Vignal, who investigated the last-mentioned of these microbes, points out that the proportions of the several enzymes it forms vary very greatly according to the nature of the fluid in which it finds itself. The fermentation in these cases is seen to be strictly subordinate to the metabolism. We have noticed in preceding chapters that the secretion of particular enzymes by the higher animals is very largely influenced by the character of their diet. We may again mention in this connection the researches of Vassilief on the influence of the latter on the relative proportions of diastase and trypsin in the pancreatic juice. In fact the production of enzymes shows a certain dependence on the physiological and alimentary conditions of the organisms which secrete them, as has been pointed out by Duclaux.

These phenomena recall also the observations of Schiff and others as to the necessity of an absorption of some nitrogenous compound or peptogen before the secretion of the gastric pepsin takes place, which we have mentioned in an earlier chapter. It also reminds us of the behaviour of an Amœba, which only forms a food vacuole with its digestive fluid round an ingested nutritious particle. Any foreign substance which is not digestible does not become enclosed in such a vacuole.

CHAPTER XXII.

THE SECRETION OF ENZYMES.

WE have seen incidentally in the case of many enzymes, particularly those formed in well differentiated glands, that a certain stimulus is necessary before secretion takes place. In the case of saliva an increased flow can be excited by reflex nervous agency, as when an afferent nerve is stimulated by some sapid substance being placed upon the tongue, or when some appetising odour reaches the olfactory membrane of the nose. The taking of food into the mouth is followed by an outpouring of both gastric and pancreatic juices.

Similar phenomena can be observed in the cases of some vegetable secretions. The leaves of Drosera and of Dionæa do not pour out their enzymes unless they are stimulated by contact with the body to be digested. Clautriau has observed that if a pitcher of Nepenthes is severed from the plant its contents will not digest albumin, though another exactly similar one left in its normal attachment will do so rapidly. He has suggested that the secretion is controlled by the plant by something comparable to a nervous influence.

Apart however from actual nervous influences there is evidence to show that the absorption of nutritive materials has a very potent influence on the formation and discharge of the secretion. This has been shown by many writers, among whom may be mentioned Heidenhain, whose experiments on the secretion of gastric juice in a portion of the stomach isolated from the rest by Thiry's method, point to a material increase in the quantity poured out, in consequence of the absorption of food by the mucous membrane of the main portion of the organ. Schiff also has called attention to the increase in the flow which follows absorption of various nutritive bodies.

This form of stimulus has been observed to be also very advantageous in the case of various vegetable organisms. The formation of diastase in the pollen grains of Zamia was shown by the writer to be directly dependent on the absorption of sugar.

The observations already alluded to, which were made by Lauder Brunton and MacFadyen on the enzymes secreted by the bacilli they examined, also bear upon this point. They found that the character of the medium in which the microbes were cultivated determined whether the enzyme produced was diastase or trypsin. The absorption of a carbohydrate in the one case or a proteid in the other was apparently antecedent to the production of the ferment.

A curious case of the influence of abundant nutrition on the secretion of enzymes has been noticed by Dubourg. Certain yeasts will not either invert or ferment cane sugar under ordinary conditions. If however they are cultivated in a liquid containing abundant supplies of nitrogenous material together with glucose they become capable of doing both. If the yeast is withdrawn from such a culture medium while it is actively fermenting its sugar, and is washed free from all saccharine matter, it will set up a fermentation on being mixed with a solution of cane sugar. It appears in this case as if the abundant supply of nutritive matter excited the formation of the enzymes.

The apparently direct influence of the nature of the diet on the secretion will also be remembered here.

In some cases however the stimulus is of a different kind. Brown and Morris showed in their experiments on the secretion of diastase by the scutellum of the barley-embryo, that no enzyme is formed in its cells so long as the embryo is artificially supplied with solutions of nutritive carbohydrates, and is not obliged in consequence to obtain its nourishment from the starch contained in its endosperm. Wortmann showed that certain microbes produce diastase when starch grains are their only available food, but do not if sugar is offered to them with the starch. De Bary indicates a similar phenomenon in the case of *Bacillus amylobacter*, which forms no cytase so long as it can obtain glucose freely. If the supply of the sugar is stopped, it attacks cellulose, and for that purpose secretes the enzyme. Pfeffer has shown that the same thing can be noticed in the cultivation of three organisms, *Penicillium glaucum*, *Aspergillus niger*, and *Bacterium megatherium*. The production of diastase by all three diminished as the percentage of sugar was increased in the medium in which they were growing, though Aspergillus proved itself less sensitive than the others. Hérissey has shown that the secretion of emulsin by Aspergillus is similarly influenced.

In several of these cases the secretion of the enzyme appears to be provoked by diminished nutrition, approaching indeed to incipient starvation.

The process of secretion and the effect of these stimulations upon it can only be advantageously studied in the case of well differentiated glandular structures. Nor do all these offer equal facilities, but those are most favourable in which the process of secretion is both intermittent and fairly rapid when once initiated.

For many reasons the cells of the pancreas are the most suitable for a study of the secretory process. It is true that this organ elaborates several enzymes, and so far as we know they are all formed together in the same cells. Their formation is however a single process so far as can be ascertained.

Heidenhain studied the appearances presented by pancreatic cells under different conditions of nutrition. In the case of an animal (dog) which had been fasting for rather more than a day, each cell was seen to consist of two zones; one which abutted on the lumen of the alveolus, and one at the back of the cell towards the basement membrane. The inner was considerably the larger in area and was studded with fine granules. The outer was narrow and its substance was homogeneous. The nucleus of the cell was shrunken and corrugated, and was found at the border of the two zones. In the pancreas of another dog, which was killed during full intestinal digestion, the same two zones were evident, but the homogeneous outer one was much wider and the inner granular one was contracted, the granules being very much less numerous. The whole cell had suffered a diminution in size, while the nucleus had regained a spherical shape and was nearly central in position. A third pancreas, excised at the time when digestion had ceased, showed the outer zone again diminished and a marked increase in the granularity of the inner zone.

The times when the granularity was most marked showed that the latter was very probably connected with the presence of the enzyme, which appeared to be formed during the period of rest and extruded from the cell during that of activity. The change in size of the cell which accompanied the pouring out of the pancreatic fluid showed that the latter removed a considerable quantity of some soluble substance with which the cell had been distended.

The formation of the granules was synchronous with an enlargement of the cell and indicated the absorption of some material, nutritive in character, at the expense of which they were constructed.

Kühne and Lea watched the process of secretion in the pancreas of the living rabbit, and noted similar appearances to those described by Heidenhain; they associated the preparation of the enzyme with the formation of the granules, and the pouring out of the secretion with their disappearance.

If we turn to the salivary glands we find similar phenomena. In the mucous glands of the tongue of the frog we can see an outer region in which the protoplasm is clear, and a much larger inner region which is crowded with granules. In a serous gland in a state of rest the whole cell appears granular, there being no outer homogeneous zone. When the secretory nerve is stimulated, an outflow of the secretion follows, and this is attended by a diminution of the granularity. In the mucous gland the granules nearly all disappear; in the serous one they are removed from the outer part of the cell, which becomes homogeneous. As the granules are carried away the whole alveolus shrinks and the outlines of the constituent cells become more evident.

The same round of changes can be observed in the chief cells of the gastric glands, which secrete pepsin. In the various regions of the stomach of a mammal we can observe an even more definite relationship between the granules and the enzyme. In the rabbit we have an animal which shows this relationship remarkably well. In the cardiac or fundus region of its stomach the chief cells are very coarsely granular, the tubular gland appearing almost as a mass of granules. Towards the greater curvature this becomes less and less obvious; about the middle of this region the cells possess an outer homogeneous border: further on, as the pyloric region is approached the cells contain fewer and fewer granules, and the pyloric glands themselves are non-granular. This distribution corresponds exactly with that of the pepsin in the mucous membrane.

The regularity of the disappearance of the granules, beginning at the outer border of the cell and extending inwards towards the lumen, is characteristic especially of mammalian glands. In the lower vertebrates, including birds, snakes, and frogs, the formation of a non-granular outer zone cannot generally be observed, but there is instead a diminution of the number and size of the granules throughout the cell.

The reconstruction of the granules takes place with great readiness. The process appears to begin before the cell is completely emptied, and to continue during the latter part of the period of digestion, so that by the time this process is finished the cells have almost regained their original granular appearance.

In some cases, especially in the pyloric glands of the stomach, the secretion does not appear as granules, but is stored in a homogeneous form in the meshes of the protoplasm.

In comparing these different changes, which we see are in the main the same for all glands, we find that the process of secretion consists of a certain sequence of events. Starting with a cell almost entirely depleted, we have its growth at the expense of something absorbed from without; immediately following this growth there is the manufacture of the coarse granules which in the resting condition of the gland fill the cells; and lastly, at the onset of the outflow of the secretion, we have their solution and extrusion in the liquid which the gland pours out.

In some cells these processes go on so far successively that the sequence can be observed. It may be that all are proceeding simultaneously but at relatively different rates. In others the round of changes is the same, but they progress simultaneously and at rates so much alike that the sequence of events becomes very indistinct.

This difficulty of recognising a sequence of changes is still more marked in the case of vegetable secretions, which are usually very prolonged and probably hardly at all intermittent. They have been studied in a few cases by various observers, but our information is not nearly so complete as in the instances of the animal cells already described. The difference depends apparently on the character of the metabolism of the plant and the animal respectively. In the latter, digestion is an intermittent process, continually repeated; in the former the corresponding process, the utilisation of reserve materials in such bodies as seeds, tubers, &c., is a continuous and very gradual one.

The observations of Brown and Morris on the changes in the scutellar epithelium of the barley grain are among the most complete which have been made.

The epithelium covering the scutellum of the barley embryo is composed of columnar cells having a length of about ·03 to ·04 mm. and a breadth of ·01 mm. Their long axes are at right angles to the surface. The cell-walls are very thin and are not cuticularised. The cell-contents before the commencement of germination are very finely granular, and the nucleus, which is large and elliptical in shape, lies near the base of the cell, with its longer dimension across it. Within a few hours after germination begins the very fine granules in the protoplasm become much larger and coarser, and increase in number to such an extent that the nucleus, which was at first very conspicuous, is so obscured as to become almost invisible. This process is complete in about a day, and the granularity is maintained until the endosperm is almost exhausted of its reserve materials. At this point the epithelium shows a great falling off in its function as a secreting tissue; the protoplasm again becomes clear and transparent. A few granules only remain, and they are very small and highly refractive, showing very little resemblance to the coarse granulation existing during secretion. Another remarkable fact now noticeable is that the nucleus of the cell has entirely disappeared.

The continued secretion depends upon the cells being amply supplied with food. This is derived from the reserve materials of the endosperm, partly the carbohydrates on which the enzymes work and partly reserve-proteids obtained from the aleurone layer.

It is impossible to disregard the general similarity of feature between this prolonged process and the intermittent one characteristic of the pancreatic cells.

The secretory process has been studied by Gardiner in the cells of the glands which are situated on the leaf of *Dionæa*. As we have seen, when the leaf of this plant is stimulated by contact of its surface with some nutritive material, the leaf folds over and encloses the exciting body, and the glands pour out a digestive secretion which contains a proteolytic enzyme. The process of secretion is much more rapid than that of the barley embryo, and may be repeated more than once by the same gland.

Gardiner distinguishes four periods in the act of secretion—(i) one antecedent to secretion, when the cells are in the resting condition; (ii) a period during which the secretion is formed and extruded from the gland; (iii) a time when absorption of the digested material is effected; (iv) a period of recovery. In the first of these, a layer of protoplasm lines the wall of the cell, and surrounds a large central vacuole, filled with cell-sap. The protoplasm is extremely granular, especially round the nucleus, which is situated at the base of the cell. In many cases it is quite obscured by the granules. At the end of the second period the nucleus has moved to the centre of the cell and is surrounded by a layer of protoplasm, which is connected by bridles with the parietal film. The protoplasm has lost its granularity and has become clear and hyaline. The nucleus itself is much smaller than before.

Gardiner noticed also in the glandular cells of the tentacles of *Drosera* that in the resting condition of the glands the protoplasm was much more granular than it became after a period of activity or secretion. Similar results were obtained by Miss Huie.

The secretion of rhamnase by the raphe of the seed of *Rhamnus infectorius* is stated by Marshall Ward to be attended with the formation of granules. The cytase of the hyphæ of *Botrytis*, as described by the same observer, appears to be associated with a number of brilliant refringent granules which give proteid reactions. These are very apparent in the drops of fluid which the hyphæ exude when they come into contact with a fresh leaf surface; moreover they only appear when the hyphæ are secreting the enzyme.

If we pass from the more definitely glandular tissues to those which are the seats of intracellular digestion, the process of secretion is not nearly so easily traced. In a few cases it has been found to be associated with granularity of the protoplasm of the cell, but this is far from universal. Nor need we expect to find the stages so evident in these cases, for the mechanism of secretion is not nearly so well differentiated. The cell in which the process is to take place must not merely furnish the enzyme; it has also to absorb and perhaps to prepare the material on which the enzyme must work, and it has to dispose in some way or other of the products of the digestion. These various functions must necessarily push the process of secretion somewhat into the background, or at any rate obscure the appearances.

We have in Guignard's researches upon myrosin some facts brought to light which indicate an essential similarity of the process to that already described. The cells in the roots of the horse-radish and other plants which contain the myrosin can be recognised by their peculiar granular contents, which give a very marked reaction with Millon's reagent, the colour produced

26—2

being much more intense than that which is developed when this fluid comes into contact with proteids. Guignard has demonstrated a close correspondence between the degree of this granularity and the amount of enzyme in the cells.

The process of the secretion of diastase in the cells of foliage leaves is hardly at all known. Some researches carried out by A. Meyer seem to indicate that it does not arise in the ordinary protoplasm of the cell, but is formed in the chromatophore or chlorophyll grain. The same thing has been noted in connection with the parts of the plant which are not green, where the diastase appears to arise in the leucoplastids which form the grains of starch. The hypothesis is largely based upon the appearances presented by grains which have been partially dissolved by the enzyme. Meyer describes many cases in which the starch grains were attacked most energetically where the sheath formed by the stroma of the plastid was thickest. In *Iris* he frequently found the side of the grain eaten out to some extent where the leucoplast was in contact with it.

Meyer's view has been confirmed by Salter, who made a series of observations on the starch grains of *Pellionia*. He says with reference to these, that when he examined grains in process of solution, he found the action had proceeded furthest at the point where the plastid was situated. The latter had practically eaten its way, in many cases, into the substance of the starch grain.

Though in so many cases we are able to associate the secretion of enzymes with the formation of the coarse granules described, it by no means follows that the latter are the ferments. Indeed evidence lies ready to hand that this is not the case. If a pancreas is removed from a fasting dog immediately after death and at once extracted with an appropriate solvent, the resulting extract has little or no fermentative activity. If however it is kept warm for a few hours, and particularly if it is faintly acidified, the extract will be very powerful. If the extract made immediately after death is acidified and kept warm for some time, it will become active. The material extracted from the gland cells, which we have seen to be connected with, if not composed of, the granules, is not the enzyme, but something which rapidly gives rise to it when warmed with a trace of acid. This antecedent substance has been named "mother of ferment" or *zymogen*.

There appears to be for each enzyme a distinct antecedent or zymogen, in which form it is stored in the cells, ready to be transformed into the active ferment directly it is required. This change seems to take place immediately prior to, or perhaps during, its extrusion from the cell.

We see thus that we can recognize in the secretion of enzymes a gradual process, at least one antecedent body being formed. Are we able to investigate this process more closely?

In the recital of the changes in the cells, we have noticed almost always that the nucleus has appeared to play a part in the secretion, though what its function is has not so far been apparent. We may recall here the observation of Brown and Morris on the scutellar secretion of diastase. This is a process which is continuous after its first inception, and not intermittent like that of animal glands. It will be remembered that when the secretion has ceased and the granularity of the cells has disappeared, the nucleus is found to be disintegrated.

Some extended researches upon this point have been carried out by Macallum in connection with an enquiry into the distribution of iron and phosphorus in cells generally.

His experiments were carried out on the pancreas of several animals, by means of staining reagents which coloured differently the various parts of the cytoplasm and nucleus, and also of certain special reagents capable of detecting iron and phosphorus. Using eosin, he found that at a time coinciding closely with the commencement of the deposition of the zymogen granules in the cell, some material which stained conspicuously with that dye disappeared from the nucleus, and the cytoplasm simultaneously developed an affinity for the eosin. In other words, an eosinophilous substance diffused out of the nucleus into the cytoplasmic zone. Later, this same substance appeared to be removed from this region to be fixed in some way in the zymogen granules.

Macallum noticed further that as the cell was filling up

with zymogen granules the latter were largest at the border of the lumen of the tubule, while the smallest were at the edge of the granular area nearest to the nucleus. The granules appeared to increase in size after their first formation, at the expense of a substance in the protoplasmic area of the gland cell.

This relation between the nucleus and the zymogen granules is supported by a striking resemblance in chemical composition. The chromatin of the nucleus contains phosphorus in combination. When nuclei are digested by artificial gastric juice a residue containing phosphorus always remains undissolved. This is associated with the nuclein which they contain, the chromatin appearing to belong to the class of nucleo-proteids. The zymogen granules in the cells of the pancreas of *Diemyctylus*, after being freed from adherent lecithin, were also found to contain phosphorus, which however was hardly so intimately combined in them as in the nuclear chromatin. The protoplasm in which the granules lay in the inner zone of the cell, also gave the phosphorus reaction, but much less distinctly. The outer protoplasmic zone gave a reaction which was intermediate in depth, and which appeared to be due not to the cytoplasm itself, but to a substance contained in its meshes.

All these various constituents of the cell were found to contain iron in addition to phosphorus and in much the same proportions. Macallum found that in a large variety of secreting cells, glandular secretion was associated with the presence of an iron-holding cytoplasm. Among them may be mentioned the cells of the pancreas, the gastric and intestinal follicles and the parotid gland of *Amblystoma*; also the cells of the submaxillary glands, and the chief cells of the cardiac gastric glands of the dog and cat. In different stages of secretion the amount of iron-holding substance varied. When the cell was exhausted and the granules of zymogen were scanty the cytoplasm still gave the reaction, though only faintly. In fact the iron-holding area of the cytoplasm varied inversely as the granular zone.

As the result of his experiments Macallum puts forward the hypothesis that the nucleus forms out of its chromatin a material which he calls *prozymogen,* and extrudes it into the cytoplasm. Before extrusion it may be diffused in the nuclear substance, or collected into definite masses which he calls *plasmo-somata.* When extruded into the cytoplasm some of it unites with a constituent of the latter and becomes zymogen, which is soon aggregated into granules. The increased size of the granules as they lie in the inner zone is due to a further addition from the prozymogen of that region.

Macallum holds further that the chromatin of the nucleus is an iron-holding nucleo-albumin, in which the iron is attached to the nuclein.

In such cells the process of secretion appears therefore to be intimately associated with this constituent of the nucleus. At the same time the zymogen is not purely a nuclear product, the cytoplasm of the cell contributing to the later stages of its construction.

Macallum has found also that the aleurone-layer of the endosperm of the wheat grain gives the same reaction for iron as do the cells of the pancreas. We have already seen reason to attribute considerable secretory activity to this layer, parti-cularly in connection with the formation of cytase.

Macallum noticed a similar distribution of iron in some Protozoa. We have seen that it is probable that these uni-cellular organisms digest their food by the agency of enzymes which are secreted into definite vacuoles which surround the ingested particles. From his observations he holds it to be probable that there is the same relation as in the gland-cells between a zymogen and an iron-holding compound which is contained in the meshes of their cytoplasm.

The view that the secretion of the enzymes may thus be a gradual process passing through several stages has been put forward by other observers. It was suggested in 1882 by Langley in the following words[1]: "There are one or two further observations on the relation of pepsin to the gland cells which are suggested by some experiments of Brücke. Brücke pointed out that prolonged extraction with water does not take out all the pepsin from the gland cells; the remaining tissue

[1] *Journal of Physiology,* 3. 290.

when treated with dilute hydrochloric acid still gives a pepsin-containing extract. He further pointed out that, when the gastric mucous membrane is treated with dilute hydrochloric acid until it is quite broken up, the residue nevertheless when washed and again treated with dilute hydrochloric acid still gives a pepsin-containing extract. We could scarcely imagine that this could be the case if the zymogen existed in the cells in one state only. It necessitates a conception of the gland cells which seems to me indeed on general grounds almost necessary. I conceive the matter thus. The protoplasm of the gland cells does not at one swoop form zymogen as it occurs immediately previous to its conversion into pepsin, but forms certain intermediate bodies in which the zymogen radicles become more and more isolated. Since the zymogen contains the radicle of the ferment, the ferment will be obtained with greater difficulty from the imperfectly elaborated zymogen, i.e. as we ascend from the final mesostate to protoplasm, the ferment will be split off less and less readily. The last traces of ferment then which are obtained by repeated extractions, I take to arise from the splitting up of substance which was on the way to be converted into zymogen."

We may now turn our attention to the relation between zymogens and enzymes, and to the differences which may be shown to exist between them.

We have pointed out already that the evidence in favour of the granules consisting of zymogen and not of enzyme is based upon a comparison of the fermentative power of a fresh tissue before and after being warmed with a dilute acid. A more convincing proof has been offered by Langley in connection with his researches on the histology of the mammalian gastric glands. He ascertained that pepsin is rapidly destroyed when warmed gently with a 1 per cent. solution of sodium carbonate, and applied this peculiarity in testing the contents of a mucous membrane. In a very striking experiment he used an extract of the mucous membrane of the fundus and greater curvature of the stomach of a rabbit, which was powdered and macerated in water for $2\frac{1}{2}$ hours at 32° C. He divided the filtered extract into two parts, and warmed one of them at 32° C. for

50 minutes in the presence of ·06 per cent. of hydrochloric acid, neutralising it with dilute alkali at the end of that time. This would effect the conversion into pepsin of any zymogen present in the extract. At the same moment he added to the other the same amount of acid and alkali simultaneously. The two extracts are referred to by him as P and Z. They only differed in that P had been warmed with the acid before the alkali was added, while Z had not. He then added to 10 c.c. of each, an equal volume of 2 per cent. solution of sodium carbonate, and kept them at 39° C. for 15 minutes. They were then removed, neutralised, and made up to ·2 per cent. of hydrochloric acid, and some fibrin was added to each. P failed to digest more than a trace, while Z was eminently peptic.

The substance extracted from the glands had been left unaffected in Z by preliminary treatment with acid; it was not destroyed by the sodium carbonate, for it was rendered active by the subsequent acidification and digested the fibrin; the same substance in P had been acted on by the acid for some time before being subjected to contact with the alkali, and was evidently destroyed by the latter, as the neutralised and subsequently acidified liquid then possessed no peptic power. The extracted material was capable of giving rise to pepsin on warming with an acid, as shown by Z. In P this pepsin, formed during the first warming with dilute acid, was destroyed by the alkali. The substance was not pepsin at the time of its extraction, or it would have been destroyed in Z also. In the condition in which it was present in the original gastric extract, the sodium carbonate had no action on it.

By this experiment Langley showed that the granular matter thus extracted from the glands is not pepsin but an antecedent or zymogen, and further demonstrated a difference between the latter and the enzyme with regard to their behaviour with dilute alkalis.

The zymogen, being the antecedent of pepsin, has been called *pepsinogen*.

Langley showed by similar means that the gastric glands of the dog, sheep, mole, snake, frog, and newt contain pepsinogen and not pepsin.

Pepsinogen can be converted into pepsin not only by dilute acids but by dilute alkalis also. The latter must be very dilute or the pepsin is destroyed almost as fast as it is formed.

Heidenhain showed in 1875 that the secretion of the pancreas contained trypsinogen and not trypsin, and that the former could be converted into the latter by the action of acids. A watery extract of the gland slowly acquires tryptic powers.

Hammarsten showed in 1872 that the rennet enzyme of the stomach is also derived from a zymogen. Langley confirmed this observation in 1881. He prepared an extract of the mucous membrane, using ·1 per cent. solution of sodium carbonate as the solvent. When a part of this extract was made acid, warmed for 15 minutes, and subsequently made alkaline again, it rapidly caused clotting in milk, while the original extract unwarmed with acid possessed no such power. It was necessary to work with a faintly alkaline solution, as a very little acidity causes precipitation of the casein, as already shown. The alkalinity had however to be very slight, as rennet zymogen is rapidly destroyed by as little as ·5 per cent. of sodium carbonate. There is not the difference in this respect between the zymogen and the enzyme as is the case with pepsin and its antecedent.

Lörcher has recently obtained similar results. He used a glycerin extract of the dried gastric mucous membrane, which he found able to coagulate in 17 minutes 20 times its volume of milk to which 1 part to 10000 parts of hydrochloric acid had been added. When he allowed the acid and the gastric extract to stand together for 2 hours before adding them to the milk, coagulation took place in 2 minutes.

The evidence for the existence of the zymogen of salivary diastase is not so complete, though analogy with other secretions points to its existence, and it is strongly suggested by the appearances in the cells of the salivary glands already described. A certain amount of direct evidence however is not wanting. Goldschmidt has demonstrated its existence in the saliva of the horse. Certain facts which were ascertained by

the writer point to its presence in human saliva also.   It is not
easy to determine it in the latter, as the secretion contains so
much ready-formed diastase.   Some saliva was secreted, freed
from mucin, and considerably diluted and ·2 per cent. of potassic
cyanide was added to the liquid.   This antiseptic was found to
be very valuable in preserving saliva, such a preparation having
been found to retain its activity unimpaired for several months.
The saliva so prepared was divided into two portions; one of
them was kept for 21 days in an incubator, the temperature
of which was maintained at 38° C., while the other remained at
the laboratory temperature.   At certain intervals 2 c.c. of each
were digested with 20 c.c. of 1 per cent. starch-paste and the
products of hydrolysis were titrated with Fehling's fluid.

The following table gives the result of the experiment.

Time of exposure to 38° C.	Cupric oxide reduced by products of action of 2 c.c. of the warmed saliva on 20 c.c. of starch-paste	Cupric oxide reduced by products of action of 2 c.c. of unwarmed saliva on 20 c.c. of starch-paste
days	grms.	grms.
1	·0565	·045
3	·0697	·045
8	·067	·062
9	·0706	·0708
15	·066	·095
21	·032	·094

From these experiments, it is probable that the saliva
contained, in addition to a certain amount of diastase, a small
quantity of zymogen, which exposure to a temperature of
38° C. converted into the enzyme almost completely in three
days.   Possibly a little remained unchanged till the ninth day.
After 15 days at this temperature the enzyme began to change,
losing its hydrolytic power, which at the end of the twenty-first
day was much less than at the beginning of the observation.   At
the temperature of the laboratory, which would be about 18° C.
the zymogen did not begin to be converted into the enzyme till
after the third day, and it then progressed gradually for the
rest of the time it was under observation.   There was no
destruction of the enzyme at that temperature during the

whole of the period, so that the curve of its activity rose somewhat higher than did the corresponding curve of the warmed extract, in which, during part of the experiment, both processes were at work.

Some experiments made by the writer upon the action of light upon dilute saliva also point to the existence of zymogen in the secretion.

A preparation of saliva made in the way already described, and preserved by the same antiseptic as in the experiments quoted, was exposed for several hours to the rays of the electric arc at such a distance that the heating effects of the lamp were imperceptible.

By the use of certain screens the effect of different parts of the spectrum were examined separately. The screens were prepared according to the directions of Landolt[1], given in his work on rotation-dispersion. The coloured liquids employed for this purpose enabled the rays of the spectrum to be divided into seven sets, which may be called roughly the infra-red, red, orange, green, blue, violet, and ultra-violet. The red rays included rays of wave-length $710—645\mu\mu$; the orange $645—585\mu\mu$, the green $585—500\mu\mu$, the blue $500—430\mu\mu$, and the violet those of the visible spectrum beyond $430\mu\mu$.

After the exposure, during which the temperature of the saliva was registered by a maximum thermometer, the diastasic power of the preparation was ascertained by allowing 5 c.c. to digest with 20 c.c. of soluble starch in 1 per cent. solution. A control preparation was in all cases examined, which was composed of the same saliva suspended with the other in front of the lamp but protected from the light by an opaque screen. The table on page 389 shows the effect of the action of the light.

The results must be received with a certain amount of caution, as with such different screens it was not at all easy to get the intensity of the transmitted light the same in each case, though as much care as possible was taken. Slight differences in intensity in any one case were found however

---

[1] Landolt. "Methode zur Bestimmung der Rotationsdispersion mit Hülfe von Strahlenfiltern." *Ber. d. deut. chem. Gesell.* 1894, p. 2872.

not to have a very marked effect during the time of the exposure.

Band examined	Cupric oxide in grms. reduced by solution exposed to band, after correction for infra-red rays	Cupric oxide in grms. reduced by unexposed solution	Increase or diminution of amount of CuO reduced. Diminution indicated by − sign	Increase or diminution of diastase per cent.	Mean increase or diminution per cent.
Red 720—640 μμ	·0557	·0354	+ ·0203	+ 57·4	
	·0565	·0393	+ ·0172	+ 43·7	
	·0921	·0568	+ ·0353	+ 62·1	
	·0518	·0343	+ ·0175	+ 51·0	+ 53·5
Orange 640—585 μμ	·0338	·0323	+ ·0015	+ 4·6	
	·0296	·0282	+ ·0014	+ 4·9	+ 4·75
Green 585—500 μμ	·0285	·0333	− ·0048	− 14·4	
	·0107	·0129	− ·0022	− 17·0	− 15·7
Blue 500—430 μμ	·0282	·0242	+ ·0040	+ 16·5	
	·0711	·0568	+ ·0143	+ 25·1	+ 20·8

	Cupric oxide reduced by solution exposed to infra-red rays	Cupric oxide reduced by unexposed solution	Gain	Increase of diastase per cent.	Mean
Infra red	·0337	·0323	+ ·0016	+ 5·0	
	·0323	·0282	+ ·0047	+ 16·6	+ 10·8

Less careful separation of the rays of light showed that the red end of the spectrum also increased the amount of diastase in a freshly-prepared extract of malt.

From these experiments it seems probable that saliva contains a zymogen which can be converted into diastase with considerable readiness by exposure to the red rays of the spectrum. The infra-red, orange and blue rays have a similar but less pronounced effect.

The other rays, particularly those of the violet and ultra-violet regions, have the power of destroying diastase almost as effectually as temperatures above 70° C.

The zymogens of the vegetable enzymes have not been very closely studied up to the present time. Their existence was first established by Vines in his experiments on *Nepenthes*. He treated some pitchers of this plant with dilute acetic acid (1 per cent.) for 24 hours before extracting them with glycerine, and at the same time extracted other similar pitchers with glycerine without preliminary treatment with acid; the first extract possessed greater proteolytic powers than the second, leading him to infer that a zymogen was present in the glands, which was converted into an enzyme by the acid.

The writer's experiments on the antecedent of the enzyme in the resting seed of the Lupin also indicate a similar condition in the cells, though its identification is not so easy, as the acid treatment usually adopted for zymogen conversion is not available, the digestion having to be a prolonged one and to be conducted in an acid medium. The zymogen was ascertained to exist by a modification of the treatment adopted by Langley in the case of the pepsinogen of the stomach, based upon the destruction of the enzyme by dilute alkalis, which do not decompose the zymogen.

Inulase can be more easily shown to exist as a zymogen in the resting tuber of the artichoke. In the writer's experiments some pieces of full-grown tubers were kept at a temperature of 35° C. for 24 hours. An extract then prepared from them was found to convert inulin into sugar, while an extract made from other pieces of the same tubers without warming, was inert. When some of this latter extract was subsequently warmed for a time with a solution of acid-albumin in ·2 per cent. hydrochloric acid, some enzyme was developed in it, though less than was obtained by warming the tubers alone before extraction as described. Treatment with acid alone was useless in the case of inulase, as the quantity needed to convert the zymogen was sufficient to destroy any ferment liberated from it.

The lipase and rennet of the castor-oil seed have also been shown by the action of acid to exist in the zymogen condition until the onset of germination, the former of them being convertible into the ferment also by the prolonged action of water without acidification.

Brown and Morris mention that the quantity of diastase secreted by the epithelium of the scutellum of the barley grain is increased 20 per cent. by the presence of very dilute formic acid. Baranetzky found that a freshly-prepared extract of the leaves of *Melianthus major* was incapable of hydrolysing starch, but that after standing a few days it possessed diastasic powers. He noted the same thing in the case of the tubers of the potato.

Experiments made by Reychler, and by Lintner and Eckhardt, point to the existence of a zymogen in the cells of the grain of wheat. They found that the action of a dilute acid upon the gluten of wheat gave rise to a diastasic enzyme. The experiments do not seem however to be quite conclusive, as the gluten might have contained a little of the diastase of translocation, which we have seen is present in the ripe endosperm, though in very small amount.

Frankhauser observed that during the germination of barley small quantities of formic acid could be detected in the grains. This may be regarded as important in discussing the increase of diastasic power attending germination, as such an acid would probably transform a zymogen into ferment.

The existence of a zymogen which is the antecedent of the fibrin ferment or thrombase has been suggested by many observers, but a satisfactory proof of it can hardly yet be said to be forthcoming.

The most definite statements on the subject which have so far been published are those of Pekelharing, who claims to have been able to prepare a substance from oxalated plasma which in many respects resembles a globulin, but is devoid of fibrino-plastic properties; after contact with calcium chloride it acquires the power of inducing coagulation of fibrinogen solutions. Pekelharing regards this as a zymogen; he says that on combustion it yields an ash which contains little or no calcium, while thrombase itself is rich in that metal. Pekelharing holds this substance, like Halliburton's thrombase, to be a nucleo-proteid. He found it possible to precipitate it from plasma by the addition of acetic acid. According to him, a number of bodies, first examined by Wooldridge, and named by him *tissue-fibrinogens*, which are capable of causing coagulation of the blood in the living blood-vessels, will yield thrombase on treatment with calcium chloride. Pekelharing considers that these bodies contain the zymogen.

He explains the action of peptone in inhibiting coagulation of the blood by the hypothesis that this proteid has a strong affinity for calcium compounds, and that when it is injected into the blood it prevents the interaction between those which the blood contains, and the zymogen which is also present. Consequently thrombase is not formed and the blood does not coagulate when shed. He quotes in support of this view a statement of Munk that soaps which combine with calcium compounds produce similar symptoms to those caused by peptone when injected into the blood-vessels. All the deleterious effects which follow the injection of peptone are obviated by injecting a solution of some salt of calcium at the same time.

Pekelharing suggests that intravascular clotting is caused by the rapid conversion of zymogen into ferment. He says that peptone will restrain coagulation in intravascular plasma if added so rapidly that the zymogen has not time to combine with the calcium salt to form the ferment.

This view is not however in accordance with the fact which has been noticed by more than one observer, that injection of very strong solutions of thrombase into the blood-vessels of a living rabbit is not followed by intravascular clotting.

According to Pekelharing the conversion of the zymogen into thrombase is effected by an actual combination with the calcium salt. He says further that fibrin itself is a calcium compound, and that the main action of the ferment appears to be the transference of the calcium to the fibrinogen.

Hammarsten endorses Pekelharing's view as to the relation of the zymogen to the thrombase, but he opposes the theory that fibrin is a calcium compound of fibrinogen, as he has found that both fibrin and fibrinogen contain the same amount of calcium.

The points of difference between zymogens and the enzymes to which they give rise have not been very completely examined. Langley and Edkins have made a comparison between pepsinogen and pepsin, and have found as might almost have been expected that they have much in common, such differences as exist being rather those of degree only. Their behaviour with regard to dilute alkalis has already been mentioned. Both are ultimately destroyed by alkalis and by alkaline salts, but the destruction of pepsinogen is much slower than that of pepsin. In the absence of acids, which rapidly convert the zymogen into the ferment, pepsinogen is fairly stable. In neutral and faintly alkaline solutions its conversion is very slow, and in a glycerine extract it may remain unchanged for years.

When carbon dioxide is passed through a solution containing pepsinogen the latter is destroyed with some rapidity, but little remaining when the stream of gas has been maintained for an hour. The rate of destruction is increased by the presence of a small quantity of magnesium sulphate, about ·1 per cent. being the most efficient proportion. Traces of acetic acid or of sodium carbonate have the same effect. Peptone on the other hand greatly delays the destruction, and when about ·25 per cent. is present, prevents it altogether. Other proteids have a similar effect, but they are not so potent in this direction as peptone.

Carbon dioxide has much less effect on pepsin.

Pepsinogen, like pepsin, is rapidly destroyed by heating its solution to about 55°—60° C.

CHAPTER XXIII.

THE CONSTITUTION OF ENZYMES.

THE knowledge which we have as to the constitution or composition of enzymes is exceedingly scanty. This is owing in great measure to two causes. In the first place they are apparently very unstable bodies and undergo decomposition with great readiness. The methods of preparation which have been employed to isolate them show a considerable variety of detail, but they seem to agree in this, that they are all accompanied by a great loss of ferment power, the enzymes being to a large extent destroyed during the preliminary extraction or the attempts at subsequent purification. In the second place a difficulty has to be encountered which is very far reaching, and goes beyond manipulative details. We have no criterion of their purity, and it is consequently impossible to say whether any of the processes so far adopted has prepared a really isolated product.

From the similarity of their action to that of the living protoplasm, which we have examined in a previous chapter, the idea is at once suggested, that they are very possibly either proteid in character, or at any rate not very different from proteids. Their instability, the changes which they undergo on heating, and their general behaviour with dilute acids and alkalis, also favour this hypothesis.

So far as various observers have been able to prepare them free from admixture with known substances, they have, with only a few doubtful exceptions, been found to give the general reactions for proteids. Their solutions at any rate assume the yellow coloration on treatment with nitric acid and ammonia which is known to physiologists as the *xanthoproteic* reaction.

Reactions which are peculiar to them, and so may be relied on as distinctive, are altogether absent, and the only evidence we can obtain of the presence of any of them in a solution is the power of the latter to bring about the changes characteristic of the particular enzyme it is suspected to contain.

Such characteristic reactions have been sought, and not infrequently various observers have announced their discovery, only however to find that further researches have demonstrated the unsatisfactory character of the suggested method.

Among the writers who have claimed to have ascertained definite reactions for the enzymes may be mentioned Weisner. He says that a characteristic colour reaction may be obtained by heating them with a solution of orcin in alcohol in the presence of strong hydrochloric acid. Nor does he stop there; he says that different enzymes give distinctive colours, and that they can be discriminated by his method, diastase in particular giving a bluish-violet.

Guignard made investigations in the same direction, and states that by the action of hydrochloric acid alone, the enzymes can be distinguished as a class and many of them identified. He boiled 1 centigramme of different enzymes with 1 c.c. of pure hydrochloric acid, and says that he found diastase yield a red which gradually turned brownish; emulsin gave a violet, papaïn an orange-red, and trypsin a greenish yellow.

These colour reactions have been found however to be characteristic of many other bodies besides the enzymes. Reinitzer has shown that dextrin, maltose, and lactose, all give similar colours when treated in the way suggested. Weisner's colour reactions have been proved to be shared by a very large number of substances, many of which occur especially in such vegetable tissues as also yield enzymes. The orcin reaction is shared by nearly all carbohydrates, and appears to be due to the production of furfurol during the treatment, which is necessarily somewhat drastic. Udransky obtained the same reaction from various proteids. Probably in Weisner's preparations the colours were given by some of these bodies which had not been separated from the enzymes, and were not characteristic of the latter at all.

Guignard again describes the action of hydrochloric acid alone on various proteids, and shows that they give colour reactions much like those which he has attributed to the enzymes.

Guignard has called attention to certain reactions yielded by the cells which contain myrosin. When a section in which these cells existed was heated with a small quantity of Millon's reagent (which is a mixture of the two nitrates of mercury and an excess of nitric acid), the cells in question rapidly assumed a more or less vivid rose colour, while the protoplasm of the surrounding cells was stained much more slowly, and so great a depth of tint was never attained. The contents of the myrosin-containing cells further became precipitated in granular form under the action of the reagent, so that they stood out quite distinctly among the rest.

Striking however as this reaction is, it cannot be regarded as undoubtedly characteristic of the enzyme, for it might equally well be due to some other constituent of the cells. It may suffice to point out where the myrosin is, but it may be attributed to something accompanying the ferment with as great a plausibility as to the latter itself.

It was ascertained by Schönbein in 1868, that when diastase is mixed with a little of the resin of guaiacum and a little peroxide of hydrogen added, a somewhat vivid blue coloration is produced. Other observers have extended this observation to other enzymes, and by many workers it has been assumed to be an unfailing test for the various members of the group, though it has not been thought to give distinctive reactions for them individually. Lintner showed in 1886 that the reaction is more delicate when an alcoholic solution of the guaiacum is employed instead of the resin.

The reaction, like the others described, has been ascertained to be by no means characteristic of diastase. It has long been one of the tests for blood-stains, and has been used in the examination of urine for the presence of albuminous matter. A careful examination of its validity as a reaction for enzymes was made in 1897 by Pawlewski, who found that a similar coloration to that given by diastase was caused by peptone, and by albumin and other native proteids, as well as by gelatin. Even the addition of hydrogen peroxide alone to guaiacum tincture gave a blue colour, which appeared at once if the solution was warm, but took a little time to manifest itself at a lower temperature. Other oxidising agents than peroxide of hydrogen, such as ozone, chlorine, nitric acid, and permanganate of potassium gave a blue colour with the tincture.

As a special test for diastase or any other enzyme the reagent is useless, for no conclusion can be drawn as to the nature of the body present, even when the blue tint appears.

If we survey all these reactions which have been put forward as means of identifying enzymes, we see that they do not give any evidence which can weigh very heavily against the hypothesis of their being proteid in nature. Weisner's preparations of some of the ferments were probably not very pure, being mixed with carbohydrate as well as proteid matters.

It has been suggested that the zymogens are proteids, and that the enzymes arise from them by a process of decomposition consequent on oxidation. Heidenhain thought that the zymogen consisted of the ferment in combination with an albuminoid body. Many observers have advocated the view that the enzymes are themselves proteids, Loew in particular considering them as allied to the peptones. In support of the hypothesis we may advance the fact that when an extract containing an enzyme has all its proteids removed by precipitation, the filtrate possesses little ferment power, the latter being diminished in many cases in proportion as the proteid is thrown out of solution.

Too much stress must not be laid however on this observation, as it is constantly found that when a precipitate, even of inert matter, is caused in a solution containing an enzyme, the latter is generally carried down with the precipitate. We have seen that this property is taken advantage of in several methods of purifying solutions of the enzymes, particularly in Brücke's process of preparing pepsin.

A more striking fact is that the temperatures at which so many enzymes are destroyed correspond very closely to the points at which proteids occurring with them are coagulated.

Most of the methods of preparation which have been described are such as would throw proteids out of solution. Two very prominent means of precipitating enzymes we have seen are the addition of excess of alcohol to the solution, and its saturation by neutral salts. These methods precipitate the proteids, and the enzymes are thrown down with them if any are present. The experiments do not show whether the proteid and the enzyme are identical, or whether the latter is merely mechanically associated with the former. It is remarkable however that prolonged exposure to strong alcohol coagulates the proteids, and that the same treatment gradually destroys the power of the enzymes.

In some cases observers have found the association of a particular enzyme and a certain proteid to be so constant and so close that they have been led to the view that the two are identical.

Osborne has studied the association of diastase with proteids in the cereals, wheat, rye, and barley. In them all he has found an albumin which he has named *leucosin*, besides a globulin and a proteose. All ferment extracts which he prepared from any of these cereals contained a mixture of these proteids, but when they were separated by various methods the greatest diastasic power clung to the leucosin. The proteose was found to be associated with a certain proportion of the enzyme, but much less than the leucosin. The quantity of leucosin and the fermentative activity also varied together, though not with very great exactness.

As the result of his experiments Osborne came to the view that the diastase is either the leucosin or a compound of the latter with some other body, presumably the proteose ; and that this compound breaks up on being heated and yields coagulated albumin, some free albumin being coagulated at the same time.

The view cannot be considered proved, as certain of his solutions which contained only the proteose possessed a certain, though relatively feeble, diastasic power.

Chittenden has shown that the bromelin and the rennet ferment which are both present in the juice of the pineapple are probably associated with two definite proteids, though he does not assert that either enzyme is identical with such proteid. Bromelin, the proteolytic enzyme, always occurs in close connection with a proteid which exhibits most of the characters of a proto-proteose. It is precipitated from a neutral solution by saturation with either ammonium or magnesium sulphate, or with sodium chloride. It is soluble in water and consequently is not precipitated by dialysis. It is further not coagulated by long contact with strong alcohol, and its aqueous solution is very incompletely precipitated by heat. In its behaviour towards nitric acid and heat and in some other ways it does not show the typical proteose reactions, and it contains a smaller amount of nitrogen than most proteids.

The rennet enzyme is more closely associated with a body whose reactions agree in the main with those of hetero-proteose. It is less soluble in water, requiring for complete solution a little neutral salt. It is more completely precipitated by heat than the proto-proteose described.

The closeness of this association in both cases may possibly indicate that these proteids are actually the enzymes in question, but the identity cannot be considered conclusively proved.

Lintner prepared diastase in what he considered to be a pure condition, and from the results of analyses of his preparation he concluded it to be a proteid, but not one of the ordinary class. He found that it contained only two-thirds as much nitrogen as, and less carbon than, the latter. It differed also in not giving the biuret reaction. Lintner hence suggests that the enzymes constitute a special class of proteids.

Wroblewski has come to the same conclusion.

Buchner associates his new zymase with a proteid in the yeast which is coagulated at 45° C. At this temperature the zymase also is destroyed. When this proteid has disappeared from the extract in consequence of the action of the proteolytic enzyme of the yeast, the power of producing alcohol is lost. Buchner does not however claim that the proteid is the zymase.

None of the researches quoted can be considered to go very

far in the direction of proving the proteid nature of the enzymes. All can apparently be explained equally well upon the hypothesis that the enzyme in any case is closely associated with the proteid. We have remarked several times that the soluble ferments are very easily carried down by a precipitate formed in the liquid in which they are dissolved. When we consider further that there is every reason to believe that an extremely small quantity of one of them is capable of causing the transformation of an enormous amount of material, we may well believe that such an association with a proteid would be detected only with very great difficulty by any method of destructive analysis. At the same time it is not impossible that they may be proteids which have not yet been isolated, as the very small quantity present mixed with other proteids might very well escape separation.

It is very difficult to accept the view that they are the proteids with which their particular activities have been associated as already described. In many cases these can be prepared without possessing any power of setting up the fermentations.

The variety of proteids with which they have been found militates still more forcibly against this view. According to Osborne, diastase is associated with a particular albumin, *leucosin*, and to a less extent with a proteose. Chittenden's enzymes occur together with a proto-proteose and a hetero-proteose respectively. Halliburton and Pekelharing associate thrombase with a body which was at first considered a globulin, though now it is held to be a nucleo-proteid. Many writers have connected other enzymes with forms of coagulable proteids, supporting their contention by the observation already quoted that the enzymic power is lost at the temperature of heat-coagulation, or precipitation, of the proteid.

It seems altogether improbable that some enzymes should belong to one group of proteids, and some to another. It is much more reasonable to predict uniformity of composition for them, and to agree with Lintner that they must form a separate class by themselves.

The connection between loss of fermentative power and

heat-coagulation cannot be supported in the case of proteoses, which are not coagulated by even the temperature of boiling water. We have no evidence so far that a proteose is changed at all by heating it to such a point as would destroy an enzyme.

Osborne's definite identification of the diastase of cereals with leucosin is open to still further objection. Leucosin is a proteid which has only a limited range of occurrence, and many tissues which yield active diastasic extracts do not contain it. Leucosin can therefore represent only one form of diastase, and as other forms exist there must be other diastasic proteids, if the enzyme has a proteid constitution. When we consider the wide distribution of the ferment and the very varied and variable composition of the proteids in different plants, this view almost forces us to the conclusion that diastase is not at all of constant composition, a view which it is very difficult to accept.

From Osborne's experiments it appears that there is no very satisfactory ratio between the amount of coagulable albumin and the diastasic power; though the latter is higher the more albumin is present, there is no numerical relation between them. This we should hardly expect to be the case if the leucosin were itself the diastase.

This lack of correspondence between the quantity of proteid and the diastasic power is very striking when we compare malt-extract and saliva. The latter is the more active in hydrolysing starch, but the former contains a far larger percentage of proteids. The difference in behaviour between the two fluids does not correspond to this difference in composition unless we admit a great variation of hydrolytic power between unit quantities of the two, which seems unlikely.

Other observers also have noticed that the ferment power, particularly in weak extracts, does not vary directly with the amount of proteid present. The writer investigated this point with some care in the case of vegetable rennet prepared from the castor-oil seed. The amount of proteid was determined by regular dilution of a standard extract; and the result showed that the coagulating action on milk diminished much more rapidly than the quantity of proteid. In a typical experiment it was found that ·5 c.c. of a particular extract clotted

5 c.c. of milk in four minutes, while ·25 c.c. took eighty-one minutes and ·125 c.c. took five-and-a-half hours.

Reychler's experiments have been considered to point to proteid either as an antecedent of diastase or as the enzyme itself. He found that by extracting the gluten of flour with a dilute acid he obtained an amylolytic solution. His results are open however to other interpretations than the one suggested. His dilute acid may have extracted some of the residual diastase of the endosperm from the flour, or some zymogen may have been mixed with and incorporated in the gluten, and have been converted into diastase by the solvent employed.

Recently Wroblewski has claimed to have shown that diastase is proteid in nature. As he prepared it he found that the greater part of his material consisted of carbohydrate matter, which yielded arabinose when boiled with acids. If freed from this, the active constituent was found to have all the properties of a proteid.

From what has already been advanced, it is evident that in discussing this point very little reliance can be placed upon analyses of supposed pure products. So far however as these analyses go they do not support the proteid theory very satisfactorily. The difficulty of admitting that what was analysed was the actual enzyme is very great.

The purest trypsin which Kühne was able to prepare had the following percentage composition:—

Carbon	47·22—48·09.
Hydrogen	7·15— 7·44.
Nitrogen	12·59—13·41.
Oxygen	31·31—29·20.
Sulphur	1·73— 1·86.

This differs from the analyses of the proteids in the percentages of the carbon and nitrogen especially, Henniger's peptone containing 52·28 of the former and 16·38 of the latter in 100 parts.

According to Würtz and Bouchut papaïn agrees much more closely with proteids, containing 52·48 per cent. of carbon and 16·59 per cent. of nitrogen.

Loew has concluded that malt diastase, and both the diastase and the trypsin of the pancreas, have a percentage
composition very similar to that of proteids. On the other
hand Lintner's purest diastase only contained 9·9 per cent.
of nitrogen.

We may oppose to contentions based upon quantitative
analyses of this nature, certain considerations suggested by the
reactions of solutions of certain of the enzymes, when these
have been subjected to such processes of purification as have
been devised.

Pepsin as prepared by Brücke in the way described at
page 177, was stated by him to give none of the reactions
characteristic of proteids and to be precipitable only by the
acetates of lead. It yielded no trace of opalescence on the
addition of tannic acid, though this reagent is capable of
detecting one part of proteid in 100,000 parts of solvent.

Hammarsten states that after purification his preparation
of rennet did not show the xanthoproteic reaction; was not
precipitated by alcohol, nitric acid, tannin, iodine, or normal
lead acetate; was precipitated by basic lead acetate.

Dastre has shown that the enzymes are not completely
insoluble in alcohol, which is inconsistent with their being
coagulable proteids.

Foster showed in 1867 that a diastasic solution could be
prepared from natural deposits of urates, with or without
previous washing in alcohol, and that this solution gave no
proteid reactions.

Other enzymes as we have seen can be prepared from the
urine.

Possibly in all these cases the quantity of enzyme was too
small to give the reactions, though sufficient to exert a certain
hydrolytic power. It must however be remembered that the
xanthoproteic reaction is exceedingly delicate.

Some serious objections to the view that enzymes are
proteids, can be based upon the action of light upon them. So
far investigations on this point have been carried out only in
the case of diastase, which was submitted to a close experimental
examination by the writer a few years ago.

It was established by Brown and Morris in their paper on the diastase of foliage leaves, that there is a diminution in the amount of enzyme these organs contain after a period of brilliant illumination. This is particularly noticeable when comparative estimations are made from leaves gathered respectively in the early morning and in the evening. The latter always show a diminution of the amount of diastase, and the diminution shows a certain relation to the amount of bright sunshine which the leaves have received during the day.

The writer examined leaves in the living condition, which were gathered in the early morning and exposed to sunlight for several hours, one half of each being shaded by a covering of blackened paper. The sunlight was found to have a powerfully destructive influence on the enzyme. The same result was obtained when the light was allowed to pass through a watery extract of the leaves, instead of through the living tissue. The diastases of malt-extract and of saliva were similarly affected. The light source also was varied, a very powerful electric-arc lamp being sometimes employed instead of sunlight. The effect was the same in this case also.

The deleterious effects of the illumination were chiefly due to the violet and ultra-violet rays of the spectrum.

Light has not so far been ascertained to have any deleterious influence on proteids. It is very improbable that it ever exercises such an influence, for in the experiments described the writer ascertained that the addition of a coagulable proteid to the diastasic solution before exposure had a very great protective effect, and up to a certain point, the degree of protection increased *pari passu* with the amount of proteid added. The proteid itself did not appear to be affected by the illumination.

The difference of behaviour of enzymes and proteids towards alcohol may also receive some attention. When coagulable proteids are exposed to prolonged contact with this reagent they pass into a condition which is almost if not quite indistinguishable from heat-coagulation. Enzymes are not destroyed by alcohol with the facility with which they are decomposed by heat. We have seen that one argument in

favour of the hypothesis of their proteid constitution is that
they are destroyed at the temperature at which the heat-
coagulation of the proteid takes place. If the argument is
valid and the two bodies identical, the coagulation of the
proteid induced by alcohol ought also to destroy the enzyme.
But it does not do so, or at any rate a diminution of enzymic
power does not take place with the same speed as the coagulation of the proteid. In many of the methods employed
for the extraction of enzymes it is usual, as we have already seen,
to free them from large quantities of proteid matter, by precipitating both together by means of alcohol; after allowing the
precipitate to stand under the spirit for a long time, a strongly
enzymic extract can be prepared from the coagulated proteid.

No extraction of an enzyme from proteid coagulated by
heat is possible. Apparently therefore we can draw a distinction between the two, from their behaviour with alcohol.

At the same time it is possible to push this argument too
far. Some proteids, such as peptones and certain albumoses,
are not coagulated by alcohol any more than they are by heat.
If the enzymes should prove to be proteids belonging to either
of these classes, alcohol would not destroy them. The action of
alcohol however seems to establish the fact that they cannot be
coagulable proteids.

Though enzymes are capable of prolonged resistance to the
action of alcohol this reagent does eventually destroy them.
They show great differences in their power of withstanding it,
some being destroyed in a few hours, others resisting decomposition for months.

A very careful examination of the composition, properties,
and reactions of invertase was made in 1889 by O'Sullivan and
Tompson, and their results were published in the following
year. The source of the invertase which they used was the
liquor obtained from pressed yeast by allowing the latter to
stand for some weeks at a temperature of about 20° C., and
subsequently filtering.

Alcohol of 70 per cent. concentration was added to a
quantity of this yeast-liquor, very slowly and with frequent
pauses. When a precipitate appeared, the solution was allowed

to stand for two hours, and then a further quantity of the same alcohol was added, and so on till no further precipitate was formed. The liquid was then computed to contain 47 per cent. of spirit. The precipitate was so obtained in a finely-divided state and with but little impurity. The solution was allowed to stand over the precipitate for two days, when the latter was removed by decantation and washed with alcohol of 47 per cent. concentration. It was next thrown on to a filter and again washed with the same alcohol. The washed precipitate was then transferred to a beaker, and water added in quantity equal to that of the original yeast liquor. Part of the precipitate remained insoluble and was found to consist of a peculiar proteid, to which the authors gave the name *yeast albuminoid*. The invertase dissolved, and the solution was found to have an inverting power but little inferior to that of the original yeast. The authors judged the invertase to be nearly pure, and estimated that about 12 per cent. had been decomposed in the process of preparation, so that what impurities were present were only the products of its own decomposition. It contained a certain amount of ash, which was principally a mixture of potassium and magnesium phosphates. A separation of the ash from the invertase was found to be possible by means of dialysis, and the authors came to the conclusion that it did not enter into the composition of the ferment. They do not however appear to have freed the latter from it completely, but to have reduced the amount present to about ·45 per cent., calculated on the weight of the solid matter.

So prepared and purified, invertase was found to be a white substance, soluble in water and yielding a clear, slightly viscous, solution which did not become turbid on boiling. It was insoluble in alcohol of ·94 sp. gr. and could be precipitated from its solution by adding alcohol to this degree of concentration. The invertase was found to be destroyed to some extent by this precipitation, and the more so the purer it was. If allowed to stand under the alcohol the ferment power was soon lost. When quite pure the solutions could not be precipitated by alcohol without causing the destruction of the enzyme.

Invertase gave no colour with Millon's reagent in the cold

but on heating the solution a pink tint was developed. The solution of the enzyme had an optical activity of $(\alpha)_j = +80°$. Destructive analysis showed it to contain 46·41 per cent. of Carbon, 6·63 per cent. of Hydrogen, and only 3·69 per cent. of Nitrogen. In respect of the latter constituent it differed in a remarkable manner from a proteid.

A study of the enzyme obtained from yeast-liquor in this manner showed that when gradually and continuously decomposed, chiefly by fractionation with alcohol and acid, it gave rise to a series of bodies, which the authors termed the *invertan* series. They prepared and examined seven members of this series, and described their properties under the designations of α, β, γ, δ, ϵ, ζ, and η invertan respectively, invertase itself being the second member.

The two members at the extreme ends of the series, α and η invertan, were found to be very stable, but the intermediate ones were very liable to spontaneous decomposition. The bodies to which each of the latter gave rise under such circumstances were invariably α invertan, and the member next below the one dissociated. α invertan itself yielded on decomposition ζ invertan and yeast albuminoid. Omitting α invertan, each succeeding member was found to possess more stability than the one above it.

All the members of the series were non-crystallisable, undialysable, and colourless; α invertan was insoluble in water, but the others dissolved readily, forming clear but viscous solutions which did not become opalescent on boiling. Alkaline or neutral solutions were transformed by the gradual addition of alcohol into a milk-like fluid which would not become clear on filtration, but from which a precipitate rapidly fell on the addition of acid.   The α invertan was precipitated in flocculent masses, but the other members separated in the form of heavy transparent syrups.

Their composition, so far as the Carbon, Hydrogen, and Nitrogen are concerned, is given in the following table, in which they are compared also with the yeast albuminoid found in the original yeast-liquor.

Substance.	Constituents per cent. on substance free from ash.		
	C	H	N
Yeast albuminoid	54·06	7·35	14·53
α Invertan	48·03	6·65	8·35
β ,, (*invertase*)	46·41	6·63	3·69
γ ,,	45·62	6·55	3·15
δ ,,	46·50	6·82	2·43
ϵ ,,	44·45	6·36	2·07
ζ ,,	44·73	6·40	1·61
η ,,	—	—	1·05

From their analyses of these bodies the authors came to the conclusion that the first six were compounds of yeast albuminoid with the lowest member, η invertan, and that the latter was probably a combination of the same proteid with a carbohydrate. They did not isolate the latter, but from their analyses they calculated that it contained 43·22 per cent. of Carbon and 6·28 per cent. of Hydrogen.

Following up this hypothesis they computed that the lowest member of the series, η invertan, contained 18 parts by weight of this carbohydrate to 1 part by weight of proteid, and that the highest, α invertan, contained 3 parts of the carbohydrate to 4 of the proteid. The other members of the series they held to be formed by the union of these two substances according to the general formula $\eta + \alpha_n$, where η represents three times η-invertan and α represents α-invertan. Invertase on this hypothesis is $\eta\alpha_5$; it split up into α- and γ-invertan according to the formula $\eta\alpha_5 = \eta\alpha_4 + \alpha$. $\eta\alpha_4$ represents γ-invertan, and this could be further transformed by the elimination of α into $\eta\alpha_3$ which is δ-invertan.

According to this view of the composition of invertase the latter is not a proteid, but as it contains proteid in its composition it must give reactions characteristic of the proteid group. It yields the pink colour on boiling with Millon's reagent and it has the power of combining with copper salts. The latter property, which is also possessed by proteids, is shared by the other members of the invertan series with the exception of the α-body. Several such copper compounds can therefore be formed, which have a similar appearance but which contain different percentages of copper oxide.

Several observers have recently advanced the view that the enzymes are nucleo-proteids. This has been urged by Halliburton in the case of thrombase; as already noticed, he associated this ferment with a body to which he gave the name "cell globulin β," and which he prepared from lymphoid cells and tissues, such as lymphatic glands and thymus. He extracted it subsequently from the stromata of the red corpuscles of blood. When a quantity of this material was submitted to artificial gastric digestion it yielded an insoluble residue of nuclein. Pekelharing has come to the same conclusion as Halliburton as to the presence of nuclein in thrombase, but he suggests that the nucleo-proteid is probably the zymogen of the enzyme and that the latter is a calcium compound of the nucleo-proteid. Pekelharing states that he has obtained thrombase from nucleo-proteid by Schmidt's method of preparation. Both he and Halliburton have shown that the ferment prepared from serum by Schmidt's method gives a residue of nuclein when submitted to artificial gastric digestion, and that on analysis it is found to contain rather more than 1 per cent. of phosphorus.

Pekelharing found that his purified pepsin also contained phosphorus when prepared in the manner described in a previous chapter, and freed from associated bodies by dialysis, and he has advanced the view that this also is a nucleo-proteid. In O'Sullivan and Tompson's researches on invertase also, the ash of the preparation was found to contain phosphorus, as already mentioned. Lintner ascertained too that the purest diastase he could prepare contained a considerable quantity of phosphorus in its ash.

This hypothesis is considerably strengthened by the observations of Macallum, to which allusion was made in a previous chapter. The nucleus has been shown by him to initiate the process of secretion, and to excrete some material into the cytoplasm, which there undergoes further changes and ultimately enters into the composition of the zymogen if it does not actually form the principal part of it. The source of this excreted material is the chromatin of the nucleus. Macallum has shown how intimately phosphorus is associated with zymogen from its inception in the nucleus till the time when the granules are fully formed.

The chief constituent of the nuclear chromatin appears to be a nucleo-proteid. It is consistent with this that the excreted material may be formed of the same substance, and that the changes it undergoes in the cytoplasm may not alter its composition in any important respect. It may there become associated with proteids of various kinds, which may exist side by side with it in the granules, for there is no evidence that these are zymogen alone. Indeed a consideration of their dimensions and abundance seems to lead to the opinion that they are mixtures of various substances. In the pancreatic cells, the granules may indeed contain more than one zymogen, though at present we have no means of testing whether this is the case.

According to the investigations of Jacobson several enzymes possess, besides their specific powers, the property of liberating oxygen from hydrogen-peroxide. Among them may be mentioned diastase, emulsin, and the ferments of the pancreas. Many organs and tissues of the body have the same power.

These organs can also cause the conversion of salicyl aldehyde into salicylic acid. According to their degree of activity some of the tissues have been arranged in the following order:—blood, spleen, liver, pancreas, thymus, brain, muscle, ovary, oviduct. We have seen that the blood and certain of the tissues contain an oxidase which can destroy sugar.

Spitzer has recently investigated these various animal fluids and tissues, and he finds that the property in question is associated with nucleo-proteids. He prepared them from several organs by the methods adopted by different observers, and ascertained that they possessed the oxidising properties of the tissues themselves.

Spitzer has come to the conclusion that all these nucleo-proteids contain iron as well as phosphorus, his results agreeing in this respect with the observations of Macallum. These experiments support the view that the enzymes are members of the nucleo-proteid group.

The hypothesis of the nucleo-proteid nature of enzymes is not contradicted by the considerations advanced in the earlier portion of this chapter. Indeed it explains many things which present considerable difficulty to an acceptance of the view that they are proteid simply. If we adopt it, their resemblance in composition to the latter group of substances will cause us to cease to feel surprise at their close association with proteids, and at the variety of proteids with which such association has been noticed. The differences which have been noticed between the mass of a supposed enzyme used in an experiment, and its zymolytic activity, are explained if we have but a trace of the nucleo-proteid associated with a much larger quantity of the proteid accompanying it.

An observation of Pekelharing's tells rather against this view of the constitution of enzymes. He says that nucleo-proteids are insoluble in dilute acids, and can be readily precipitated from their solutions by faint acidification. This is certainly not the case with many of the proteolytic enzymes, nor with saliva and other diastasic preparations.

CHAPTER XXIV.

THE MODE OF ACTION OF ENZYMES.
THEORIES OF FERMENTATION.

BEFORE discussing the views at present held as to the manner in which enzymes carry out their specific functions, it will be well to trace the steps by which they have been reached, and to review the theories which have been advanced from time to time to explain the phenomena of fermentation.

We have already seen that it is only within comparatively recent years that there has been brought forward a general rather than a restricted view of this process. To the older writers this term generally conveyed the idea of alcoholic fermentation only, or at most such processes were included as involved the evolution of a stream of gas. The processes of putrefaction were held to be akin to this, though the idea was based only upon general relationships, and did not include the view that organic or living bodies played a leading part in both operations. The early theories of fermentation consequently sought to explain only the phenomena which we now associate with the formation of alcohol.

The subject is mentioned in writings dating from early times, and ferments are spoken of frequently, but we have very little to guide us in forming an opinion of what the writers understood by the term. Basilius Valentinus, a German alchemist, gives us the view that was current at the beginning of the 15th century. In his opinion alcohol existed in the wort, or extract of germinated barley, before fermentation, but it was held in combination by various impurities which masked its special properties. Fermentation was a process of purification in which the yeast or ferment, whose nature was unknown, communicated to the liquid an internal inflammation or disturbance, by which the initially turbid and discoloured liquid was so improved and freed from its impurities that the alcohol was enabled to show its true properties. The great growth of yeast which accompanied the process, being deposited at the bottom of the vessels when the fermentation was over, was thought to be composed of the impurities, and by some writers was spoken of as *fœces vini*, or the excrement of the fermenting liquor.

Libavius, about the end of the 16th century, held that the ferment was a body which transformed the fermenting substance into something of the same nature as itself, the chief agent in the change being found in the heat or *zeal* of the ferment.

A distinction between fermentation and effervescence, which was for a long time thought to be a similar phenomenon, was drawn by de la Boë, who pointed out that in the former case decomposition, and in the latter combination, were involved.

Lemery held less accurately that the chief difference between the two processes was one of relative rapidity, fermentation being the slower and more complicated operation of the two. He considered that the spirit produced in alcoholic fermentation was an oil refined by the essential salts which the must of wine contained. The deposit of yeast accompanying the process was composed of the grosser elements of the must, which became separated during the action of the salts, and took the form partly of a sediment, and partly of a scum floating on the surface.

No intelligible view of the nature of the action appears among the writers of this period until Willis in 1659, and Stahl some forty years later, put forward the first physical hypothesis. They suggested that a ferment was a body possessing an internal movement which it transmitted to the fermentable material. Stahl extended the idea to include putrefaction, of which he held alcoholic fermentation to be a particular case. He considered all fermentable material to be composed of particles formed by a loose combination of salt, earth and oil; under the influence of the internal movement set up by the ferment the various heterogeneous particles were separated from each other, and were subsequently recombined in various ways, gradually forming compounds of increased stability, containing indeed the same constituents but in different proportions.

This idea of a physical movement or vibration shows a great advance upon any earlier hypothesis, and has occupied a leading position in later speculations based upon accurate chemical knowledge. It held its place much as Stahl stated it, until the beginning of the present century.

In 1810 Gay-Lussac put forward the view that fermentation was set up in the must of wine and other saccharine fluids by the presence of oxygen. He did not consider it necessary that the gas should be present during the whole process, but held that it could not be started without contact with oxygen or air.

About 1837 the physical conception of fermentation was superseded by the vitalistic theory of Cagniard de Latour, which was based upon his discovery of the true nature of yeast, a discovery independently made about the same time by Schwann at Jena and by Kützing at Berlin. He suggested as the result of his observations that the yeast-cells disengaged carbon dioxide and formed alcohol by some effect of their own vital processes. This idea had previously occurred twenty years before to Erxleben, but he laid no stress upon it in his published writings.

Astier also even earlier (1813) foreshadowed the vitalistic theory. Though he held that the globules of the ferment were animal in nature, he said that they were generated and nourished at the expense of the sugar, and that they gave rise to a disturbance of equilibrium among the elementary particles of that body.

Schwann's views, put forward simultaneously with de Latour's, are quoted by Lafar[1] in the following words:—"Vinous fermentation must therefore be regarded as the decomposition

---

[1] Lafar. *Technical Mycology*, Eng. Ed. p. 16.

occasioned by the sugar-fungus extracting from the sugar and a nitrogenous body the materials necessary to its nutrition and growth, whereby such elements of these bodies (probably among other substances) as are not taken up by the plant unite by preference to form alcohol."

Kützing, whose name is also associated with the investigation into the nature of yeast and with the consequent theory of its action, was the discoverer of the nature of the organism causing acetous fermentation of alcohol, and in the expression of his views he groups the two fermentations together. Lafar quotes from his writings the following passages: "It is well-known that chemistry explains vinous fermentation by the reaction of the so-called gluten on the amylum (starch) and sugar. I must firmly maintain that the explanation does not give me a clear idea of the process, and I am inclined to doubt whether others are more fortunate in this respect. It is however certain that the entire process of alcoholic fermentation is dependent on the formation of yeast, and the acid fermentation on the growth of the vinegar plant.......Along with the increased growth of these organisms, the reproductive impulse also increases, and concurrently, their reaction on the liquid present.......In so far as fermentation is synonymous with a reciprocal reaction of organic and inorganic bodies on the constituents of a given liquid which may be regarded as forming the nutrient medium of the organic product, so is it necessarily synonymous with every organic vital function; wherefore organic life = fermentation. On the other hand, such processes as lead to the production of vinegar from alcohol by the use of platinum black, or other similar methods, cannot be compared with fermentation, being purely chemical, whilst fermentation is an organo-chemical process, as is also the life process of any organic body[1]."

The upholders of the older view combatted the new theory with considerable vigour. Liebig, who was the most conspicuous champion of the physical or physico-chemical hypothesis, somewhat modified Stahl's statement, bringing it into accordance with the discoveries that had been made in the field of

[1] Lafar. *loc. cit.* p. 17.

chemistry in the years between them. He drew a more definite distinction between putrefaction and fermentation than Stahl had done, while admitting that they presented many features in common. He states his opinion in the following sense—" The yeast of beer, and in general all animal and vegetable matter undergoing putrefaction, transmit to other bodies the state of decomposition in which they are themselves; the vibration which by disturbance of their equilibrium has been imparted to their own particles is communicated equally to the particles of the bodies which are in contact with them." The difference between putrefaction and fermentation consisted in that in the former case the decomposition was transmitted by the decomposing matter itself, so that it continued to proceed even if the original cause became inactive; while in fermentation the sugar was incapable by itself, even while in a state of incipient decomposition, of transmitting the movement or vibration to still undecomposed substance. Hence the necessity for the ferment not only to originate, but to continue, the decomposition. While however admitting the necessity of the presence of the yeast, Liebig did not at all agree that its action was related to its vegetative powers, as was claimed by the vitalists. He regarded it simply as a source of nitrogenous matter, by the decomposition of which fermentation was maintained, just as putrefaction is connected with the decomposition of similar albuminous material.

Liebig's views found many sympathisers, for his hypothesis seemed to account for not only alcoholic fermentation and putrefaction but certain other fermentations which had hitherto been unexplained, particularly the formation of lactic and butyric acids from sugar. Though we know now that these processes are the work of organic beings, at that time the latter had not been observed, and the only explanation that seemed possible was that they resulted from the action of decomposing nitrogenous matter in the sugar solutions. Some of the writers of this school, among them Fremy, held that the character of a particular fermentation varied according to the degree to which the decomposition of the albuminous exciting body had proceeded. At first it would set up alcoholic fermentation, but as its decomposition became more advanced this would be replaced by lactic or butyric fermentation, and so on.

The vitalistic theory of fermentation put forward by de Latour, Schwann, and Kützing was supported by Pasteur, to whose brilliant researches reference has already been made. He carried it much further than its original propounders, and sought to show the reason why the vegetative life of the yeast is associated with such a complete decomposition as takes place. In the course of his work he very greatly strengthened his position by the discovery that the formation of alcohol and carbon dioxide is not the only transformation taking place in the sugar, but that about four per cent. of the latter is disposed of in another manner, giving rise to succinic acid and glycerine. He showed too that other alcohols than the ordinary one are formed in varying quantities during the metabolic activity of the yeast.

As the outcome of his researches Pasteur was led to the view that the larger decomposition of the sugar was the expression of the effort of the yeast to obtain a supply of oxygen, and to him therefore the phenomenon became one of the intramolecular respiration of the yeast plant. As already mentioned this view was strengthened by the researches of Lechartier and Bellamy on alcohol-production in fruits kept in an atmosphere of carbonic dioxide.

Other later observers, particularly Schützenberger, controverted Pasteur's intramolecular respiration theory, and argued that the sugar was an alimentary substance for the yeast, a view which is at present held.

The controversy between the physical and the physiological schools was maintained with considerable vigour. Liebig largely modified his views to bring them into harmony with the facts established by Pasteur, while still maintaining that fermentation is essentially vibration or movement among the molecules of the fermenting substance. He went so far as to admit that the cause of the vibrations was the living organism rather than any decomposing nitrogenous matter.

Efforts were made to harmonise the two views somewhat

on the lines of Liebig's latest opinion. According to Pasteur the action was essentially intracellular; the decompositions took place in the cells of the yeast plant, in direct relation to their metabolic processes, and the products of the fermentation left the cells as excreta. The advocates of Liebig's opinion, while admitting that the living cells set up the vibrations they contended for, suggested that the decomposing forces themselves proceeded beyond the cells and attacked the fermenting matter in the medium around them. The most noteworthy advocate of this view was Naegeli, who held that in fermentation there was a transference of vibrations or movements in the material composing the living substance of the yeast to the fermenting material outside, the living substance itself remaining chemically unchanged. By such transferred vibrations the equilibrium of the molecules of the fermenting substance became destroyed and so its decomposition took place. Naegeli computed what he thought to be the distance through which such vibrations might be transmitted, and concluded that the sphere of action of each cell extended to a distance of from 20 to 50μ from its centre.

While this controversy was going on a new discovery, or rather series of discoveries, was being made, the influence of which soon began to be felt, and which was destined to very largely modify and extend the physiological hypothesis of the vitalists. The work of Payen and Persoz in 1833, and of Schwann in 1836, led to the recognition of a class of substances capable of setting up fermentations of very much the same order as the alcoholic fermentation of yeast, marked by very similar features of their action, but yet not depending on the presence of any living organism. These bodies, which we know now as enzymes, were in all cases derived from living cells, and their action was readily held to be very similar to that of the yeast cell and other organisms which were beginning to be known. They were ascertained to act as ferments, and the term *unorganised ferment* was given them, to indicate at once their mode of action and the distinction existing between them and the living cells of yeast.

The theory of fermentation in favour clearly had no room

for these new bodies, and for a time no effort was made to bring them into harmony with the popular views. Indeed efforts were made to show that though at first sight similar, they were really fundamentally distinct from the organised ferments. In his Theory of Fermentation, published in 1879, Naegeli dwells particularly on this point. He says, "Fermentation by means of fungi only takes place in immediate contact with the protoplasm and so far as its molecular action extends." He finds a point of difference between the two classes of ferments in the fact that the enzymes convert plastic material from the passive into the active condition, so that it is rendered suitable for nutrition, while the products of fermentation by means of fungi are without exception less nutritious compounds, and such fermentation destroys especially the most nutritious substances. "The contrast appears most striking in the case of carbohydrates and proteid substances; while the action of unorganised ferments (enzymes) produces from them glucoses and peptones, fermentation by means of fungi breaks up these compounds into alcohol, mannite, lactic acid, and into leucin, tyrosin, &c. In some cases several fermentations follow one another; their products then become less nutritious step by step. We may say generally that the yeast-fungi render the medium in which they occur chemically less suitable for nutrition by every process of fermentation which they effect."

Sachs, writing in 1882, supports the views of Naegeli. In his "Lectures on the Physiology of Plants" he says, "In fermentation by means of unorganised ferments the chemical transformation proceeds smoothly and completely; dextrin is entirely transformed into grape-sugar, cane-sugar entirely into inverted sugar, and albuminates entirely into peptones. In the alcoholic fermentation, on the other hand, the products of which alone have hitherto been quantitatively determined, only the greater part of the sugar is broken up into alcohol and carbon dioxide, while according to Pasteur about 5 per cent. of the sugar is decomposed by the way into glycerine, succinic acid, and carbon dioxide. It is likewise certain that in the lactic acid fermentation, not all the sugar is decomposed into

THE MODE OF ACTION OF ENZYMES. [CH.

lactic acid. Carbon dioxide especially appears to be a by-product of all fermentations due to the action of fungi and of all putrefactive processes. The alleged difference between fermentation by means of unorganised ferments and that due to the action of fungi would be unintelligible if both processes were due to the same cause. Every difficulty vanishes, on the other hand, if fermentation by means of fungi is caused not by a contact substance (ferment) but by living protoplasm. We then apprehend that while the ferment as a simple chemical compound alters another chemical compound in a simple and equable manner, so that all molecules suffer the same kind of decomposition, an organised substance with its various molecular movements and molecular forces produces more complicated decompositions."

Both Naegeli and Sachs lay emphasis on the fact that at the time they wrote enzymes which could effect fermentation in the absence of the cells had not been extracted from any of these fungi. Naegeli also denied that any of the intracellular fermentative changes in any organism, whether animal or vegetable, were due to unorganised ferments, and attributed them always to the fermentative activity of the protoplasm.

This view was however controverted almost from the first. In 1858 Moritz Traube made the origin and influence of these enzymes the basis of a new theory of fermentation, which all subsequent investigations have gone far to support and which is continually growing into favour. This view is that all fermentation is not instigated so much by the organisms themselves as by enzymes which are formed as products of their vitality, and are either excreted by them to work on substances outside them, or retained in the cells in which they are secreted to effect intracellular digestion. This hypothesis is of great value as again bringing all fermentative phenomena into line with one another, and in showing clearly that alcoholic fermentation is not something *sui generis*, but only one mani-festation of a very wide-spread power. The fact that it can be excited by an enzyme extracted from the yeast cell strongly supports the theory, already greatly helped by the extraction of enzymes from so many other micro-organisms.

Berthelot was one of the first chemists to accept this view, being led to endorse it by the consideration, based on experiment, that in certain cases production of alcohol took place without the intervention of yeast at all. The experiments of Lechartier and Bellamy and of Pasteur on this point have already been referred to. The weight of the opinion of Hoppe-Seyler has also been given in favour of this theory, reasons for the acceptance of which have already been advanced in one of the earlier chapters of this volume.

Before turning to consider how far we can explain the action of the enzymes it seems desirable to examine a little more closely the objections to this view which Naegeli and Sachs based upon the nature of the products of enzymic action as contrasted with those arising from the activity of the organised ferments or Fungi. The peculiar features of the latter which appeared to them radical find their representatives among the metabolic phenomena of other plants than Fungi, and are really the expression of the vital activity of plants in conditions which are common among the lowly forms, but which occur abnormally among the higher ones. As we have seen, the marked feature of the life of these plants is the fact that the products of their activity become less and less nutritious as the life goes on, or as the authors quoted put it, as the fermentation proceeds. When we consider that the fermentation in question is the expression of the striving of the cell to procure nourishment from the environment in which it finds itself, this ceases to be a matter for surprise and becomes only what might be expected. But under certain conditions the higher plants behave similarly, and they ought therefore to be considered as ferments also, if this feature is a specific mark of fermentation. Boehm and de Luca have shown that if any part of a living plant is insufficiently supplied with oxygen, hydrogen and sometimes marsh-gas are evolved from it. Boussingault and Schulz have observed similar phenomena. These exhalations are ordinary features of the putrefaction set up by micro-organisms. From plants containing mannite, hydrogen is given off, while according to de Luca acetic acid is formed from malic acid in the fruits, flowers, and leaves of the Privet. In the decomposition of proteids again Boehm found ammonia exhaled. The condition under which these results are obtained, viz. the absence of oxygen, is the normal condition of many of the microbes, they being anærobic in their mode of life. When oxygen is present we find the same agreement. The result of the action of invertase is the same whether that action is brought about by living yeast or by the enzyme extracted from a higher plant. The decomposition set up by trypsin in the formation of proteoses, peptones, and amides is similar to that induced by some of the proteolytic bacteria which have not at present been found to secrete an enzyme. Sachs claims as a peculiarity of all fermentation set up by Fungi that carbon dioxide appears as a by-product. This is not to be wondered at, for Fungi like all other living beings exhale this gas in the course of their respiration. The quantity exhaled can in many cases be very greatly increased by the simple process of allowing the plant an insufficient supply of oxygen.

The difference drawn between the two so-called classes of ferments is therefore unsupported by evidence and breaks down as soon as it is carefully scrutinised.

The exact mode of action of the enzymes is still to a large extent a matter of speculation and must remain so until we arrive at some definite conclusion as to their exact nature and constitution. This, as we have seen, is beyond our grasp at present, for even if we accept the view that they belong to the group of nucleo-proteids, we have no conception either of their chemical composition or of the manner in which their groups of molecules are arranged. There are however features of their action which are more or less clearly established.

We have seen that in many cases this action is one of hydration, especially in the cases of the simplest transformations. To recall several of these which have been discussed in preceding chapters, we have among others the reactions expressed by the equations:

Invertase $C_{12}H_{22}O_{11} + H_2O = C_6H_{12}O_6 + C_6H_{12}O_6$.

Emulsin $C_{20}H_{27}NO_{11} + 2H_2O = C_6H_5COH + HCN + 2C_6H_{12}O_6$.

Lipase $C_{57}H_{104}O_6 + 3H_2O = 3C_{18}H_{34}O_2 + C_3H_5(HO)_3$.

Urease $CON_2H_4 + 2H_2O = (NH_4)_2CO_3$.

The oxidases exhibit another process of a similar character, but instead of inducing hydration they set up oxidation. In the case of the action of laccase upon hydroquinone we have the reaction expressed by the equation

$$2C_6H_4(OH)_2 + O_2 = 2H_2O + 2C_6H_4O_2.$$

The enzyme differs from those of the other group by transferring oxygen to the fermenting body instead of water.

These two classes of reaction suggest that the enzyme is concerned with the transference of some molecule to the fermenting body which it could not otherwise combine with, and a fairly simple chemical hypothesis might be framed upon observations such as those quoted. But it would be unsafe to generalise so hastily, for other decompositions due to enzymes are not apparently of the same order. One noteworthy instance at least is known in which decomposition of the fermenting substance takes place apparently without the intervention of either water or oxygen. Buchner's zymase splits up sugar according to the equation

$$C_6H_{12}O_6 = 2CO_2 + 2CH_3CH_2OH.$$

We must not however conclude that even the first reactions quoted do more than express the final results of the actions of the particular enzymes in question. We have no ground for supposing the latter to be of so simple a nature as the equations would indicate, but on the contrary there are many reasons for entertaining the opposite view. We know that the action of diastase, which is capable of being represented by almost as simple an expression, is a progressive one and that many stages of the process have been demonstrated.

Certain observations of O'Sullivan and Tompson on the behaviour of invertase lead to the conclusion that the hydrolysis of cane-sugar by the enzyme is complex rather than simple. In studying the effect of a rise of temperature on the enzyme they ascertained that there was a great difference in its power of resisting decomposition by heat according to the conditions in which it was placed with regard to cane-sugar. When it was heated in the absence of the latter, the enzyme was almost all destroyed at 50° C.; while if the sugar was present this effect was not produced till the temperature was 75° C., a difference of 25° C. The authors advance as an explanation the view that the invertase enters into combination with the sugar as a stage in the hydrolysis and that the compound so formed can resist the heat more successfully than can the invertase alone. They hold further that the new compound is unstable and becomes decomposed when its molecule meets with another molecule of cane-sugar.

Whether we endorse their explanation or not, their observation on the influence of cane-sugar on the temperature of destruction must lead us to believe that the reaction is complicated in some manner by the invertase entering into it.

This observation does not stand alone. The writer has observed a similar difference of behaviour in the case of vegetable rennet, though in the opposite direction. This enzyme is weakened and destroyed at a lower temperature if heated in the presence of caseinogen than when it is similarly treated in the absence of the proteid.

Buchner's zymase, again, is more stable in the presence than in the absence of sugar.

Chittenden has observed that if neutralised pine-apple juice, containing bromelin, is heated to 60° C. in the absence of any proteoses or peptones, the enzyme is rapidly destroyed, whereas this temperature is the one at which it is most active if proteids are present during the warming.

Biernacki found similarly that albumoses or peptones raised the temperature at which trypsin was destroyed by 5° C. or more, and that while pepsin in the absence of peptone was destroyed in acid solutions by a temperature of 55° C. it was active after being heated to 70° C. if peptone was present.

We may also note in this connection the possible significance of the inhibitory effects of traces of acids or alkalis in the solution in which an enzyme is working. The minute trace of the reagent required to cause inhibition seems to correspond to the extremely small amount of the enzyme usually present, and it does not seem a wild speculation to suppose that it may affect the action of the latter by entering into some combination with it, the resulting body not being capable then of uniting with the substance which the enzyme would ordinarily transform. That acids and enzymes do in some cases combine we know from a consideration of the relations between pepsin and hydrochloric acid, the ferment being absolutely inoperative without hydrochloric or some other acid. Biernacki has shown that pepsin in the absence of an acid is destroyed at a temperature 5° C. lower than that which is required for similar destruction in its presence. Chittenden found that bromelin on the other hand would withstand a higher temperature in neutral than in acid solutions. In this case the compound, if such exists, is less stable than the enzyme alone.

As we have seen the decompositions set up by enzymes are very varied. In some cases the action appears to be hydrolytic, in others oxidative, and in others disruptive without previous hydrolysis or oxidation.

In endeavouring to form a hypothesis which shall be capable of accounting for all these transformations there seems to be no necessity to draw any distinction in this respect between (i) the fermentative activity of protoplasm, (ii) the action of enzymes which act only intracellularly, and which have only a transient existence, being decomposed as soon as their immediate duty is discharged, and (iii) the action of those which are excreted from the cells, and are much more stable, being capable of preservation by the various methods already described. The difference between these three agents is probably one of degree only and not of kind.

It seems clear however that any hypothesis which shall be satisfactory must go back in some way to a physical or physicochemical theory, although such a view need not have much or indeed anything in common with that which gave place to the physiological one. Admitting that the problem is physiological as regards the functions of the enzymes and their position with regard to the metabolism of the organism, we must recognize that the changes which we call metabolic, and in which these enzymes take part, are fundamentally connected with chemical or physical processes going on in the cells.

We cannot now be satisfied with the statement that fermentation is connected with the life of the organism and is a symptom or manifestation of that life. We must pass beyond this to enquire into the nature of the processes by which it assists to maintain that life and takes part in that metabolism. Such a physical or chemical hypothesis will have very little in common with that of Liebig beyond perhaps embracing the view that molecular vibration plays a part in the transformations, if indeed any hypothesis based upon vibration will survive examination.

Before discussing the question of a physical as distinct from a chemical process the recent researches of Emil Fischer demand some attention. They are based upon the great development of stereo-chemistry during the last decade, and develop the idea that in the reactions of various chemical bodies the molecular configurations of those taking part must show a certain correspondence.

This view appears to have been suggested by the anomalous behaviour of different species of Yeast or Saccharomyces when brought into contact with different sugars. Hansen's very complete researches into the biology of the Yeast plant enabled him to determine at least ten definite species of this genus, whose life-history and capabilities of inducing fermentation he ascertained with extreme accuracy. Of these ten species, six hydrolyse both cane-sugar and maltose, forming hexoses as described in a preceding chapter. They then further carry on alcoholic fermentation at the expense of these hexoses. *Saccharomyces Marxianus* cannot attack maltose, but can hydrolyse cane-sugar. *S. membranæfaciens* does not invert either sugar and is not capable of setting up alcoholic fermentation. With a particular species of Yeast again there are conspicuous differences among the sugars as to their power of undergoing fermentation. Of the biose polysaccharides, cane-sugar, maltose and trehalose are capable of producing alcohol, but they must first be hydrolysed to hexoses. Of the latter glucose and fructose are much the most readily fermented; galactose is very refractory, but is said by Bourquelot to be fermentable when a small quantity of either glucose, fructose, or maltose is present also, but not under other conditions. Lactose again is said by Duclaux and by Kayser to be capable of fermentation by three species of yeast.

Fischer was led from a consideration of such facts as these to suggest that these ferments can only attack those sugars which possess a molecular configuration corresponding to their own. Researches which he carried out, using various enzymes, enabled him to put forward this view as applying to fermentation in general, whether set up by living protoplasm or the enzymes it secretes. His first investigations were directed towards the behaviour of glucase when brought into contact with various sugars and glucosides.

When a solution of a sugar in methyl alcohol is saturated with hydrochloric acid it loses the power of reducing Fehling's fluid, and a crystallisable compound having the formula $C_6H_{11}O_6CH_3$ is formed as the result of a reaction represented by the equation $C_6H_{12}O_6 + CH_3OH = C_6H_{11}O_6CH_3 + H_2O$. The same reaction takes place with all the alcohols which are capable of dissolving sugar, and the compounds formed correspond to natural glucosides. The sugar entering into the reaction need not necessarily be glucose, so that series of such artificial methyl-glucosides may be prepared. These bodies are not altered by Fehling's solution, phenylhydrazine or caustic potash, but when boiled with dilute acid they take up water and are decomposed into sugar and alcohol.

Theory indicates the existence of two glucosides, both of which are formed by the action of hydrochloric acid on a solution of d glucose in methyl alcohol. According to Fischer they have respectively the formulæ

$$H-C-O-R^1 \quad \text{and} \quad {}^1R-O-C-H$$

CHOH	CHOH
O⟨ CHOH	O⟨ CHOH
CH	CH
CHOH	CHOH
CH₂OH	CH₂OH

[1] R denotes CH_3

Fischer calls them respectively α and β methyl-glucoside. He acted upon these new compounds (which only differ in the configuration of the one asymetric carbon atom) with a preparation of glucase made by digesting yeast with 15 times its weight of water for 15 hours at 30—35° C. and filtering. The α body was decomposed by the enzyme after 20 hours action. The quantity of glucase solution was one-twentieth part of the solution of the glucoside, and the temperature was 30—35° C. About half of the glucoside was transformed into glucose and alcohol.

The β body was tested in the same way, but remained altogether unaffected.

Fischer prepared the α- ethyl-glucoside and found that glucase decomposed it as it did the corresponding methyl one.

When the artificial glucoside was prepared with l glucose, glucase was found to be without action on it. When d glucose was replaced by galactose again, no decomposition was effected by glucase.

When ordinary d fructose (levulose) was used instead of d glucose, the resulting glucosides of the α type were decomposed by glucase. Invertase could not effect these decompositions.

Fischer attributes these differences of behaviour to the difference of configuration of the sugar molecules in the several cases. This view gains some support from the fact that each of the biose polysaccharides is hydrolysed by a special enzyme, cane-sugar by invertase, maltose by glucase, milk-sugar by lactase, and others by other enzymes, as described in Chapter IX. of this volume. If the reason is a correspondence in configuration between the sugar and the enzyme it would apply equally to the failure of glucase to decompose the methyl-galactoside, as lactose, which yields galactose and glucose on hydrolysis, is unaffected by glucase.

Fischer carried his observations further and experimented on the action of glucase on the aromatic glucosides. He found it unable to attack many of these, but to be capable of setting up a decomposition of amygdalin, which affected the sugar radicle of the complex molecule only. Instead of acting like

emulsin and forming bitter almond oil and hydrocyanic acid
with two molecules of glucose, it was only capable of splitting
off one molecule of glucose, leaving a new glucoside containing
the other molecule of the hexose.

Fischer studied the action of emulsin as well as glucase on
the artificial glucosides. He found that like glucase, emulsin
attacked the glucosides of d glucose, but it was incapable of
decomposing those of galactose and similar sugars. It was
also without effect upon the methyl l glucosides. Its action
was however easily distinguishable from that of glucase, for
it only decomposed the glucosides possessing the β constitution, while glucase attacked only the a group.

Fischer found that emulsin was without action on methyl-
d-mannoside, methyl-sorboside or methyl-galactoside. It would
not decompose either the a or β form of methyl-glucoside
prepared from l glucose.

Myrosin showed similar differences; it split up the a methyl-
d-glucoside but not the β one.

Glucase prepared from yeast did not attack methyl-sorboside,
nor methyl-mannoside prepared from reserve cellulose.

On the basis of these researches of Fischer we may arrive
at the following view, which was in the main suggested by him.
The enzymes being probably nucleo-proteid in nature, are like
true proteids optically-active substances, and on account of this
their molecules have an asymmetric structure. As the work
described shows that the action of enzymes is a selective one,
it follows that those substances with which they come into such
contact as will enable them to provoke decompositions must
have a configuration corresponding to their own. The two
molecules may be taken to have such a mutual relationship as
that existing between a key and the lock for which it is constructed, the shape of the key enabling it only to unfasten the
particular lock to the arrangement of whose wards it corresponds.

The configurations of the two bodies, the sugar and the
enzyme, fitting as it were into each other, the disruption of the
former by the latter becomes conceivable. Fischer lays stress
especially on the behaviour of the a and β methyl-glucosides

with glucase. The two glucosides have the same composition, are formed from the same alcohol and the same sugar, and differ only in the configuration of a single carbon-atom which is rendered asymmetric by the introduction of the methyl-group into the sugar.

The same relationship or correspondence of form may well exist between other enzymes than those decomposing sugars and the bodies whose decompositions they effect. The latter are for the most part asymmetric substances.

We can now return to the consideration of the question whether the actual processes of fermentation can best be explained on a physical or a chemical hypothesis.

The original position of Liebig demanded a vibratory movement of the molecules of the body exciting fermentation, and he associated such a movement with the decomposing nitrogenous matter in the solution. This was modified partly by himself and partly by his successors to embrace the idea that the vibrations were set up by living cells or by something excreted from them. Fischer's hypothesis does not appear inconsistent with this view; indeed it supplies a reason for the selective fermentations set up by different excreted bodies or enzymes, showing that only such bodies can be thrown into increased molecular vibration as have a similar configuration to the enzyme.

But a view based upon molecular vibration does not seem satisfactory, in the face of what is now known about the nature of enzymes, little as it is. They are not continually undergoing decomposition as was the case with Liebig's organic nitrogenous material. On the other hand they are fairly stable substances, even when in aqueous solution. Moreover in the light of the observations noted already, there is reason to believe that they enter into some form of chemical combination with the bodies which they hydrolyse.

Berzelius long ago rejected Liebig's view of molecular vibration in favour of a theory that the action was one of contact, comparable to the effect which spongy platinum produces on water holding oxygen in solution. Not only spongy platinum, but many other bodies, such as finely powdered charcoal, or

manganese dioxide, or finely precipitated silver, when added to
water saturated with oxygen cause the latter to be given off in
bubbles. Many of the enzymes will produce the same effect,
having this property in common, as well as possessing the
specific powers already described. Schoenbein, who first noticed
this peculiarity, thought it was especially connected with their
fermentative power. Jacobson has shown that this is not the
case, for if gently and gradually heated their solutions lose this
property before they part with the power of exciting their
specific actions.

The suggestion does not however throw any light upon the
mode of action which can be associated with them. Nor is
there such a similarity between enzymes and a body like
spongy platinum as would allow us to regard their behaviour
as identical.

A hypothesis has been advanced that fermentation is due
to electric hydrolysis; that as the result of electrical action
in the liquid the elements of water are withdrawn from
the sugar and added on again in a different order, so that
there is going on in the liquid an electric hydration and
dehydration. The suggestion has been brought forward by
Armstrong, who has pointed out that in adopting this view
it is not necessary to assume that there is ever any actual
separation of the elements of water from the molecule and
their subsequent readmission, but that the action may be
brought about by a kind of electrolytic change taking place
locally within the molecule. On this hypothesis the part
played by the enzyme or the organism might well be to
complete the galvanic circuit and so to allow the electrolytic
current to pass. This is not out of harmony with Fischer's
views, as a close approximation of the two sufficient to form a
pathway for the current might well depend upon their possessing
a corresponding configuration. This hypothesis demands however that enzymes shall be electrolytes, and of this so far
there is no evidence.

A physical hypothesis of a different nature has been
advanced by de Jager. Starting with Naegeli's view that
fermenting yeast-cells emit vibrations which pass out of the

cells and decompose the sugar in the solution surrounding them, de Jager suggests that the enzymes may be regarded not as substances at all, but as the vibrations themselves, that is as properties of substances rather than material bodies. In this way he compares them to other physical forces, such as light, electricity, and magnetism. Fermentation so would come to consist of chemical transformations set up by physical forces and not in any sense dependent on the chemical action of a molecular substance.

This theory has been recently taken up and elaborated by Maurice Arthus, who has very ingeniously compared the properties of enzymes with those of various physical forces. He lays stress particularly on two facts: 1st, that the composition of the enzyme so far has eluded the investigations of the most eminent chemists, and that all their results have only shown that these bodies if they exist cannot be separated from impurities. 2nd, that the amount of any of them which effects a decomposition is infinitesimal in comparison with the amount of material which it can decompose. He regards the "carrying" theory of Bunsen as totally inadequate to account for this enormous disproportion. Further he endeavours to show that their various so-called "properties" are paralleled by properties of the different physical agencies with which he is disposed to rank them.

Though the theory is ingenious it seems impossible to reconcile it with the very definite processes of formation or secretion which we have seen can be observed in both animal and vegetable tissues. The different reactions which they present, their respective powers of solubility and precipitation, their varying capability of dialysis, &c., also point much more strongly to their being definite material entities.

The view that the action of enzymes is chemical rather than physical in its character has met with considerable support. The changes which so many of them set up we have seen are hydrolytic, the fermenting body taking up water through their instrumentality and subsequently undergoing decomposition. Most of the changes so effected can be brought about in the laboratory by ordinary chemical reagents,—starch can be hydrolysed to sugar by dilute mineral or organic acids; the same bodies will produce glucose and fructose from cane-sugar; fats can be decomposed by alkalis or by superheated steam; peptones can be formed by heating proteids to a high temperature in a Papin's digester. Moreover when the changes set up by the enzymes are investigated carefully, they show considerable correspondence in detail with changes which are distinctly chemical. In the course of the observations upon invertase to which allusion was made in the last chapter, C. O'Sullivan and Tompson showed that the rate of inversion of cane-sugar which it brings about may always be represented by a definite time curve which " is practically that given by Harcourt as being the one expressing a chemical change of which no condition varies, excepting the diminution of the changing substance[1]." It thus obeys the law formulated by Harcourt that in the case of a substance undergoing such change under such conditions, the amount of change occurring at any moment is directly proportional to the quantity of the substance. J. O'Sullivan also has found that the hydrolytic action of yeast at the ordinary temperature follows the same course as that of a simple chemical interchange.

But if we adopt the view that we are dealing with an ordinary chemical reaction, leading in these cases to hydration and subsequent decomposition we have still to enquire what is the exact nature of that reaction. Any explanation of the latter must be such as to include also those more complex changes which involve the actual disruption of the molecule, and which we find in the cases of myrosin and Buchner's zymase.

In considering the question of the chemical changes set up by enzymes, the first thing that strikes an observer is the extremely small amount of one of them that is needed to bring about the transformation of an enormous amount of the body which it attacks. Thus O'Sullivan and Tompson showed in one of their experiments that a sample of invertase induced inversion of 100,000 times its own weight of cane-sugar. Moreover it was not destroyed or even greatly injured by its activity. The decomposition of an enzyme during the course of its activity

[1] O'Sullivan and Tompson, 'On Invertase.' *Jour. Chem. Soc.*, 1890, p. 926.

has been shown by Tammann to be very slow and gradual, and
to be influenced by differences of temperature in the same way
as the activity itself.   These considerations lend a good deal of
plausibility to a suggestion which has been made by Bunsen and
by Hüfner that enzymic action may be similar to the behaviour
of nitric oxide in the manufacture of sulphuric acid or of sulphuric acid in the preparation of ether.   In the former process
a stream of sulphur dioxide mixed with oxygen in the presence
of water is presented to nitric oxide, and the further oxidation
of the sulphur dioxide is effected with great rapidity.   A small
proportion of the oxide of nitrogen will effect the combination of
an almost indefinite amount of sulphurous anhydride and oxygen.
Nitric oxide in the presence of oxygen immediately becomes
nitrogen peroxide, and the latter when mixed with sulphurous
anhydride and a large quantity of water gives up oxygen to the
sulphurous anhydride, which becomes sulphuric acid, while the
original amount of nitric oxide is again formed.   This process
continues so long as any sulphurous anhydride remains in the
presence of oxygen uncombined with it.

The action of sulphuric acid in the manufacture of ether from
alcohol is of a similar kind.   An interaction of the acid with the
alcohol takes place, which can be expressed by the equation

$$C_2H_5HO \quad + H_2SO_4 = C_2H_5HSO_4 \quad + H_2O,$$
Alcohol   Sulphuric acid   Ethyl sulphate   Water

and the ethyl sulphate reacts with another molecule of alcohol
in a manner indicated by the further equation

$$C_2H_5HSO_4 + C_2H_5HO = (C_2H_5)_2O \quad + H_2SO_4.$$
Ethyl sulphate   Alcohol   Ether   Sulphuric acid

On this view we may suppose the enzyme to attach itself to
the body on which it acts, with or without the elimination or
addition of water.   The new compound meeting with a further
molecule of the same body will then be decomposed with the
formation of the body found to result from the action of the
enzyme, the latter being again liberated unchanged or perhaps
reformed.

This view must remain at present hypothetical, as no one
so far has isolated a compound of the enzyme and the body it

attacks. But as we saw in the last chapter, O'Sullivan and
Tompson have found reason to believe that such a combination
exists, their view being based on the effect of heat on invertase
in the presence and in the absence of cane-sugar in the solution
under examination. Matthieu and Hallopeau were led by some
of their observations to the view that in the action of pepsin in
the stomach the chlorine of the hydrochloric acid enters into
combination with the albuminoid substances, and even with the
peptone. If this conclusion is well-founded the acid may be an
intermediary between the pepsin and the proteid, for we know
it is in very close association, if not union, with the former.

If the hypothesis is a valid one, the behaviour of enzymes
must fall into line with other chemical actions. One of the
fundamental features of the latter is that if two compounds
capable of chemical interaction are present in the same solution
the resulting interchange will not be complete unless one of
the products formed is removed as fast as it appears. A certain
amount of all four possible bodies will result, the relative
quantities depending on various conditions, and the chemical
interchange will only cease when a definite point of dynamical
equilibrium between them is reached. This point must depend
among other things upon the relative amounts of the reacting
bodies in the solution and will therefore vary accordingly.

The behaviour of sulphuric acid and alcohol will illustrate
this point. So long as alcohol is in excess the ethyl sulphate
first formed will combine with it as described, giving rise to
ether. But if water is in excess a reaction will occur, which
will lead to the decomposition of the ethyl sulphate into the
compounds from which it was formed,

$$C_2H_5HSO_4 \ + H_2O = C_2H_5HO \ + H_2SO_4.$$

Ethyl sulphate   Water   Alcohol Sulphuric acid

In the process therefore all four bodies, alcohol, sulphuric
acid, ether, and water will be found together, and the ultimate
equilibrium point will depend on the relative proportions of the
alcohol, the water, and the sulphuric acid.

We must therefore expect that in the action of enzymes
the continuous and complete decomposition of the original

substance on which the enzyme works cannot take place unless the ultimate product formed is removed as fast as it appears. If this is not done the action should go on rapidly at first, but gradually more slowly until the equilibrium point is reached, when the action should cease. This is exactly what has been found in many cases. Tammann has shown that a considerable inhibition of the action of emulsin on amygdalin is caused by the accumulation of the products of the decomposition, and that this is not due at all to a destruction of the enzyme. Brown and Morris have shown that it is almost impossible to completely hydrolyse starch by means of diastase, generally nearly 20 per cent. of dextrin remaining. Lea has found that the hydrolysis can be completed if the sugar is removed as fast as it is formed.

Again, if the relative amounts of the interacting bodies regulate the course of the action, an enzyme should be capable of effecting a certain amount of a reverse action, when the final instead of the initial body is presented to it. Certain investigations made recently have shown that this also is one of the features of the behaviour of enzymes.

An extended research upon the action of maltase or glucase upon malt-sugar was made by Croft Hill in 1897 and 1898. The enzyme was prepared from dried yeast with appropriate antiseptic precautions and the experiments were safeguarded by the use of toluene in all the digestions. Hill found at the outset that the presence of glucose had a marked inhibitory effect upon the hydrolysis of maltose, and this effect was greater as the sugar solutions were more concentrated. When the enzyme was digested with a 40 per cent. solution of glucose at 30° C. a slow conversion of the latter into maltose was detected. As the action proceeded its amount was measured by observing the rotation of a ray of polarised light in a polarimeter and by ascertaining the cupric-oxide-reducing power of the solution. After five days 3·25 per cent. of the glucose had undergone conversion, which was increased to 10 per cent. after 28 days, and to 14·5 per cent. after 70 days. Beyond this point no change occurred, so that an equilibrium point for these two sugars was reached when there was 85·5 per cent. of glucose and 14·5 per cent. of maltose present. The quantitative determinations were

checked by treatment of the contents of the flask with phenyl-
hydrazine acetate, when two osazones were formed, having the
properties of glucosazone and maltosazone respectively.

In a subsequent experiment, in which a mixture of the
two sugars was used, consisting of 75 per cent. of glucose and
25 per cent. of maltose, the concentration of the total sugar
being as before 40 per cent., the same equilibrium point was
reached, the digestion being continued 56 days.

With a smaller percentage of sugars in the digesting liquid
the equilibrium point was not the same. Using a 20 per cent.
concentration Hill found that on the conclusion of the experiment 90·5 per cent. of the total sugar was glucose and 9·5 per
cent. maltose; with a 10 per cent. concentration the equilibrium
point was reached when the proportion of glucose to maltose
was 94·5 : 5·5. In more dilute solutions the transformation of
glucose was still less—with 4 per cent. concentration the
ratio became 98 : 2, and with 2 per cent. 99 : 1. Hill found
therefore that with very dilute solutions of maltose hydrolysis
to glucose by the enzyme was almost complete: with a concentration of 20 per cent. a reverse action could be detected
which was well marked when the concentration was 40 per cent.

These experiments of Croft Hill afford considerable support
to the view that the action of glucase is a chemical one, and
that the transformations which it effects do not differ in any
essential feature from ordinary chemical changes.

Croft Hill's results have been challenged by Emmerling,
who claims that the sugar produced from the glucose is isomaltose and not the original sugar.

Hanriot states that he finds lipase capable of setting up a
reversing action and forming fat from fatty acid and glycerine
when appropriate proportions are present. The same observation has been made by Kastle and Loevenhart.

The action of other enzymes needs investigation from this
point of view. The decompositions which many of them set up
are more complex than those just described. The hydrolysis
effected by diastase for instance we have seen is progressive and
not complete in one operation, dextrins of different degrees
of complexity making their appearance simultaneously with

maltose, as the starch molecule is gradually broken down. Trypsin also does not effect the conversion of proteid into peptone and amides by a single operation but by a succession of stages, each characterised by the appearance of definite intermediate bodies. The complexity of the decompositions in these cases, however, does not negative the idea that the course of action is the same as in the cases already examined.

Fischer's hypothesis of the dependence of the action upon the configuration of the enzymes may throw some light upon their destruction by heat. It has generally been considered that they are entirely broken up, possibly by hydrolysis, at the temperature which we have called their maximum point. This view is supported by the observation that when perfectly free from water they may be heated to a much higher point without destruction. Pavy has shown that this is the case when the water is removed by alcohol instead of by desiccation. He found liver diastase active after being boiled in absolute alcohol. The process appears thus as if it might be due to hydrolysis. On the other hand it may be that the heating causes only such a change in their configuration as to cause their molecules no longer to correspond to those of the bodies they decompose. Tammann has shown that an enzyme loses its activity at a low temperature, though much more slowly than at a high one. As the temperature at which it is kept is raised it shows a greater proneness to become inactive. This is consistent with the view that their so-called decomposition is only a change in configuration.

We are acquainted with numerous cases in which simply heating an optically active substance changes its configuration. We may mention especially the classical experiment of Jungfleisch in which ordinary tartaric acid is transformed by this means into racemic acid, which is optically inactive. Many other instances of this phenomenon are to be met with in Fischer's researches on the sugar series.

BIBLIOGRAPHY.

THE following list, though not exhaustive, will be found to include the more important works that have a bearing upon the subject matter of this book. The figures in brackets after the names of Authors refer to the pages on which particular reference is made to the papers indicated.

ABELOUS et BIARNÉS (329). Sur le pouvoir oxydant du sang et des organes. *Compt. Rend. de la Soc. Biol.* S. x. 1 (1894). 536, 799.

ABELOUS et GÉRARD (331). Sur la présence dans l'organisme animal d'un ferment soluble, réduisant les nitrates. *Compt. Rend.* 129 (1899). 56.

ABELOUS et GÉRARD (332). Sur la présence dans l'organisme animal d'un ferment soluble réducteur. Pouvoir réducteur des extraits d'organes. *Compt. Rend.* 129 (1899). 164.

ABELOUS et GÉRARD (332). Transformation de la nitrobenzine en phényl-amine ou aniline par un ferment réducteur et hydrogénant de l'organisme. *Compt. Rend.* 130 (1900). 420.

ABELOUS et HEIM (44, 123, 198). Note sur l'existence de ferments digestifs dans les œufs de Crustacées. *Compt. Rend. Soc. Biol.* S. IX. 3 (1892). 273.

ALBERT und BUCHNER (360). Hefepresssaft und Fällungsmittel. *Ber. d. deut. chem. Gesell.* 33 (1900). 266, 971.

ALCOCK, Miss (189). The digestive processes of Ammocœtes. *Proc. Camb. Phil. Soc.* 7 (1891). 252.

ARTHUR, J. C. (95). The movement of protoplasm in cœnocytic hyphæ. *Annals of Botany* 11 (1897). 501.

ARTHUS (327). Glycolyse dans le sang et ferment glycolytique. *Arch. de Physiol.* (5) 3 (1891). 425.

ARTHUS (327). Glycolyse dans le sang. *Compt. Rend.* 114 (1892). 605.

ARTHUS (278). Fibrinogène et Fibrine. *Compt. Rend. Soc. Biol.* S. x. 1 (1894). 306.

ARTHUS et HUBER. Fermentations vitales et fermentations chimiques. *Compt. Rend.* 115 (1892). 839.

ARTHUS et PAGÈS. Recherches sur l'action du lab et la coagulation du lait. *Arch. de Physiologie* (5) 2 (1890). 335.

464   BIBLIOGRAPHY.

ARTHUS et PAGÈS (273). Nouvelle théorie chimique de la coagulation du sang. *Arch. de physiol. norm. et path.* (5) 2 (1890). 739.

ARTHUS, M. (456). Nature des Enzymes. *Thèse pour le Doctorat en Médecine.* Paris 1896.

ASTACHEWSKY (69). Reaction des Parotisspeichels beim gesunden Menschen. *Centr. f. d. med. Wissensch.* 16 (1878). 257.

ASTIER (4, 438). Expériences faites sur le sirop et le sucre de raisin. *Ann. de Chimie* t. LXXXVII. (1813). 271.

ATKINSON (119, 345). Note on the action of the new diastase *Eurotin* on starch. *Chemical News* 1880. 169.

BABCOCK and RUSSELL (207). Unorganised ferments in milk : a new factor in the ripening of cheese. *14th Ann. Rep. Agricultural Exp. Station, University of Wisconsin* (1897). 161.

BABCOCK, RUSSELL and VIVIAN (207). Properties of galactase, a digestive ferment of milk. *15th Ann. Rep. Agricultural Exp. Station, University of Wisconsin* (1898). 77.

BABCOCK, RUSSELL and VIVIAN (207). Distribution of galactase. *15th Ann. Rep. Agricultural Exp. Station, University of Wisconsin* (1898). 87, 93.

BABCOCK, RUSSELL, VIVIAN and HASTINGS (207, 232). The action of galactase and other proteolytic ferments on milk and cheese. *16th Ann. Rep. Agricultural Exp. Station, University of Wisconsin* (1899).

BAEYER, A. Ueber die Wasserentziehung und ihre Bedeutung für das Pflanzenleben und die Gährung. *Ber. d. deut. chem. Gesell.* (1870). 63, also *Journ. Chem. Soc. Trans.* (1871). 331.

BAGINSKY (262). Ueber das Vorkommen und Verhalten einiger Fermente. *Zeit. physiol. Chem.* 7 (1882). 209.

BARANETZKY (16, 65, 69, 415). Die stärkeumbildenden Fermente. 1878.

BARR (142). Melibiase. *Chem. Zeit.* (1895). 19. 1873.

BARTH. Zur Kenntniss des Invertins. *Ber. d. deut. chem. Gesell.* XI. (1878). 474.

DE BARY. Ueber die Myxomyceten. *Bot. Zeit.* 16 (1858). 357 et seq.

DE BARY (102). Sur la fermentation de la cellulose. *Bull. Soc. Bot. de France* 26 (1879). 25.

DE BARY (90). Ueber einige Sclerotinien und Sclerotien-Krankheiten. *Bot. Zeit.* 1886. 377 et seq.

DE BARY. Lectures on Bacteria. Eng. Trans. Oxford 1887.

BAUER. Ueber eine aus Aepfelpektin entstehende Zuckerart. *Landw. Versuchsstat.* 43. 191.

BAUMANN und BÖMER (180). Ueber die Fällung der Albumosen durch Zinksulfat. *Zeit. f. Untersuch. d. Nahr- u. Genussmittel* 1 (1898). 106.

BÉCHAMP. Sur l'acide acétique et les acides gras volatils de la fermentation alcoolique. *Compt. Rend.* 56 (1863). 969, 1086, 1231.

BÉCHAMP. Sur la fermentation alcoolique. *Compt. Rend.* 57 (1863). 674.

BÉCHAMP (113). Sur des nouveaux ferments solubles. *Compt. Rend.* 59 (1864). 496.

BÉCHAMP. Recherches sur la théorie physiologique de la fermentation alcoolique par la levure de bière. *Compt. Rend.* 75 (1872). 1036.

BÉCHAMP. *Mem. Académie Sciences* 28, 269, 347, 352.

BÉCHAMP. Existe-t-il une digestion sans ferments digestifs des matières albuminoïdes ? *Compt. Rend.* 118 (1894). 1157.

BÉCHAMP. *Ann. de Chim. et de Phys.* 13 (1867). 103.

BEIJERINCK (171). On Indigo-fermentation. *Botan. Zeit.* 2 Abt. 58 (1900). 188, abs. from *Proc. K. Akad. Wetensch. Amsterdam* 2 (1900). 120.

BEITLER. Ueber das Chlorproteinochrom. *Ler. d. deut. chem. Gesell.* 31 (1898). 1604.

BENDERSKY. Ueber die Ausscheidung der Verdauungsfermente aus dem Organismus bei gesunden und kranken Menschen. *Virchow's Archiv* 121 (1890). 554.

BENJAMIN. Beiträge zur Lehre von der Labgerinnung. *Virchow's Archiv* 145 (1896). 30.

BERNARD, Claud. De l'assimilation du sucre de canne. *Mem. Soc. Biol. Paris* 1849. 114.

BERNARD, Claud (236). Recherches sur les usages du suc pancréatique dans la digestion. *Compt. Rend.* 28 (1849). 249.

BERNARD, Claud (39). Critique expérimentale sur le mécanisme de la formation du sucre dans le foie. *Compt. Rend.* 85 (1877). 519.

BERNARD, Claud. Leçons sur les phénomènes de la vie. Paris 1878-9.

BERNARD, Claud (358). La fermentation alcoolique. Contributed by Berthelot to *Rev. Scientifique*, July 20, 1878, page 49.

BERTHELOT (6, 112). Sur la fermentation glucosique du sucre de canne. *Compt. Rend.* 50 (1860). 980.

BERTHELOT (350, 357). Remarques sur la formation de l'alcool et de l'acide carbonique et sur l'absorption de l'oxygène par les tissus des plantes. *Compt. Rend.* 128 (1899). 1366.

BERTRAND (311). Sur le latex de l'arbre à laque. *Compt. Rend.* 118 (1894). 1215.

BERTRAND (311). Recherches sur le latex de l'arbre à laque du Tonkin. *Bull. Soc. Chim.* (3) (1894). 11, 717.

BERTRAND (312). Sur la laccase et sur le pouvoir oxydant de cette diastase. *Comp. Rend.* 120 (1895). 266.

BERTRAND (313, 314). Sur la recherche et la présence de la laccase dans les végétaux. *Compt. Rend.* 121 (1895). 166.

BERTRAND (318). Sur une nouvelle oxydase ou ferment soluble oxydant, d'origine végétale. *Bull. de la Soc. Chim.* (3) xv. (1896). 793.

BERTRAND (312). Sur les rapports qui existent entre la constitution chimique des composés organiques, et leur oxidabilité sous l'influence de la laccase. *Compt. Rend.* 122 (1896). 1132.

BERTRAND (312, 319). Sur la présence simultanée de la laccase et de la tyrosinase dans le suc de quelques champignons. *Compt. Rend.* 123 (1896). 463.

BERTRAND (313). Sur l'intervention du manganèse dans les oxydations provoquées par la laccase. *Compt. Rend.* 124 (1897). 1032.

BERTRAND (314). Sur l'action oxydante des sels manganeux et sur la constitution chimique des oxydases. *Compt. Rend.* 124 (1897). 1355.

BERTRAND et MALLÈVRE (290 et seq.). Recherches sur la pectase et sur la fermentation pectique. *Morot's Journ. de Botanique* 8 (1894). 340.

BERTRAND et MALLÈVRE (290 et seq.). Sur la pectase et sur la fermentation pectique. *Compt. Rend.* 119 (1894). 1012.

BERTRAND et MALLÈVRE. Nouvelles recherches sur la pectase et la fermentation pectique. *Compt. Rend.* 120 (1895). 110.

BERTRAND et MALLÈVRE (293). Sur la diffusion de la pectase dans le règne végétal et sur la préparation de cette diastase. *Morot's Journ. de Botanique* 10 (1896). 37.

BERTRAND et MALLÈVRE (294). Sur la diffusion de la pectase dans le règne végétal, et sur la préparation de cette diastase. *Compt. Rend.* 121 (1895). 726.

BEYERINCK. Recherches sur la contagiosité de la maladie de gomme chez les plantes. *Archives Néerlandaises* 19 (1884). 43.

BEYERINCK (145). Die Lactase, ein neues Enzym. *Cent. f. Bact.* 1890. Bd. 6. 44.

BEYERINCK. Sur l'aliment photogène et l'aliment plastique des Bactéries lumineuses. *Archiv Néerlandais* 24 (1891). 369.

BEYERINCK. Ueber Nachweis und Verbeitung der Glucase, das Enzym der Maltose. *Cent. f. Bact.* 1895. II. Abth. Bd. I. 229, 265, 329.

BEYERINCK. Ueber die Arten der Essigbakterien. *Cent. f. Bakt.* II. Abt. 1898. 209.

BIAL (34, 132). Ueber die diastatische Wirkung des Blut- und Lymphserums. *Pflüger's Archiv* 52 (1892). 137.

BIAL (34, 132). Weitere Beobachtungen über das diastatische Ferment des Blutes. *Pflüger's Archiv* 52 (1893). 156.

BIAL (41). Ueber die Beziehungen des diastatischen Fermentes des Blutes und der Lymphe zur Zuckerbildung in der Leber. *Pflüger's Archiv* 55 (1894). 434.

BIEDERMANN. Beiträge zur vergleichenden Physiologie der Verdauung. *Pflüger's Archiv* 72 (1898). 105.

BIEDERMANN und MORITZ (102). Ueber ein celluloselösendes Enzym im Lebersecret der Schnecke (*Helix pomatia*). *Pflüger's Archiv* 73 (1898). 219.

BIEDERMANN und MORITZ. Ueber die Function der sogenannten "Leber" der Mollusken. *Pflüger's Archiv* 75 (1899). 1.

BIERNACKI. Das Verhalten der Verdauungsenzyme bei Temperaturerhöhungen. *Zeit. Biol.* 28 (1891). 49.

Biffi. Zur Kenntniss der Spaltungsprodukte des Caseïns bei der Pankreasverdauung. *Virchow's Archiv* 152 (1898). 130.

Biffin (248). A fat-destroying Fungus. *Ann. of Botany* 13 (1899). 375.

Biondi. Beiträge zur Lehre der fermentativen Prozesse in den Organen. *Virchow's Archiv* 144 (1896). 373.

Biot (261). Sur l'existence dans le sang des animaux d'une substance empéchant l'action de la présure sur le lait. *Compt. Rend.* 128 (1899). 1359.

Boehm (367). *Sitzber. d. k. Akad. d. Wiss. in Wien* 71 (1875).

Boehm. Ueber Stärkebildung aus Zucker. *Bot. Zeit.* 1883. 33, 49.

Bömer. *Zeit. Analyt. Chem.* 1895. 562.

Bokorny (339). *Allg. Brauer und Hopfen Zeit.* 36 (1896). 1573, 1591.

Bondonneau. Dextrine pure du malt. *Bull. Soc. Chem.* 23 (1875). 98.

Bondonneau. De la saccharification des matières amylacées. *Bull. Soc. Chim. de Paris* 25 (1876). 2 and *Compt. Rend.* 81 (1875). 972, 1210.

Bouchardat (33). Note sur la fermentation saccharine ou glucosique. *Compt. Rend.* 20 (1845). 107.

Bouchardat et Sandras (16, 33). Des fonctions du pancréas et de son influence dans la digestion des féculents. *Compt. Rend.* 20 (1845). 1085.

Bouchut (224). Sur un ferment digestif contenu dans le suc de figuier. *Compt. Rend.* 91 (1880). 67.

Boudourg (43). Recherches sur la valeur physiologique des tubes pyloriques de quelques Téléostéens. *Compt. Rend.* 128 (1899). 745.

Bouffard (320). Sur la cassage des vins. *Compt. Rend.* 118 (1894). 827.

Bouffard (322). Observations sur quelques propriétés de l'oxydase des vins. *Compt. Rend.* 124 (1897). 706.

Bouffard et Semichon (326). Contribution à l'étude de l'oxydase des raisins. *Compt. Rend.* 126 (1898). 423.

Bourquelot (190, 198). Recherches sur les phénomènes de la digestion chez les mollusques céphalopodes. *Thèse pour le Doctorat ès Sciences.* Paris 1884.

Bourquelot. Recherches sur les propriétés physiologiques du maltose. *Journ. de l'Anat. et de la Physiol.* 1886. 193.

Bourquelot. Les fermentations. Paris 1893.

Bourquelot (230). Ferments solubles sécrétés par l'*Aspergillus niger* et le *Penicillium glaucum*. *Bull. Soc. Biol.* 1893. 653.

Bourquelot (153). Présence et rôle de l'émulsine dans quelques champignons parasites des arbres ou vivants sur le bois. *Bull. Soc. Biol.* 1893. 804.

Bourquelot (137). Sur un ferment soluble nouveau dédoublant le tréhalose en glucose. *Compt. Rend.* 116 (1893). 826.

Bourquelot (81, 84). Inulase et fermentation alcoolique indirecte de l'inuline. *Compt. Rend.* 116 (1893). 1143.

Bourquelot (151). Présence d'un ferment analogue à l'émulsine dans les champignons. *Compt. Rend.* 147 (1892). 383 and *Bull. Soc. Mycol. de France* x. 1° fasc. (1894). 49.

BOURQUELOT (84). Inulase et fermentation alcoolique indirecte de l'inuline. *Compt. Rend. Soc. Biol.* S. IX. 5 (1893). 481.

BOURQUELOT (137). Transformation du tréhalose en glucose dans les champignons par un ferment soluble : la tréhalase. *Bull. de la Soc. Mycol. de France* T. IX. 3° fasc. (1893). 189.

BOURQUELOT (23, 81, 121, 130, 137, 141, 151, 230). Les ferments solubles de l'*Aspergillus niger*. *Bull. de la Soc. Mycol. de France* T. IX. 4° fasc. (1893). 230.

BOURQUELOT. Sur la présence de l'éther méthylsalicylique dans quelques plantes indigènes. *Journ. pharm. et chim.* (5) 30 (1894). 433.

BOURQUELOT (131). Maltase et fermentation alcoolique du maltose. *Journ. pharm. et chim.* (6) 2 (1895). 97.

BOURQUELOT. Sur la consommation du maltose par les êtres vivants. *Compt. Rend. Soc. Biol.* S. x. 2 (1895). 474.

BOURQUELOT (143). Sur l'hydrolyse de raffinose (mélitose) par les ferments solubles. *Journ. pharm. et chim.* 6 ser. 3 (1896). 390.

BOURQUELOT (166). Sur la présence dans le *Monotropa hypopythis* d'un glucoside et sur le ferment soluble hydrolysant de ce glucoside. *Compt. Rend.* 122 (1896). 1002 and *Journ. pharm. et chim.* 3 (1896). 577.

BOURQUELOT (319). Des composés oxydables sous l'influence du ferment oxydant des champignons. *Compt. Rend.* 123 (1896). 315.

BOURQUELOT (319). Action du ferment soluble oxydant des champignons sur les phénols insolubles dans l'eau. *Compt. Rend.* 123 (1896). 423.

BOURQUELOT (319). Nouvelles recherches sur le ferment oxydant des champignons II. et III. Son action sur les phénols et sur quelques dérivés éthérés des phénols. *Journ. pharm. et chim.* (6) 4 (1896). 241, 440.

BOURQUELOT. Ferments solubles oxydants et médicaments. *Journ. pharm. et chim.* (8) 4 (1896). 481.

BOURQUELOT. Les ferments solubles. Paris 1896.

BOURQUELOT (319). Sur la présence générale dans les champignons d'un ferment oxydant agissant sur la tyrosine. *Bull. de la Soc. Mycol. de France* XIII. (1897). 2° fasc. 65.

BOURQUELOT (127). Sur la physiologie du gentianose: son dédoublement par les ferments solubles. *Compt. Rend.* 126 (1898). 1045.

BOURQUELOT. Sur les pectines. *Journ. pharm. et chim.* (6) 9 (1899). 563.

BOURQUELOT. Sur les pectines. *Compt. Rend.* 128 (1899). 1241.

BOURQUELOT et BERTRAND (315). La laccase dans les champignons. *Compt. Rend.* 121 (1895). 783.

BOURQUELOT et BERTRAND (316). Sur la coloration des tissus et du suc de certains champignons au contact de l'air. *Journ. pharm. et chim.* 63 (1896). 177.

BOURQUELOT et BERTRAND (315). Les ferments oxydants dans les champignons. *Bull. de la Soc. Mycol. de France* T. XII. 1896. 1e fasc. 18.

BOURQUELOT et BERTRAND. Sur la coloration des tissus et du suc de certains champignons au contact de l'air. *Bull. de la Soc. Mycol. de France* T. XII. 1896. 1e fasc. 27.

BOURQUELOT et GLEY (133). Action du sérum sanguin sur la matière glycogène et sur le maltose. *Soc. de biologie* séance du 30 mars 1895.

BOURQUELOT et GLEY (140). Digestion du tréhalose. *Compt. Rend. Soc. Biol.* S. x. 2 (1895). 555.

BOURQUELOT et GRAZIANI (121). Sur quelques points relatifs à la physiologie du *Penicillium Duclauxi*. *Bull. de la Soc. Mycol. de France* T. VIII. 3e fasc. 1891. 147.

BOURQUELOT et HÉRISSEY. Sur les propriétés de l'émulsine des champignons. *Compt. Rend.* 121 (1895). 693.

BOURQUELOT et HÉRISSEY (23, 122, 140, 230). Les ferments solubles de *Polyporus sulphureus*. *Bull. de la Soc. Mycol. de France* T. x. 4e fasc. (1895). 235.

BOURQUELOT et HÉRISSEY (153). Action de l'émulsine de l'*Aspergillus niger* sur quelques glucosides. *Bull. de la Soc. Mycol. de France* T. XI. 3e fasc. (1896). 199.

BOURQUELOT et HÉRISSEY (143). Sur l'hydrolyse du mélézitose par les ferments solubles. *Journ. pharm. et chim.* (6) 4 (1896). 385.

BOURQUELOT und HÉRISSEY (143). Ueber die Hydrolyse der Melezitose durch lösliche Fermente. *Journ. pharm. et chim.* (6) 4 (1896). 385 and *Chem. Centr.* 1897. 1, 30.

BOURQUELOT et HÉRISSEY (107). Sur l'existence, dans l'orge germée, d'un ferment soluble, agissant sur la pectine. *Compt. Rend.* 127 (1898). 191.

BOURQUELOT et HÉRISSEY (127). De l'action des ferments solubles sur les produits pectiques de la racine de gentiane. *Journ. pharm. et chim.* (VI.) 8 (1898). 145.

BOURQUELOT et HÉRISSEY. Recherche et présence d'un ferment soluble protéolytique dans les champignons. *Compt. Rend.* 127 (1898). 666 and *Bull. Soc. Mycol. de France* 15 (1899). 66.

BOURQUELOT et HÉRISSEY (107). Sur la composition de l'albumen de la graine de caroubier. *Compt. Rend.* 129 (1899). 228.

BOURQUELOT et HÉRISSEY (107). Germination de la graine de Caroubier; production de mannose par un ferment soluble. *Compt. Rend.* 129 (1899). 614.

BOURQUELOT et HÉRISSEY (107). Sur les ferments solubles produits pendant la germination par les graines à albumen corné. *Compt. Rend.* 130 (1900). 42.

BOURQUELOT et HÉRISSEY (107). Sur l'individualité de la séminase, ferment soluble sécrété par les graines de légumineuses à albumen corné pendant la germination. *Compt. Rend.* 130 (1900). 340.

BOURQUELOT et HÉRISSEY (107). Les hydrates de carbone de réserve des graines de Luzerne et de Fenugrec. *Compt. Rend.* 430 (1900). 731.

BOURQUELOT et HÉRISSEY (107). Sur la présence de séminase dans les graines à albumen corné au repos. *Compt. Rend.* 131 (1900). 903.

BOURQUELOT et NARDIN. Sur la préparation de Gentianose. *Comptes Rend.* 126 (1898). 280.

BOUSSINGAULT (367). *Boussingault Agronomie* t. III. 1864.

BOUTRON et FREMY. Recherches sur les semences de moutardes noires et blanches. *Compt. Rend.* 9 (1839). 817.

BOUTRON et FREMY. Recherches sur la fermentation lactique. *Ann. de chim. et phys.* (3) t. 2 (1841). 257.

BOUTROUX. Sur la fermentation panaire. *Compt. Rend.* 113 (1891). 203.

BRACONNOT (297). Recherches sur un nouvel acide universellement répandu dans tous les végétaux. *Ann. de chim. et phys.* 28 (1825). 173.

BRACONNOT. Nouvelles observations sur l'acide pectique. *Ann. de chim. et phys.* 30 (1825). 96.

BRASSE (17). Sur la présence de l'amylase dans les feuilles. *Compt. Rend.* 99 (1884). 878.

BRÉAUDAT (170, 326). Sur le mode de formation de l'indigo dans les procédés d'extraction industriels. Fonctions diastasiques des plantes indigofères. *Compt. Rend.* 127 (1898). 769.

BRÉAUDAT (326). Nouvelles recherches sur les fonctions diastasiques des plantes indigofères. *Compt. Rend.* 128 (1899). 1478.

BREFELD (349). Ueber Gährung. *Landwirth. Jahrbuch* v. 1876.

BRISSEMORET et JOANNE. Sur le ferment digitalique. *Journ. pharm. et chim.* (VI.) 8 (1898). 481.

BROWN, A. J. (376). The chemical actions of pure cultivations of *Bacterium aceti*. *Journ. Chem. Soc. Trans.* 1896. 172.

BROWN, A. J. (376). An acetic ferment which forms cellulose. *Journ. Chem. Soc. Trans.* 1896. 432.

BROWN, H. T. (98). On the search for a cytolytic enzyme in the digestive tract of certain grain-feeding animals. *Journ. Chem. Soc. Trans.* 1892. 352.

BROWN and ESCOMBE (29, 96). On the depletion of the endosperm of *Hordeum vulgare* during germination. *Proc. Roy. Soc.* Vol. 63 (1898). 3.

BROWN and HERON (51). Contributions to the history of starch and its transformations. *Journ. Chem. Soc. Trans.* 1879. 596.

BROWN and HERON (44, 129). Some observations on the hydrolytic ferments of the pancreas and small intestine. *Proc. Roy. Soc.* 1880. 393.

BROWN and MORRIS (51, 71). The non-crystallisable products of the action of diastase on starch. *Journ. Chem. Soc.* 47 (1885). 527.

BROWN and MORRIS (53). The amylodextrin of W. Nägeli and its relation to soluble starch. *Journ. Chem. Soc.* 55 (1889). 449.

BROWN and MORRIS (18, 25, 73, 95, 397, 401, 414). On the germination of some of the Gramineae. *Journ. Chem. Soc. Trans.* 1890. 458.

BROWN and MORRIS. On the analysis of a beer of the last century. *Trans. Laboratory Club* III. 4 Feb. 1890.

BROWN and MORRIS (20, 59, 71, 116, 428). A contribution to the chemistry and physiology of foliage leaves. *Journ. Chem. Soc. Trans.* 1893. 604.

BROWN and MORRIS (54). On the action of diastase in the cold on starch-paste. *Journ. Chem. Soc. Trans.* 1895. 309.

BROWN and MORRIS (57). On the isomaltose of C. J. Lintner. *Journ. Chem. Soc. Trans.* 1895. 709.

BROWN, MORRIS and MILLAR. Experimental methods employed in the determination of the products of starch-hydrolysis by diastase. *Journ. Chem. Soc. Trans.* 1897. 72.

BROWN, MORRIS and MILLAR. Relations of specific rotatory powers and cupric-reducing powers of starch-hydrolysis by diastase. *Journ. Chem. Soc. Trans.* 1897. 115.

BROWN and PICKERING. Thermal phenomena attending the change in rotatory power of freshly prepared solutions of certain carbohydrates. *Journ. Chem. Soc. Trans.* 1897. 758.

BROWN and PICKERING. Thermochemistry of carbohydrate hydrolysis. *Journ. Chem. Soc. Trans.* 1897. 783.

BRÜCKE. *Wien. Acad. Ber.* (3) 65. 126.

BRÜCKE (7, 187, 427). Beiträge zur Lehre von der Verdauung. *Sitzungsber. d. k. Akad. d. Wiss. in Wien* 43 (1861). 601.

BRUNTON, Lauder, and MACFADYEN (24, 31, 73, 232). The Ferment-action of Bacteria. *Proc. Roy. Soc.* 46 (1889). 542.

BUCHANAN (269). Contributions to the physiology and pathology of the Animal Fluids. *London Medical Gazette* 18 (1835–6). 50.

BUCHANAN (269, 278). On the coagulation of the blood and other fibriniferous liquids. *London Medical Gazette* 1 (New ser. 1845). 617.

BUCHNER, E. (358, 423). Alcoholische Gährung ohne Hefezellen. *Ber. d. deut. chem. Gesell.* 30 (1897). 117 and 30 (1897). 1110.

BUCHNER, E. (326). Ueber zellenfreie Gährung. *Ber. d. deut. chem. Gesell.* 31 (1898). 568.

BUCHNER und RAPP. Alcoholische Gährung ohne Hefezellen. *Ber. d. deut. chem. Gesell.* 30 (1897). 2668 ; 31 (1898). 209, 1084, 1090, 1581 ; 32 (1899). 127.

BÜSGEN. Aspergillus Oryzae. *Ber. d. deut. bot. Gesell.* 3 (1885). LXVI.

BUFALINI. *Ann. di chim. e di farmac.* (4) T. X. 1889. 207.

BULL, Buckland W. Einige Beobachtungen über Emulsin und dessen Zusammensetzung. *Ann. d. Chem. und Phys.* 69 (1849). 145.

BUNGE. Textbook of Physiological and Pathological Chemistry. (English Translation by Wooldridge.) *London*, 1890.

472   BIBLIOGRAPHY.

BUSSEY (155). Note sur la formation de l'huile essentielle de moutarde. *Compt. Rend.* 9 (1839). 719.

BUTKEWITSCH (222). Ueber das Vorkommen proteolytischer Enzyme in gekeimten Samen und über ihre Wirkung. *Ber. d. deut. bot. Gesell.* 18 (1900). 185.

CAMUS (248). Formation de lipase par le *Penicillium glaucum. Compt. Rend. Soc. Biol.* 49 (1897). 192.

CAMUS (248). De la lipase dans les cultures d'*Aspergillus niger. Compt. Rend. Soc. Biol.* 49 (1897). 230.

CAMUS et GLEY (286). Action coagulante du liquide prostatique sur le contenu des vésicules séminales. *Compt. Rend.* 123 (1896). 194.

CAMUS et GLEY. Action du sérum sanguin sur quelques ferments digestifs. *Compt. Rend. Soc. Biol.* S. x. 4 (1897). 825.

CAMUS et GLEY (261). A propos de l'action empéchante du sérum sanguin sur la présure. *Compt. Rend.* 128 (1899). 1416.

CARLES (327). Valériane et oxydase. *Journ. de pharm. et chim.* (VI.) 12 (1900). 148.

CAZENEUVE (322). Sur le ferment soluble oxydant de la casse des vins. *Compt. Rend.* 124 (1897). 406.

CAZENEUVE (323). Sur quelques propriétés du ferment de la casse des vins. *Compt. Rend.* 124 (1897). 781.

CHEVASTELON (81). Contribution à l'étude des hydrates de carbon. *Thèse pour le Doctorat ès Sciences.* Paris 1894.

CHEVASTELON (81). Sur l'inuline de l'ail, de la jacinthe, de l'asphodèle et de la tubereuse. *Journ. Pharm.* (6) 2 (1895). 1, 83.

CHITTENDEN (210, 234, 262, 423). On the proteolytic action of Bromelin, the ferment of Pine-apple juice. *Journ. of Physiol.* xv. (1894). 249.

CHITTENDEN. Human Saliva. *Proc. Amer. Physiol. Soc.* 1898. 3.

CHITTENDEN and AMERMAN (193, 206). A comparison of natural and artificial gastric digestion. *Journ. of Physiol.* xiv. (1893). 483.

CHITTENDEN and ELY (68). Influence of Peptones and certain inorganic salts on the diastatic action of Saliva. *Journ. of Physiol.* iii. (1880). 327.

CHITTENDEN and GOODWIN. Myosin-Peptone. *Journ. of Physiol.* xii. (1891). 34.

CHITTENDEN and GRISWOLD (69). *Amer. Chem. Journ.* iii. (1881). 305.

CHITTENDEN and HARTWELL (193). The relative formation of proteoses and peptones in gastric digestion. *Journ. of Physiol.* xii. (1891). 12.

CHITTENDEN and MEARA (183). A study of the primary products resulting from the action of superheated water on coagulated egg-albumin. *Journ. of Physiol.* xv. (1894). 501.

CHITTENDEN and MENDEL (181, 193). Proteolysis of crystallized globulin. *Journ. of Physiology* xvii. (1894). 48.

CHITTENDEN, MENDEL and McDERMOTT. Papaïn-Proteolysis. *Amer. Journ. Physiology* (1898). 1, 255.

CHITTENDEN and PAINTER. Casein Digestion. *Stud. Lab. Physiol. Chem. Yale Univ.* 2 (1887). 156.

CHITTENDEN and SMITH. On the primary cleavage products formed in the digestion of gluten-casein of wheat by pepsin-hydrochloric acid. *Journ. of Physiology* IX. (1890). 410.

CHITTENDEN and SOLLEY (194). The primary cleavage products formed in the digestion of gelatin. *Journal of Physiology* XII. (1891). 23.

COHN. Zur Kenntniss des bei der Pancreasverdauung entstehenden Leucins. *Ber. d. deut. chem. Gesell.* 27 (1894). 2727.

COHN. Zur Kenntniss des bei Pancreasverdauung entstehenden Leucins. *Zeit. f. physiol. Chem.* 20 (1895). 203.

COHNHEIM (45). Zur Kenntniss der zuckerbildenden Fermente. *Virchow's Archiv* 28 (1863). 241.

COHNSTEIN und MICHAELIS (238). Ueber die Veränderung der Chylusfette im Blute. *Pflüger's Archiv* 65 (1897). 473.

CONN (264). Isolirung eines Lab-Fermentes aus Bakterienkulturen. *Centr. f. Bakt.* 12 (1892). 223.

CORNU (327). Sur les ferments oxydants de la vigne. *Journ. pharm. et chim.* (VI.) 10 (1899). 342.

COURMONT et DOYON. La substance toxique qui engendre le tétanos résulte de l'action, sur l'organisme récepteur, d'un ferment soluble fabriqué par le bacille de Nicolaier. *Compt. Rend.* 116 (1893). 593.

CREMER. Ueber die Umlagerungen der Zuckerarten unter dem Einflusse von Ferment und Zelle. *Zeit. Biol.* 31 (1895). 183.

CROSS and BEVAN (86, 103). Cellulose. London 1895.

CUBONI. Recherches sur la formation de l'amidon dans les feuilles de la vigne. *Arch. Ital. de Biol.* 7 (1886). 209.

CUISINIER (132). Sur une nouvelle matière sucrée diastasique et sa fabrication. *Monit. scientif.* 1886. 718.

CZUPEK (107). Zur Biologie der holzbewohnenden Pilze. *Ber. d. deut. bot. Gesell.* XVII. (1899). 166.

DACCOMO and TOMMASI (224). On a proteolytic ferment in *Anagallis arvensis. Rev. de Therap.* 59, 470.

DANILEWSKI (195). Ueber specifisch wirkende Körper des näturlichen und künstlichen pancreatischen Saftes. *Virchow's Archiv* 25 (1862). 279.

LE DANTEC. Recherches sur la digestion intracellulaire chez les Protozoaires. 1e et 2e parties. *Ann. de l'Inst. Pasteur* 4 (1890). 776 and 5 (1891). 163.

DARWIN (229). Insectivorous plants. London 1888.

DASTRE (427). Solubilité et activité des ferments solubles, en liqueurs alcooliques. *Compt. Rend.* 121 (1895). 899 and *Bull. de l'Acad. des Sciences* 24 (1895). 899.

DASTRE (35). Recherches sur la glycogène de la lymphe. *Compt. Rend. Soc. Biol.* S. X. 2 (1895). 242.

474   BIBLIOGRAPHY.

DASTRE et BOURQUELOT (128).   De l'assimilation du maltose.   *Compt.
Rend.* 98 (1884). 1604.

DASTRE et FLORESCO (194).   Liquéfaction de la gélatine.   Digestion saline
de la gélatine.   *Compt. Rend.* 121 (1895). 615.

DASTRE et FLORESCO (275).   Contribution à l'étude du ferment coagulateur
du sang.   *Compt. Rend.* 124 (1897). 94.

DAVIS (217).   Papaïn.   *Pharmaceutical Journal* 3 ser. 24 (1893–4). 207.

DAVY, E. (375).   On some combinations of Platinum.   *Phil. Trans. Roy.
Soc.* 1820. 108.

DENIS (267).   Mémoire sur le sang.   1859.

DESCHAMPS (253).   *Dingler's Polytechnisches Journal* 1840. No. 78 p.
445.

DESMAZIÉRES (4).   Recherches microscopiques et physiologiques sur le
genre Mycoderma.   *Ann. des Sciences Naturelles* t. x. 4 (1827). 4.

DETMER (242).   Vergleichende Physiologie des Keimungsprocesses der
Samen.   Jena 1880.

DETMER.   Ueber Fermentbildung und fermentative Prozesse.   Jena 1884.

DEVAUX (350).   Asphyxie spontanée et production d'alcool dans les tissus
profonds des tiges ligneuses poussant dans les conditions naturelles.
Compt. Rend. 128 (1899). 1346.

DICKINSON (271, 276).   Note on leech extract and its action on blood.
Journal of Physiology XI. (1890). 566.

DIENERT.   Sur la fermentation du galactose.   *Compt. Rend.* 128 (1899).
569, 617.

DIENERT.   Sur la sécrétion des diastases.   *Compt. Rend.* 129 (1899). 63.

DONATH (115).   Ueber den invertirenden Bestandtheil der Hefe.   *Ber. d.
deut. chem. Gesell.* 8 (1875). 795.

DOTT.   The digestive action of papaïn and pepsin compared.   *Pharm.
Journ.* 3 ser. 24 (1893–4). 758.

DUBOIS, R. (226).   Sur le prétendu pouvoir digestif du liquide de l'urne
des Népenthes.   *Compt. Rend.* 111 (1890). 315.

DUBOURG (131).   Recherches sur l'amylase de l'urine.   *Thèse pour le
Doctorat ès Sciences.*   Paris 1889.

DUBOURG.   De la fermentation der saccharides.   *Compt. Rend.* 128 (1899).
440.

DUBRUNFAUT.   Mémoire sur la saccharification.   *Société d'agriculture de
Paris.*   1823.

DUBRUNFAUT (6, 15).   Ueber Verwandlung des Stärkemehls in Zucker
durch Malz.   *Erdm. Journ. Tech. Chem.* IX. 156.

DUBRUNFAUT.   Note sur quelques phénomènes rotatoires et sur quelques
propriétés des sucres.   *Compt. Rend.* 23 (1846). 40.

DUBRUNFAUT.   Note sur la chaleur et le travail mécanique produits par la
fermentation vineuse.   *Compt. Rend.* 42 (1856). 945.

DUCLAUX (121).   Chimie biologique.   Paris 1883.

DUCLAUX.   *Thèses présentées à la faculté des sciences.*   Paris 1865.

DUCLAUX. Sur la saccharification. *Ann. de l'Institut Pasteur* 1895. 56, 170.

DUCLAUX (61). Traité de Microbiologie. Tome II. Paris 1899.

DUDDERIDGE. The manufacture of Pepsin and determination of its proteolytic power. *Pharm. Journ.* 3 ser. 23 (1892–3). 588.

DUMAS. Recherches sur la fermentation alcoolique. *Ann. de chim. et phys.* 5ᵉ ser. t. 3 (1874). 57.

DUNSTAN and HENRY (169). The nature and origin of the poison of *Lotus Arabicus. Proc. Roy. Soc.* LXVII. (1900). 224.

EBSTEIN und GRÜTZNER. Ueber Pepsinbildung im Magen. *Pflüger's Archiv* 8 (1874). 122.

EDELBERG (279). Ueber die Wirkungen des Fibrinfermentes im Organismus. *Arch. J. exper. Path. u. Pharmakol.* 12 (1880). 283.

EDKINS. The changes produced in casein by the action of pancreatic and rennet extracts. *Journ. of Physiology* XII. (1891). 193.

EDMUNDS (255). Note on Rennet and on the Coagulation of Milk. *Journ. of Physiology* XIX. (1896). 466.

EFFRONT (66). Sur des conditions chimiques de l'action des diastases. *Compt. Rend.* 115 (1892). 1324.

EFFRONT. Sur la formation de l'acide succinique et de la glycérine dans la fermentation alcoolique. *Compt. Rend.* 119 (1894). 169.

EFFRONT (67). Sur l'amylase. *Compt. Rend.* 120 (1895). 1281.

EFFRONT. Sur un nouvel hydrate de carbone, la caroubine. *Compt. Rend.* 125 (1897). 38.

EFFRONT (105). Sur une nouvelle enzyme hydrolytique "la caroubinase." *Compt. Rend.* 125 (1897). 116.

EFFRONT. Sur la caroubinose. *Compt. Rend.* 125 (1897). 309.

EFFRONT (326). Action de l'oxygène sur la levure de bière. *Compt. Rend.* 127 (1898). 326.

EFFRONT. Les enzymes et leurs applications. Paris 1899.

ELFERT. Ueber die Auflösungsweise der secundären Zellmembranen der Samen bei ihrer Keimung. *Bibliotheca Bot.* 30 Hft. 1894.

ELFVING. Studien über die Pollenkörner der Angiospermen. *Jenaische Zeit.* 13 (1879).

EMMERLING (348). Butylalcoholische Gährung. *Ber. d. deut. chem. Gesell.* 30 (1897). 451.

EMMERLING (461). Synthetische Wirkung der Hefenmaltase. *Ber. d. deut. chem. Gesell.* 34 (1901). 600.

ENGELMANN (203). *Hermann's Handbuch d. Physiologie.*

ETZINGER (193). Ueber die Verdaulichkeit der leimgebenden Gewebe. *Zeit. f. Biologie* Bd. X. 92.

EVANS (75). *Journ. Fed. Inst. of Brewing* 1899.

EVES, Miss (39, 63). Some experiments on the liver ferment. *Journ. of Physiology* V. (1884). 342.

EWALD (195). Versuche über die Wirksamkeit künstlicher Verdauungs-
Präparate. *Zeitsch. f. klin. Med.* I. (1880). 231.

FABRONI (3). Mémoire sur les fermentations présenté à l'Académie de
Florence 1787.

FANO. Das Verhalten des Peptons und Tryptons gegen Blut und Lymphe.
Du Bois-Reymond's Arch. Physiol. 1881. 277.

FERMI (233). Weitere Untersuchungen über die tryptischen Enzyme der
Mikroorganismen. *Cent. f. Bact.* 1891. Bd. 10. 401.

FERMI. Beitrag zum Studium der von den Mikroorganismen abgeson-
derten diastatischen und Inversionsfermente. *Cent. f. Bact.* 1892.
Bd. XII. 714.

FERMI und MONTESANO. Ueber die Dekomposition des Amygdalins durch
Mikroorganismen. *Cent. f. Bakt.* 15 (1894). 722.

FERMI und PERNOSSI. Ueber die Enzyme. *Centr. f. Bact.* 1894. Bd. 15.
229.

FERNBACH (121). Recherches sur la sucrase, diastase inversive du sucre
de canne. *Thèse pour le Doctorat ès Sciences.* Paris 1890.

FERNBACH. Sur le dosage de la sucrase. *Ann. de l'Inst. Pasteur* 3 (1889).
473, 531; 4 (1890). 1.

FERNBACH. Sur l'invertine ou sucrase de la levure. *Ann. de l'Inst.
Pasteur* 4 (1890). 641.

FERNBACH (168). Sur la tannase. *Compt. Rend.* 131 (1900). 1214.

FERNBACH et HUBERT (222). Sur la diastase protéolytique du malt.
Compt. Rend. 130 (1900). 1783.

FERNBACH et HUBERT (70). De l'influence des phosphates et de quelques
autres matières minérales sur la diastase protéolytique du malt.
Compt. Rend. 131 (1900). 293.

FICK. Ueber die Wirkungsart der Gerinnungsfermente. *Pflüger's Arch.*
45 (1889). 293.

FISCHER, E. Synthese einer neuen Glucobiose. *Ber. d. deut. chem.
Gesell.* 23 (1890). 3689.

FISCHER, E. (136, 145, 450). Einfluss der Configuration auf die Wirkung
der Enzyme. *Ber. d. deut. chem. Gesell.* 27 (1894). 2985, 3479 and 28
(1895). 1429.

FISCHER. Spaltung von Trehalose. *Ber. d. deut. chem. Gesell.* (1895). 1433.

FISCHER, E. Ueber die Isomaltose. *Ber. d. deut. chem. Gesell.* 28 (1895).
3024.

FISCHER. Bedeutung der Stereochemie für die Physiologie. *Zeit. f.
physiol. Chem.* 26 (1898-9). 60.

FISCHER und LINDNER (135, 142). Ueber die Enzyme von *Schizo-Sac-
charomyces octosporus* und *S. Marxianus.* *Ber. d. deut. chem. Gesell.*
28 (1895). 984.

FISCHER und LINDNER (122, 142). Ueber die Enzyme einiger Hefen.
Ber. d. deut. chem. Gesell. 28 (1895). 3034.

FISCHER und LINDNER. Verhalten der Enzyme gegen Melibiose, Rohrzucker und Maltose. *Ber. d. deut. chem. Gesell.* 28 (1895). 3035.

FISCHER und NIEBEL (140, 141, 144). Ueber das Verhalten der Polysaccharide gegen einige thierische Sekrete und Organe. *Chem. Centr.* 1896. 1, 499.

FITZ (347). On *Mucor racemosus* and alcoholic fermentation. *Soc. chim. de Berlin* Jan. and Feb. 1873.

FITZ (347). Ueber die Gährung des Glycerins. *Ber. d. deut. chem. Gesell.* 8 (1876). 1348.

FITZ. Ueber alkoholische Gährung. *Ber. d. deut. chem. Gesell.* 9 (1876). 1352.

FITZ (370). Ueber Schizomycetengährungen. *Ber. d. deut. chem. Gesell.* 10 (1877). 276.

FITZ (370). Ueber Spaltpilzgährungen. *Ber. d. deut. chem. Gesell.* 11 (1878). 1898; 12 (1879). 479; 13 (1880). 1309.

FLÜGGE. Microorganismen. 1896. III. Aufl. 202.

FOSTER (35, 73, 427). Notes on Amylolytic ferments. *Journal of Anatomy and Physiology* 1 (1867). 107.

FRANKFURT (119). Zur Kenntniss der chemischen Zusammensetzung des ruhenden Keims von Triticum vulgare. *Landw. Versuchsst.* 47 (1896). 449.

FRANKHAUSER (415). "What is diastase?" *Ann. Agrin.* 12 (1885). 340.

FRANKLAND, P. F. and G. C. The Nitrifying Process and its specific ferment. *Phil. Trans. Roy. Soc. B.* 1890. 107.

FREDERICQ (197). La digestion des matières albuminoïdes chez quelques invertébrés. *Arch. de Zool. Expérimentale* 7 (1878). 39.

FREDERICQ. *Bull. de l'Akad. Roy. de Belgique* 46 (1878).

FREDERICQ (281). Note sur le sang de l'Homard. *Bull. de l'Acad. Roy. de Belgique* (2) 47 (1879). No. 4.

FREDERIKSE. Einiges über Fibrin und Fibrinogen. *Zeit. f. physiol. Chem.* 19 (1894). 143.

FREMY. Premiers essais sur la maturation des fruits. *Bull. de trav. de la Soc. Pharm. de Paris* 26 (1840). 368.

FREMY (287). Mémoire sur la maturation des fruits. *Ann. chim. et phys.* (3) 24 (1848). 1.

FREMY. Recherches sur les fermentations. *Compt. Rend.* 75 (1872). 976 and 1060.

FRIEDBURG (254). The active principle of rennet, the so-called Chymosin. *Pharmaceutical Journal* 5 Jan. 1889. 526.

v. FÜRTH. Ueber die Eiweisskörper des Muskelplasmas. *Arch. f. exper. Path. u. Pharm.* 36 (1895). 231.

v. FÜRTH. Ueber die Einwirkung von Giften auf die Eiweisskörper des Muskelplasmas, und ihre Beziehung zu Muskelstarre. *Arch. f. exper. Path. u. Pharm.* 37 (1896). 389.

GAD (239).   Zur Lehre von der Fettresorption. *Archiv f. Anat. u. Physiol.* (1878). 181.

GAMGEE (269).   Some old and new experiments on the Fibrin-ferment. *Journ. of Physiology* II. (1879). 145.

GARDINER, W.   On the changes in the gland cells of *Dionæa muscipula* during secretion. *Proc. Roy. Soc.* 36 (1884). 180.

GARDINER, W.   On the phenomena accompanying stimulation of the gland cells in the tentacles of *Drosera dichotoma.* *Proc. Roy. Soc.* 39 (1885). 229.

GARDINER, W. (98).   The Histology of the cell-wall, with special reference to the mode of connexion of cells. *Proc. Roy. Soc.* 62 (1897). 100.

GAY-LUSSAC (3, 438).   Sur l'analyse de l'alcool et de l'éther sulfurique. *Ann. de chimie* t. XCV. 318.

GAYON (121).   Sur l'inversion et sur la fermentation alcoolique du sucre de canne par les moisissures. *Compt. Rend.* 86 (1878). 52.

GEDDES (281).   On the coalescence of Amœboid cells into Plasmodia and on the so-called coagulation of Invertebrate Fluids. *Proc. Roy. Soc.* 30 (1879). 252.

GEDULD (132).   *Wochenschrift für Brauerei* 8. 620 ; abs. in *Journ. Soc. Chem. Industry* 1892. 627.

GÉRARD (151).   Présence dans le *Penicillium glaucum* d'un ferment agissant comme l'émulsine. *Compt. Rend. Soc. Biol.* 1893. 651 and *Journ. de Pharm.* (5) 28. 11.

GÉRARD (154).   Sur le dédoublement de l'amygdaline dans l'économie. *Journ. de pharm. et chim.* (6) 3 (1896). 233.

GÉRARD (248).   Sur une lipase végétale extraite du *Penicillium glaucum. Compt. Rend.* 124 (1897). 370.

GÉRARD.   Transformation de la créatine en créatinine par un ferment soluble déshydratant de l'organisme. *Compt. Rend.* 132 (1901). 153.

GERBER (10, 351).   Maturation des fruits charnus. *Ann. de Sc. Nat.* Ser. VIII. T. 4. p. 1.

GERET und HAHN (231).   Zum Nachweis des im Hefepresssaft enthaltenen proteolytischen Enzyms. *Ber. d. deut. chem. Gesell.* 31 (1898). 202, 2335.

GESSARD.   Sur la tyrosinase. *Compt. Rend.* 130 (1900). 1327.

GILLESPIE (173, 182).   On the gastric digestion of proteids. *Journ. Anat. and Physiol.* 27 (1893). 195.

GOEBEL (226).   *Pflanzenbiologische Schilderungen* 11 (1893). 186.

GOLDSCHMIDT (410).   Zur Frage : Ist in Parotidenspeichel ein Ferment vorgebildet vorhanden oder nicht? *Zeitsch. f. physiol. Chem.* 10 (1886). p. 273.

GOLDSCHMIDT.   Zur Frage : Enthält die Luft lebende auf Stärke verzuckernd wirkende Fermente? *Zeit. f. physiol. Chem.* 10 (1886). 299.

GOLDSCHMIDT.   Zur Frage : Ist das Speichelferment ein vitales oder chemisches Ferment? *Zeit. f. physiol. Chem.* 10 (1886). 294.

GONNERMANN (24). *Chem. Zeit.* 19 (1895). 1806.

GONNERMANN (118). *Bied. Centr.* 28 (1899). 550.

v. GORUP-BESANEZ (16, 219). Ueber das Vorkommen eines diastatischen und peptonisirenden Fermentes in den Wickensamen. *Ber. d. deut. chem. Gesell.* 7 (1874). 1478.

v. GORUP-BESANEZ (226). *Sitzber. d. phys. med. Soc. zu Erlangen* 1874. 75 and 76.

v. GORUP-BESANEZ und GRIMM. Peptonbildende Fermente im Planzenreich. *Ber. d. deut. chem. Gesell.* 8 (1875). 1510.

v. GORUP-BESANEZ und WILL. Fortgesetzte Beobachtungen über peptonbildende Fermente im Pflanzenreiche. *Ber. d. deut. chem. Gesell.* 9 (1876). 673.

v. GORUP-BESANEZ und WILL (226). *Sitzber. d. phys. med. Soc. zu Erlangen* 1876.

GOUIRAND (320). Sur la présence d'une diastase dans les vins cassés. *Compt. Rend.* 120 (1895). 887.

GREEN, Reynolds (272). On certain points connected with the coagulation of the blood. *Journ. of Physiology* VIII. (1887). 354.

GREEN, Reynolds (219, 414). On the changes in the proteids in the seed which accompany germination. *Phil. Trans.* 178 (1887). B. 39.

GREEN, Reynolds (79, 414). On the germination of the Jerusalem Artichoke (*Helianthus tuberosus*). *Ann. of Botany* Vol. I. 1888.

GREEN, Reynolds (222, 243, 263). On the germination of the seeds of the Castor-oil plant (*Ricinus communis*). *Proc. Roy. Soc.* 48 (1890). 370.

GREEN, Reynolds (224). On the occurrence of vegetable trypsin in the fruit of *Cucumis utilissimus*. *Annals of Botany* 6 (1892). 195.

GREEN, Reynolds. On vegetable ferments. *Ann. of Botany* 1893. Vol. 7, p. 83.

GREEN, Reynolds (22, 59, 73, 120). On the Germination of the Pollen grain and the Nutrition of the Pollen tube. *Phil. Trans.* 185 (1894). B. 385.

GREEN, Reynolds (412). The action of light on diastase. *Phil. Trans.* B. Vol. 188 (1897). 167).

GREEN, Reynolds (289). The cell-membrane. *Science Progress* Vol. VI. 344.

GREEN, Reynolds. On the supposed alcohol-producing enzyme in yeast. *Ann. of Bot.* XI. (1897). 555.

GREEN, Reynolds. The alcohol-producing enzyme of yeast. *Annals of Botany* XII. (1898). 491.

GREENWOOD, Miss (202, 230). On the digestive process in some Rhizopods. *Journ. of Physiology* VII. (1885). 253 and VIII. (1887). 263.

GREENWOOD, Miss. On digestion in Hydra. *Journ. of Physiology* IX. (1888). 317.

GREENWOOD, Miss (190). On the Food Vacuoles in Infusoria. *Phil. Trans.* 185 (1894). B. 355.

GREENWOOD (Miss) and SAUNDERS (Miss). On the rôle of acid in proto-
zoan digestion. *Journ. of Physiol.* XVI. (1894). 441.

GRIESSMEYER (50). Ueber das Verhalten von Stärke und Dextrin gegen
Jod und Gerbsäure. *Liebig's Annalen der Chim. u. Pharm.* 160
(1871). 40.

GRIFFITHS (43). Physiology of the Invertebrata. London, Reeve and Co.
1892.

GRIS. Recherches sur la germination. *Ann. des Sc. Nat.—Botanique* (5)
2 (1865). 5.

GRÜSS. Ueber den Eintritt von Diastase in das Endosperm. *Ber. d.
deut. bot. Gesell.* 11 (1893). 286.

GRÜSS (100). Ueber die Einwirkung der Diastase auf Reservecellulose.
Ber. d. deut. bot. Gesell. 12 (1894). (60).

GRÜSS. Ueber das Verhalten des diastatischen Enzyms in der Keim-
pflanze. *Pringsheim's Jahrb.* 26 3 (1894). 379.

GRÜSS. Die Diastase in Pflanzenkörper. *Ber. d. deut. bot. Gesell.* 13
(1895). 2.

GRÜSS. *Wochenschr. f. Brauerei* (1895). 1257.

GRÜSS. *Bibliotheca bot.* (1896). Hft. 39.

GRÜSS. *Landw. Jahrb.* (1896). Bd. 25.

GRÜSS. Ueber Lösung von Cellulose durch Enzyme (Cytase). Abs. in
Chem. Centr. (1896). Bd. I. 313.

GRÜSS. Ueber die Secretion des Schildchens. *Pringsheim's Jahrb.* 30
(1897). 645.

GRÜSS. Studien über Reservecellulose. *Bot. Centr.* LXX. (1897). 242.

GRÜTZNER. Notizen über einige ungeformte Fermente des Säugethier-
organismus. *Pflüger's Archiv* 12 (1876). 302 and 14 (1877). 285.

GUIGNARD (148). Sur la localisation dans les plantes des principes qui
fournissent l'acide cyanhydrique. *Compt. Rend.* 110 (1890). 477.

GUIGNARD (148). Sur la localisation dans les amandes et le laurier-
cerise des principes qui fournissent l'acide cyanhydrique. *Journal de
Botanique* 4 (1890). 3.

GUIGNARD (156, 419, 420). Sur la localisation des principes actifs des
crucifères. *Journal de Botanique* 4 (1890). 385, 412, 435 and *Compt.
Rend.* 111 (1890). 249, 920.

GUIGNARD (156). Recherches sur la localisation des principes actifs chez
les Capparidées, Tropéolées, Limnanthés, Résédées. *Journal de
Botanique* (1893). Nes 19, 20, 22, 23, 24.

GUIGNARD (156). Recherches sur certains principes actifs encore in-
connus chez les Papayacées. *Journal de Botanique* 8 (1894). 67,
85.

GUIGNARD (160). Sur quelques propriétés chimiques de la Myrosine.
Bull. Soc. Bot. de France (3) 1 (1894). 418.

GUNNING (123). Untersuchungen über den Einfluss der Hefen auf
Zuckerlösung. *Ber. d. deut. chem. Gesell.* 5 (1872). 821.

DE HAAS und TOLLENS (288, 289). Untersuchungen über Pectinstoffe. *Liebig's Annalen* 286 (1895). 278.

HABERLANDT (25, 28). Die Kleberschicht des Gras-Endospermes als Diastase ausscheidendes Drüsengewebe. *Ber. d. deut. bot. Gesell.* 8, 40.

HAHN (230). Das proteolytische Enzym des Hefepresssaftes. *Ber. d. deut. chem. Gesell.* 31 (1898). 200.

HALLIBURTON (217). Physiological Chemistry.

HALLIBURTON (281). On the blood of decapod Crustacea. *Journ. of Physiology* VI. (1885). 183.

HALLIBURTON (284). On Muscle-plasma. *Journ. of Physiology* VIII. (1887). 159.

HALLIBURTON (270, 278, 433). On the nature of fibrin ferment. *Journ. of Physiology* IX. (1888). 265.

HALLIBURTON (271). Nucleo-proteids. *Journ. of Physiology* XVIII. (1895). 306.

HALLIBURTON and BRODIE. Nucleo-Albumins and Intravascular Coagulation. *Journ. of Physiology* XVII. (1894). 135.

HALLIBURTON and BRODIE (252, 260). Action of pancreatic juice on milk. *Journ. of Physiology* XX. (1896). 97.

HALLIER (259). Gährungserscheinungen (1867). 39.

HAMBURGER (34, 133). Vergleichende Untersuchung über die Einwirkung des Speichels, des Pancreas- und Darmsaftes, sowie des Blutes auf Stärkekleister. *Pflüger's Archiv* 60 (1895). 543.

HAMMARSTEN (253, 331). Ueber die Milchgerinnung und die dabei wirkenden Fermente der Magenschleimhaut. *Maly's Berichte* II. (1872). 118.

HAMMARSTEN. Ueber den chemischen Verlauf bei der Gerinnung des Caseïns mit Lab. *Maly's Berichte* IV. (1874). 135.

HAMMARSTEN (253, 256). Zur Kenntniss des Caseïns, und der Wirkung des Labfermentes. Upsala 1877.

HAMMARSTEN (253, 256). Zur Kenntniss des Caseïns, und der Wirkung des Labfermentes. *Maly's Berichte* VII. (1877). 166.

HAMMARSTEN. Zur Lehre von der Faserstoffgerinnung. *Pflüger's Arch.* 14 (1877). 211.

HAMMARSTEN. Ueber das Paraglobulin. *Pflüger's Archiv* 17 (1878). 413 ; 18 (1878). 38.

HAMMARSTEN. Untersuchungen über die Faserstoffgerinnung. *Nov. Acta Reg. Soc. Scientiar. Upsal.* Ser. 10. Vol. 10. Separatabdruck, Upsala 1878.

HAMMARSTEN. Ueber das Fibrinogen. *Pflüger's Arch.* 19 (1878). 563.

HAMMARSTEN. Ueber den Faserstoff und seine Entstehung aus dem Fibrinogen. *Pflüger's Archiv* 30 (1883). 457.

HAMMARSTEN (274, 416). Ueber die Bedeutung der löslichen Kalksalze für die Faserstoffgerinnung. *Zeit. f. physiol. Chem.* 22 (1896). 333.

HAMMARSTEN (274, 278). Weitere Beiträge zur Kenntniss der Fibrin-
bildung. *Zeit. physiol. Chem.* 28 (1899). 98.

HANRIOT (238). Sur un nouveau ferment du sang. *Compt. Rend.* 123
(1896). 753.

HANRIOT (241). Sur la non-identité des lipases d'origine différente.
Compt. Rend. 124 (1897). 778.

HANRIOT (461). Sur le méchanisme des actions diastatiques. *Compt.
Rend.* 132 (1901). 145, 212.

HANRIOT et CAMUS. Sur le dosage de la lipase. *Compt. Rend.* 124
(1897). 235.

HANSEN, A. (24, 224). Ueber Fermente und Enzyme. *Arbeit. des bot.
Inst. in Würzburg* III. (1887). 251, and *Bot. Zeit.* (1886). 137.

HANSEN. Die Verflüssigung der Gelatine durch Schimmelpilze. *Flora*
72 (1889). 88.

HANSTEEN (30). Ueber die Ursachen der Entleerung der Reservestoffe
aus Samen. *Flora* 79 (1894). 419.

HARCOURT, A. Vernon (457). On the observation of the course of chemical
change. *Journ. Chem. Soc. Trans.* 20, 460.

HARDY (282). The blood corpuscles of the Crustacea, together with a
suggestion as to the origin of the Crustacean fibrin-ferment. *Journ.
of Physiology* XIII. (1892). 165.

HARLAY. Caractères différentiels des produits de la digestion pepsique
et de la digestion pancréatique de la fibrine. *Journ. pharm. et
chim.* (VI) 9 (1899). 225.

HARLAY (193, 200, 218). De l'application de la Tyrosinase à l'étude des
ferments protéolytiques. *Thèse pour Doctorat de l' Université de Paris*
1900.

HARLEY. The behaviour of saccharine matter in the blood. *Journ. of
Physiology* XII. (1891). 391.

HARRIS and GOW (252). Ferment actions of the pancreas in different
animals. *Journ. of Physiology* XIII. (1892). 469.

HARTIG, R. Lehrb. d. Baumkrankheiten II. Berlin 1889.

HAYDUCK (338). *Zeit. Spirit. Ind.* (1881).

HAYEM. Observation à l'occasion du travail de M. Arthus sur le dosage
comparatif du fibrinogène et de la fibrine. *Compt. Rend. Soc. Biol.*
S. x. 1 (1894). 309.

HEDIN (199). Zur Kenntniss der Produkte der tryptischen Verdauung des
Fibrins. *Du Bois Archiv f. Physiol.* (1891). 273.

HEIDENHAIN (398, 410). Beiträge zur Kenntniss des Pancreas. *Pflüger's
Archiv* 10 (1875). 557.

HELBING and PASSMORE (217). *Pharmacological Record,* July 1893.

HELWES. Ueber Labferment im menschlichen Harn. *Pflüger's Archiv*
43 (1888). 384.

HENIGER. Recherches sur les peptones. *Compt. Rend.* 86 (1878).
1464.

HENNEBERG und STOHMANN. Ueber die Bedeutung der Cellulose-Gährung für die Ernährung der Thiere. *Zeit. f. Biol.* XXI. (1885). 613.

HÉRISSEY (145, 153, 154). Étude comparée de l'émulsine des amandes et de l'émulsine d'*Aspergillus niger. Bull. Soc. Biol.* (1896). 640.

HÉRISSEY (154). Sur la présence de l'émulsine dans les lichens. *Bull. de la Soc. Mycol. de France* 15 (1898).

HÉRISSEY (152). Recherches sur l'émulsine. *Thèse pour Doctorat de l'Université de Paris* 1899.

HÉRY (385). *Ann. de Micrographie* IV. 13.

HERZFELD (51). Ueber Maltodextrin. Halle 1879.

HERZFELD. Ueber die Einwirkung der Diastase auf Stärkekleister. *Ber. d. deut. chem. Gesell.* 12 (1879). 2120.

HEWSON (266). The Works of Wm. Hewson, F.R.S., edited with an introduction and notes by G. Gulliver, F.R.S. London, printed for the Sydenham Society, 1846.

HIEPE. On some products of starch transformation. *Coventry Brewers' Gazette* (1893). 732 ; (1894). 12, 37, 68, 99.

HILL, A. Croft (460). Reversible Zymohydrolysis. *Journ. Chem. Soc. Trans.* (1898). 634.

HIRSCHLER (199, 208). Bildung von Ammoniak bei der Prankreasverdauung von Fibrin. *Zeit. f. physiol. Chem.* 10 (1886). 302.

HJORT (230). Neue eiweissverdauende Enzyme. *Cent. f. Physiol.* X. (1896). 192.

HOFFMANN. Ueber das Schicksal einiger Fermente im Organismus. *Pflüger's Archiv* 41 (1887). 148.

HOFMEISTER. Ueber die chemische Structur des Collagens. *Zeit. f. physiol. Chem.* 2, 299.

HOFMEISTER. *Arch. f. Thierheilkunde* Bd. VII. (1881). 169 ; Bd. XI. Hfte. 1 and 2.

HOOKER (225). The carnivorous habits of plants. *Brit. Assoc. Reports* 1874 (Belfast). 102.

HOPPE-SEYLER. *Physiol. Chem.* (1877).

HOPPE-SEYLER. Ueber Unterschiede im chemischen Bau und der Verdauung höherer und niederer Thiere. *Pflüger's Archiv* 14 (1877). 395.

HOPPE-SEYLER. Gährung der Cellulose. *Ber. d. deut. chem. Gesell.* XVI. (1883). 122.

HÜFNER. Untersuchungen über ungeformte Fermente und ihre Wirkungen. *Journ. f. prakt. Chem.* N. F. Bd. 5 (1872). 372 ; 10 (1874). 1 ; 11 (1875). 43, 194.

HÜFNER. Recherche sur le ferment non organisé. *Bull. de la Soc. Chim. de Paris* (1877).

HUIE, Miss (403). Changes in the Cell-organs of *Drosera rotundifolia* produced by feeding with Egg-albumen. *Quart. Journ. Mic. Sc.* 39 (1897). 387.

31—2

484 BIBLIOGRAPHY.

JACOBSON (434, 455). Ueber ungeformte Fermente. *Inaug. Diss.* Berlin 1891.

JACOBSON. Untersuchungen über lösliche Fermente. *Zeitsch. f. physiol. Chem.* 16 (1892). 340.

JACQUEMART (301). Note sur la fermentation urinaire. *Ann. Chim. et Phys.* 3 ser. VII. (1843). 149.

JADIN (159). Localisation de la myrosine et de la gomme chez les Moringa. *Compt. Rend.* 130 (1900). 733.

DE JAGER. Erklärungsversuch über die Wirkungsart der ungeformten Fermente. *Virchow's Archiv* 121, 182.

JALOWITZ (57). *Chem Zeit.* (1895). 19, 2003.

JAQUET (330). Ueber die Bedingungen der Oxydationsvorgänge in den Geweben. *Archiv Exp. Path. Pharm.* 29 (1892). 386.

JOHANSEN (148). Sur la localisation de l'émulsine dans les Amandes. *Ann. Sci. Nat. (Bot.)* 7 ser. 6 (1887). 118.

JORGENSEN. Micro-organisms and fermentation. Trans. by Miller and Lenholm. London 1893.

JORISSEN et HAIRS (170). La Linamarine, nouveau glucoside fournissant de l'acide cyanhydrique par dédoublement. *Bull. de l'Académie royale des sciences de Belgique* 3 ser. 21 (1891). 529.

KASTLE und LOEVENHART (461). *Chem. Centr.* 72 (1901). 263.

KATZ. *Sitzungsb. d. Sächs. Ges. d. Wiss. Leipzig* (1896). 513.

KAUFMANN (40). Contribution à l'étude de ferment glucosique du foie. *Compt. Rend. Soc. Biol.* 9, 1 (1889). 600.

KAWALIER. Untersuchung der Blätter von *Arctostaphylos uva ursi. Journ. f. prakt. Chemie* 58 (1853). 193.

KEAN (94). On Rhizopus nigricans. *Bot. Gazette* xv. 173 (1890).

KELLNER, MORI und NAGAOKO (119). Beiträge zur Kenntniss der invertirenden Fermente. *Zeit. f. physiol. Chem.* 14 (1890). 297.

KIRCHHOFF (6, 15). Ueber die Zuckerbildung beim Malzen des Getreides. *Schweig's Journ.* 14 (1815). 389.

KJELDAHL (118). I. Recherches sur les ferments producteurs de sucre. II. Recherches sur la ptyaline. *Meddelelser from Carlsberg laboratory.* Vol. 1 (1878). 82.

KJELDAHL (70). *Meddel. fra Carlsberg Laboratoriet* 1879 t. 1.

KJELDAHL. *Compt. rend. des travaux du Laboratoire de Carlsberg* (1879). 129 and (1881). 189.

KLUG (191). Untersuchungen über Pepsinverdauung. *Pflüger's Archiv* 60 (1895). 43.

v. KNIERIEM. Ueber die Verwerthung der Cellulose im thierischen Organismus. *Zeit. f. Biol.* XXI. (1885). 67.

v. KNIERIEM und MAYER (375). Ueber die Ursache der Essiggährung. *Landwirth. Versuchsstat.* 16 (1873). 305.

KOCH. *Jahresb. d. Gährungsorg.* (1890). 171.

KOCH und HOSŒUS. Das Verhalten der Hefen gegen Glykogen. *Chem. Centr.* (1894). 2, 869.

KOHL. Ueber Assimilationsenergie und Spaltöffnungsmechanic. *Bot. Centr.* 64 (1895). 109.

KOSMANN (16, 118). Recherches chimiques sur les ferments contenus dans les végétaux. *Bull. de la Soc. Chim. de Paris* XXVII. (1877). 251.

KOSSEL und MATHEWS (200). Zur Kenntniss der Trypsinwirkung. *Zeit. physiol. Chem.* 25 (1898). 190.

KRABBE (19). Untersuchungen über das Diastaseferment unter specieller Berücksichtigung seiner Wirkung auf Stärkekörner innerhalb der Pflanze. *Pringsheim's Jahrb.* 21 (1890). 520.

KRAUCH (16). Beiträge zur Kenntniss der ungeformten Fermente im Pflauzenreich. *Landwirthsch. Versuchsstat.* XXIII. (1879).

KRAUCH (219). Ueber Pepton-bildende Fermente in den Pflanzen. *Landw. Versuchsst.* 27 (1882). 383.

KRAWKOW (46). Eine allgemeine Methode zur Darstellung unorganischer Fermente in reinen Wasseraufgüssen. *Journ. der russ. phys.-chem. Gesell.* 1 (1887). 387 ; and *Ber. d. deut. chem. Gesell.* Referatband 1887. 735.

KROBER. *Zeit. ges. Brauwes.* (1895). 18, 324, 334.

KRUKENBERG (190, 230). Ueber ein peptisches Enzym im Plasmodium der Myxomyceten und im Eidotter vom Huhne. *Maly's Jahresb. Thiersch.* 9 (1879). 270.

KRUKENBERG. *Kühne's Untersuchungen* II. (1882). 4, 13.

KRUKENBERG (230). Ueber ein peptisches Enzym im Plasmodium der Myxomyceten. *Kühne's Untersuchungen* II. (1882). 273.

KÜHNE (176). Ueber die Verdauung der Eiweissstoffe durch den Pankreassaft. *Virchow's Archiv* 39 (1867). 130.

KÜHNE (174). Ueber das Verhalten verchiedener organisirter und ungeformter Fermente. *Verhandl. d. naturhist.-med. Ver. Heidlbg.* 1 (1876). 190.

KÜHNE (196, 426). Ueber das Trypsin. *Verhandl. d. naturhist.-med. Ver. Heidlbg.* N. F. Bd. I. (1876). 194 and Bd. III. (1886). 463.

KÜHNE. Erfahrungen und Bemerkungen über Enzyme und Fermente. *Physiologisches Institut Heidelberg* (1878).

KÜHNE. Vereinfachte Darstellung des Trypsins. *Verhandl. d. naturhist.-med. Ver. Heidlbg.* 3 (1886). 463.

KÜHNE und CHITTENDEN (180). Ueber die nächsten Spaltungsproducte der Eiweisskörper. *Zeit. f. Biol.* 19 (1883). 184.

KÜHNE und CHITTENDEN (170). Ueber Albumosen. *Zeit. f. Biol.* 20 (1884). 11.

KÜHNE und CHITTENDEN. Ueber die Peptone. *Zeit. f. Biol.* 22 (1886). 423.

KÜHNE und LEA, Sheridan (399). Ueber die Absonderung des Pankreas. *Verhandl. naturhist.-med. Vereins Heidelberg* 1 (1877). 445.

KÜLZ und FRERICHS. Ueber den Einfluss der Unterbindung des Ductus choledochus auf den Glycogengehalt der Leber. *Pflüger's Archiv* 13 (1876). 460.

KÜLZ und VOGEL (58). Welche Zuckerarten entstehen bei dem durch thierische Fermente bewirkten Abbau der Stärke und des Glykogens? *Zeit. f. Biol.* 31 (1895). 108.

KURAÉEFF. Zur Kenntniss der Bromproteinchrome. *Zeit. physiol. Chem.* 26 (1899). 501.

KUTSCHER (184). Die Endprodukte der Trypsinverdauung. *Thèse inaugurale.* Strasbourg 1899.

KÜTZING (4, 374, 438). Organische Untersuchungen über die Hefe und Essigmutter. *Erdm. Journ. f. prakt. Chem.* XI. (1837). 381.

LABORDE. Sur la consommation du Maltose par une moisissure nouvelle, l'*Eurotiopsis Gaym.* Ast. *Compt. Rend. Soc. Biol.* S. x. 2 (1895). 472.

LABORDE. Recherches physiologiques sur une moisissure nouvelle, l'*Eurotiopsis Gayoni.* Paris 1896.

LABORDE (323). Sur l'oxydase du *Botrytis cinerea. Compt. Rend.* 126 (1898). 536.

LAFAR. Technical Mycology. Eng. Trans. by Salter. London 1898.

LAILLER. Note sur la fermentation ammoniacale de l'urine. *Compt. Rend.* 78 (1874). 361.

LANGLEY. On the destruction of ferments in the alimentary canal. *Journal of Physiology* III. (1881). 246.

LANGLEY (260, 407). On the histology of the mammalian gastric glands. *Journal of Physiology* III. (1881). 269.

LANGLEY and EDKINS (192, 408, 416). Pepsinogen and Pepsin. *Journal of Physiology* VII. (1886). 371.

LANGLEY and EVES (68). On certain conditions which influence the amylolytic action of Saliva. *Journal of Physiology* IV. (1882). 18.

LASZCYNSKI, De Verbus. *Zeit. ges. Brauw.* Abs. in *Journ. Chem. Soc.* Abs. 2 (1899). 791.

DE LATOUR, Cagniard (4, 438). Mémoire sur la Fermentation vineuse. *Ann. de chim. et phys.* 2e sér. t. LXVIII. (1838). 206.

LAVOISIER. Éléments de Chimie.

LEA, Sheridan (263). On the action of the Rennet Ferment contained in the seeds of *Withania coagulans. Journal of Physiology* V. (1883) VI. and *Proc. Roy. Soc.* 36 (1883). 55.

LEA, Sheridan (304). Some notes on the isolation of a soluble urea-ferment from the *Torula ureae. Journal of Physiology* VI. (1885). 136.

LEA, Sheridan (58, 72, 199). A comparative study of artificial and natural Digestions. *Journal of Physiology* XI. (1890). 226.

LEA, Sheridan, and GREEN, Reynolds (270). Some notes on the Fibrin-ferment. *Journal of Physiology* IV. (1883). 380.

LEBER (191). Bowman Lecture 1892. *Brit. Med. Journ.* 1643, 1357.

LECHARTIER et BELLAMY (5, 348). De la fermentation des fruits. *Comptes Rendus* 81 (1875). 1129.

LEIDY (202). Freshwater Rhizopods of N. America.

LEMERY (437). Cours de chimie. 1675.

LEPINE (328). Sur la présence normale dans le chyle d'un ferment destructeur du sucre. *Compt. Rend.* 110 (1890). 742.

LEPINE (328). Sur la production du ferment glycolytique. *Compt. Rend.* 120 (1895). 139.

LEPINE et BARRAL (327). Sur le pouvoir glycolytique du sang et du chyle. *Compt. Rend.* 110 (1890). 1314.

LEPINE et BARRAL (327). Sur la destruction du sucre dans le sang *in vitro. Compt. Rend.* 112 (1891). 146.

LEPINE et BARRAL (327). Sur l'isolement du ferment glycolytique du sang. *Compt. Rend.* 112 (1891). 411.

LEPINE et BARRAL (328). Sur les variations des pouvoirs glycolytique et saccharifiant du sang, et sur la localisation du ferment saccharifiant dans le sérum. *Compt. Rend.* 113 (1891). 1014.

LEPINOIS (326). Note sur les ferments oxydants de l'Aconit et de la Belladone. *Journ. de Pharm.* (6) 9 (1899). 49.

LEUCHS (6, 15, 33). Wirkung des Speichels auf Stärke. *Poggend. Annal.* XXII. (1831). 623.

LEUCHS. Ueber die Verzuckerung des Stärkmehls durch Speichel. *Kastner. Archiv Chemie* III. (1831). 105.

LÉVY. De la fermentation alcoolique des topinambours. *Compt. Rend.* 116 (1893). 1381.

LIBAVIUS (437). Alchymia. 1595.

LIBORIUS. Beiträge zur Kenntniss des Sauerstoffbedürfnisses der Bacterien. *Zeitsch. f. Hygiene* 1886. Bd. I. 156.

LIEBIG (5, 439). Sur les Phénomènes de la Fermentation et de la Putréfaction, et sur les causes qui les provoquent. *Ann. de chim. et de phys.* 2e sér. t. LXXI. (1839). 147.

LIEBIG. Ueber die Erscheinungen der Gährung, Fäulniss und Verwesung und ihre Ursachen. *Ann. d. Chem. u. Pharm.* 30 (1839). 250, 363.

LIEBIG (258). *Chemische Briefe* 1865. 159.

LIEBIG. Ueber die Gährung und die Quelle der Muskelkraft. *Ann. d. Chem. u. Pharm.* 153 (1870). 1, 137.

LIEBIG (375). Sur la fermentation. *Ann. de chim. et de phys.* (4) 23 (1871). 7.

LIEBIG et WOEHLER (6, 147). Sur la formation de l'huile d'amandes amères. *Ann. de chim. et phys.* 64 (1837). 185.

LILIENFELD (279). Ueber Blutgerinnung. *Zeit. physiol. Chem.* 20 (1894). 89.

LINDBERGER. Beiträge zur Kenntniss der Trypsinverdauung bei Gegenwart von freien Säuren. *Maly's Berichte* 13 (1884). 280.

LINDET (324). Sur l'oxydation du tanin de la pomme à cidre. *Compt. Rend.* 120 (1895). 370.

LING and BAKER (57, 135). The action of diastase on starch. *Journal of the Chemical Society. Trans.* 1895. 702 ; *Trans.* 1895. 739 ; *Trans.* 1897. 508.

LINNÆUS (261).   Pinguicula Vulgaris.   *Flora Laponica* 1737. page 10.

LINTNER, C. J.   Studien über Diastase.   *Zeit. f. d. ges. Brauwesen* 1886. 474.

LINTNER, C. J. (46).   Studien über Diastase.   *Journ. f. prakt. Chem.* (2) 34 (1886). 378; 36 (1887). 481.

LINTNER, C. J.   Ueber die chemische Natur der vegetabilischen Diastase. *Pflüger's Arch.* (1881). 205.

LINTNER, C. J. (56).   Ueber das Vorkommen von Isomaltose im Biere und in der Wurze.   *Zeit. f. d. ges. Brauwesen* 1891. 281.

LINTNER, C. J.   Ueber Isomaltose und deren Bedeutung für die Brauerei. *Zeit. f. d. ges. Brauwes.* 1892. 6.

LINTNER, C. J.   Ueber die Vergährbarkeit der Isomaltose.   *Zeit. f. d. ges. Brauwes.* 1892. 106.

LINTNER, C. J. (56).   Ueber die Einwirkung von Diastase auf Isomaltose. *Zeit. f. d. ges. Brauwesen* (1894). 378 and (1895). 173 and *Chem. Centr.* 1895. 1, 91.

LINTNER (135).   Ueber die Invertierung von Maltose und Isomaltose durch Hefe.   *Chem. Centr.* 1895. 1, 271.

LINTNER und DÜLL (56).   Versuche zu Gewinnung der Isomaltose aus den Produkten der Stärkeumwandlung durch Diastase.   *Zeit. f. angew. Chem.* 1892. 363, also *Zeit. f. d. ges. Brauwesen* 1892. 145.

LINTNER und DÜLL.   Ueber den Abbau der Stärke unter dem Einflusse der Diastasewirkung.   *Ber. d. deut. chem. Gesell.* 16 (1893). 2533.

LINTNER and DÜLL.   On a second achroodextrin formed by the action of diastase on starch.   *Zeit. f. d. ges. Brauwesen* 1894. 339.

LINTNER und ECKHARDT (30, 66, 415).   Studien über Diastase.   *Zeit. f. das ges. Brauwes.* 1889. 389 and *Journ. f. pr. Chem.* [2] 41 (1890). 91.

LINTNER und KRÖBER (135).   Verschiedenheit der Hefeglucase von Maïs-glucase und Invertine.   *Ber. d. deut. chem. Gesell.* 28 (1895). 1050.

LINZ.   Beiträge zur Physiologie der Keimung von *Zea Mais* L.   *Prings-heim's Jahrb.* 29 (1896). 267.

LOEW (427).   Ueber die chemische Natur der ungeformten Fermente. *Pflüger's Archiv* 27 (1882). 203 and 36 (1885). 170.

LOEW (71, 427).   On the Chemical Nature of Enzymes.   *Science.* N.S. vol. x. (1899). 955.

LÖRCHER (410).   Ueber Labwirkung.   *Pflüger's Archiv* 69 (1897). 141.

LUBAVIN (194).   Ueber die künstliche Pepsin-Verdauung des Caseins und die Einwirkung von Wasser auf Eiweisssubstanzen.   *Hoppe-Seyler's med.-chem. Untersuch.* Hf. IV. (1871). S. 463.

DE LUCA (349, 367).   Recherches chimiques tendant à démontrer la production de l'alcool dans les feuilles, les fleurs et les fruits de certaines plantes.   *Ann. de Sc. Nat.* (6) 6 (1878).

LUDWIG (316).   Ueber das Chromogen des *Boletus cyanescens*, und anderer auf frischem Bruche blau werdenden Pilze.   *Arch. de Pharm.* (2) CXLIX. 1872. 107.

LUTZ. Amygdalin und Emulsin in den Samen gewisser Pomaceen. *Rep. de Pharm.* 1897. 312.

MACALLUM (405, 433). Contributions to the Morphology and Physiology of the Cell. *Trans. Canadian Inst.* 1 (1890). 247.

MACALLUM (405). On the distribution of assimilated iron compounds in animal and vegetable cells. *Quart. Journ. Mic. Sc.* 38 (1895). 229.

MACALLUM (405). On the detection and localisation of phosphorus in animal and vegetable tissues. *Proc. Roy. Soc.* 63 (1898). 467.

MACFADYEN, MORRIS and ROWLAND (359). On expressed yeast-cell Plasma (Buchner's "zymase"). *Proc. Roy. Soc.* 67 (1900). 250.

MALFITANO (230). La protéolyse chez l'Aspergillus niger. *Ann. de l'Inst. Pasteur* 14 (1900). 60.

MALY (188). Ueber die chemische Zusammensetzung und physiologische Bedeutung der Peptone. *Pflüger's Archiv* 9 (1874). 592.

MANGIN (87). Étude historique et critique sur la présence des composés pectiques dans les tissus des végétaux. *Journ. de Botanique* 5 (1891). 400, 440.

MANGIN (87, 288, 289). Propriétés et réactions des composés pectiques. *Morot's Journ. de Botanique* 6 (1892). 206, 235, 363.

MANGIN. Recherches sur les composés pectiques. *Journ. de Botanique* 7 (1893). 37, 121, 325.

MARCANO (223). Sur la fermentation peptonique. *Compt. Rend.* 99 (1884). 811 and 107 (1888). 117.

MARCANO (209). *Bulletin of Pharmacy* v. (1891). 77.

MARLIÈRE (107). Sur la graine et spécialement sur l'endosperme du *Ceratonia siliqua*. *La Cellule* 13 (1897). 7.

MAROCHOWETZ und LAUROW. Zur Kenntniss der Chemismus der peptischen und tryptischen Verdauung der Eiweissstoffe. *Zeit. f. physiol. Chem.* 26 (1898). 513.

MARTIN, S. H. C. (215, 262). Papaïn-Digestion. *Journ. of Physiology* v. (1884). 213.

MARTIN, S. H. C. (215). The nature of Papaïn and its action on vegetable proteids. *Journ. of Physiology* vi. (1885). 336.

MARTINAND (322). Action de l'air sur le moût de raisin et sur le vin. *Compt. Rend.* 121 (1895). 502.

MARTINAND (321). Sur l'oxydation et la casse des vins. *Compt. Rend.* 124 (1897). 512.

MATHIEU et HALLOPEAU. *Arch. de Méd. expér.* No. 3. 1893.

MAYER, Ad. Die Lehre von den chemischen Fermenten. Heidelberg 1882.

MAYS, Karl. Ueber die Wirkung von Trypsin in Säuren und von Pepsin und Trypsin aufeinander. *Unters. a. d. physiol. Inst. d. Univ. Heidel.* 1880. 378.

MAZÉ (350). Signification physiologique de l'alcool dans le regne végétal. *Compt. Rend.* 128 (1899). 1608.

MAZÉ (245). Recherches sur la digestion des réserves dans les graines en voie de germination et leur assimilation par les plantules. *Compt. Rend.* 130 (1900). 424.

MEISSNER (175). *Zt. f. nat. Med.* VII. 1; VIII. 280; X. 1; XII. 46; XIV. 303; résumé by Lehmann in *Biol. Centr.* 4 (1884). 407.

v. MERING (130). Ueber den Einfluss diastatischer Fermente auf Stärke, Dextrin, und Maltose. *Zeit. f. physiol. Chem.* v. (1881). 185.

METSCHNIKOFF (201). Untersuchungen über die intracelluläre Verdauung bei wirbellosen Thieren. *Arbeit. a. d. Zool. Instit. Wien* 5 (1883). 141.

METSCHNIKOFF. Recherches sur la digestion intracellulaire. *Ann. de l'Inst. Pasteur* 3 (1889). 25.

METZLER (193). Beiträge zur Verdauung des Leims. Giessen 1860. *Schmidt's Jahrb.* Bd. 110, 153.

MEYEN (4). Pflanzenphysiologie. 3e v. 455.

MEYER, A. (404). Ueber Stärkekörner welche sich mit Jod rot färben. *Ber. d. deut. bot. Gesell.* 4 (1886). 337.

MEYER, A. Zu der Abhandlung von Krabbe "Untersuchungen über das Diastase." *Ber. d. d. bot. Gesell.* 1891. 9, 238.

MEYER, A. Untersuchungen über die Stärkekörner. Jena 1895.

MIALHE (6, 16, 33). De la digestion et de l'assimilation des matières sucrées et amiloïdes. *Comptes Rend.* 20 (1845). 954, 1485.

MIALHE. Mémoire sur la digestion et l'assimilation des matières amyloïdes et sucrées. 1846.

MIERAN (119). *Chem. Zeit.* 17. 1283.

MIGUEL (303, 307). Sur le ferment soluble de l'urée. *Compt. Rend.* 111 (1890). 397.

MITSCHERLICH (4). Chemische Zersetzung und Verbindung vermittelst Contactsubstanzen. *Poggendorff's Ann.* LV. 225.

MITTELMEIER (54). Beiträge zur Kenntniss der diastatischen Zersetzung der Stärke. *Chem. Centr.* 1895. II. 163.

MITTELMEIER. *Chem. Centr.* 1897. II. 1010.

MIURI. Inversion des Rohrzuckers. *Ber. d. deut. chem. Gesell.* (1895). 623.

MIURI (122). Ist der Dünndarm im Stande, Rohrzucker zu invertiren? *Zeit. Biol.* 32 (1895). 266.

MIX (342). On a kephir-like yeast found in the United States. *Proc. Amer. Acad. of Arts and Sciences* 26 (1891). 102.

MIYOSHI. On a chitin-dissolving enzyme found in *Empusa, Entomophthora*, &c. *Jahrb. f. wiss. Bot.* 29 (1895). 277.

MOLISCH. Ueber die sogenannte Indigogährung und neue Indigopflanzen. *Sitzber. d. k. Akad. der Wiss. in Wien.* July 1898.

MONOYER. *Thèse de médecine.* Strasbourg 1862.

MORITZ and GLENDINNING (72). Note on diastatic action. *Journ. Chem. Soc.* 1892. 688.

MORRIS, G. H. (135). Glucase. *Trans. Inst. Brew.* 6 (1894). 132.

MORRIS and WELLS. Diastase in yeast. *Trans. Instit. Brewing* v. 6.

MÜLLER. *Ann. Agrin.* 12 (1885). 481.

MÜLLER. *Landwirth. Versuchsstat.* 16. 273.

MÜLLER (302). Ueber Conservirung und Concentrirung des menschlichen Harns. *Journ. f. prakt. Chem.* 81 (1860). 467.

MÜLLER und EBSTEIN. Ueber den Einfluss der Säuren und Alkalien auf das Leberferment. *Ber. d. deut. chem. Gesell.* 8 (1875). 679.

MÜLLER und MASUYAMA (44). Ueber ein diastatisches Ferment im Hühnerei. *Zeit. Biol.* 39 (1900). 547.

MÜLLER und SCHWANN. Versuche über die künstliche Verdauung des geronnenen Eiweisses. *Müller's Archiv* 1836. 66.

MÜLLER et SCHWANN. Expériences sur la digestion artificielle. *Ann. Sci. Nat. Zool.* (2) 7 (1837). 313.

MÜNTZ (242). Sur la germination des graines oléagineuses. *Annales de chim. et phys.* (4) 22 (1871). 472.

MÜNTZ (350). Recherches sur la fermentation alcoolique intracellulaire des végétaux. *Compt. Rend.* 86 (1878). 49, and *Ann. de chim. et phys.* (5) 13 (1878). 543.

MÜNTZ (350). *Boussingault Agronomie* VI. 1878.

MÜNTZ et MARCANO. Sur la formation des terres nitrées dans les régions tropicales. *Ann. chim. et phys.* (6) 10 (1887). 563.

v. MUSCULUS (50). Remarques sur la transformation de la matière amylacée en glucose et dextrine. *Compt. Rend.* 50 (1860). 785.

v. MUSCULUS. Nouvelle note sur la transformation de l'amidon en dextrine et glucose. *Compt. Rend.* 54 (1862). 194.

v. MUSCULUS (50). De la dextrine. *Ann. chim. et phys.* (4) 6 (1865). 177.

v. MUSCULUS (303). Sur un papier réactif de l'urée. *Compt. Rend.* 87 (1874). 132.

v. MUSCULUS (304). Sur le ferment de l'urée. *Compt. Rend.* 82 (1876). 334.

v. MUSCULUS und GRÜBER. Ein Beitrag zur Chemie der Stärke. *Zeit. f. physiol. Chem.* II. (1878). 177.

v. MUSCULUS et GRÜBER (50). Sur l'amidon. *Compt. Rend.* 86 (1878). 1459 and *Bull. Soc. Chim.* 30 (1878). 54.

v. MUSCULUS und v. MERING. Ueber die Umwandlung von Stärke und Glycogen durch Diastas, Speichel, Pancreas- und Leberferment. *Zeit. f. physiol. Chem.* 2 (1878). 403.

v. MUSCULUS et v. MERING. De l'action de la diastase de la saliva et du suc panc. sur l'amidon et sur le glycogène. *Bull. de la Soc. Chim. de Paris* T. XXXI. No. 3. 105.

MUSSI (224). Cradina, a proteolytic enzyme in the Fig (*Ficus Carica*). *L'Orosi* Nov. 1890. 364. Abs. in *Pharm. Journ.* 3 ser. 21 (1890–1). 560.

NADELMANN. Ueber die Schleimendosperme der Leguminosen. *Jahrb. f. wiss. Bot.* 21 (1890). 670.

v. NAEGELI, C. (5). Theorie der Gährung. München 1879.

NÄGELI, W. Beiträge zur Kenntniss der Stärkegruppe. Leipzig 1874.

NASSE. Untersuchungen über die ungeformten Fermente. *Pflüger's Arch.* 11 (1875). 138.

NASSE (55). Bemerkungen zur Physiologie der Kohlehydrate. *Pflüger's Archiv* 14 (1877). 477.

NASSE. Fermentprocesse unter dem Einfluss von Gasen. *Pflüger's Archiv* 15 (1877). 471.

NASSE (42). Zur Anat. und Physiol. der quergestreiften Muskelsubstanz. Leipzig 1882.

NASSE und FRAMM. Bemerkungen zur Glykolyse. *Pflüger's Archiv* 63 (1896). 203.

NENCKI. Ueber die Zersetzung der Gelatine und des Eiweisses bei der Fäulniss mit Pancreas. Bern 1876. Abs. in *Jahresber. f. Thier-chemie* 6, 31.

NENCKI (200). Zur Kenntniss der pankreatischen Verdauungsproducte des Eiweisses. *Ber. d. deut. chem. Gesell.* 26 (1895). 560.

NEUMEISTER (181). Zur Kenntniss der Albumosen. *Zeit. f. Biol.* 23 (1887). 381.

NEUMEISTER. Bemerkungen zur Chemie der Albumosen und Peptone. *Zeit. f. Biol.* 24 (1888). 267.

NEUMEISTER (183). Ueber die nächste Einwirkung gespannter Wasser-dämpfe auf Proteïne. *Zeit. f. Biol.* 26 (1890). 57.

NEUMEISTER (200). Ueber die Reactionen der Albumosen und Peptone. *Zeit. f. Biol.* 26 (1890). 324.

NEUMEISTER. Physiol. Chem. 1893.

NEUMEISTER (218). Ueber das Vorkommen und die Bedeutung eines eiweisslösenden Enzyms in jugendlichen Pflanzen. *Zeit. Biol.* 30 (1894). 447.

NEUMEISTER. Bemerkungen zu Edward Buchner's Mittheilungen über "Zymase." *Ber. d. deut. chem. Gesell.* 30 (1897). 2963.

NEWCOMBE (99, 101). Cellulose-Enzymes. *Ann. of Botany* 13 (1899). 49.

ONIMUS. Phénomènes consécutifs à la dialyse des cellules de la levure de bière. *Compt. Rend.* 119 (1894). 479.

ORTLOFF. Ueber die Natur und chemische Constitution des in den Mandeln enthaltenen Emulsins. *Arch. der Pharmacie* 48. 1846.

OSBORNE, T. B. (422, 425). The Chemical Nature of diastase. *Journ. Amer. Chem. Soc.* 17 (1895). 587.

OSBORNE and CAMPBELL (425). The Chemical Nature of diastase. *Journ. Amer. Chem. Soc.* 18 (1896). 536.

OSBORNE and CAMPBELL (222). The proteids of malt. *Journ. Amer. Chem. Soc.* 18 (1896). 542.

OSTWALD. *Ber. d. Sächs. Ges. d. Wiss.* 1894. 337.

O'SULLIVAN, C. (50, 51). On the transformation-products of Starch. *Journ. Chem. Soc. Trans.* 1872. 579 and 1879. 770.

O'SULLIVAN, C., and TOMPSON (123, 429). Invertase, a contribution to the history of an enzyme or unorganised ferment. *Journ. Chem. Soc. Trans.* 57 (1890). 834.

O'SULLIVAN, J. (118). The influence of germination upon the constituents of barley. *Trans. Laboratory Club* III. 5 1890.

PARKIN (79). Contributions to our knowledge of the formation, storage and depletion of Carbohydrates in Monocotyledons. *Phil. Trans. Roy. Soc.* ser. B. 191 (1899). 35.

PASTEUR. Mémoire sur la fermentation appelée lactique. *Ann. de chim. et de phys.* 3ᵉ ser. T. 52 (1858). 407.

PASTEUR. De l'origine des ferments. *Compt. Rend.* 50 (1860). 849.

PASTEUR (4). Mémoire sur la fermentation alcoolique. *Ann. de chim. et phys.* 3ᵉ ser. T. 58 (1860). 323.

PASTEUR. Sur les ferments. *Bull. Soc. Chim.* 1861. 61.

PASTEUR. Sur la fermentation visqueuse et la fermentation butyrique. *Bull. Soc. Chim.* 1861. 30.

PASTEUR. Note sur un mémoire de M. Liebig, relatif aux fermentations. *Ann. chim. et phys.* 4 ser. T. 25 (1872). 145.

PASTEUR. Faits nouveaux pour servir à la connaissance de la théorie des fermentations proprement dites. *Compt. Rend.* 75 (1872). 784.

PASTEUR. Sur la production de l'alcool par les fruits. *Compt. Rend.* 75 (1872). 1054.

PAUTZ und VOGEL (134, 141, 144). Ueber die Einwirkung der Magen- und Darmschleimhaut auf einige Biosen und auf Raffinose. *Zeit. Biol.* 32 (1895). 304.

PAVY. Physiology of the Carbohydrates. London 1894.

PAVY (40). Hepatic Glycogenesis. *Journal of Physiology* XXII. (1898). 391.

PAWLEWSKI (421). Ueber die Unsicherheit der Guajak-Reaction auf wirksame Diastase. *Ber. d. deut. chem. Gesell.* 30 (1897). 1313.

PAYEN. Analyse de la partie corticale de l'Ailanthus glandulosa. *Ann. de chim. et phys.* (2) 26 (1824). 329.

PAYEN (298). *Recueil des savants étrangers* (9) (1846). 148.

PAYEN. Note sur la racine charnue du cerfeuil bulbeux. *Compt. Rend.* 43 (1856). 769.

PAYEN. Réaction de la Diastase sur la substance amylacée dans différentes conditions. *Ann. chim. et phys.* (4) 4 (1865). 286.

PAYEN et PERSOZ (6, 15, 50). Mémoire sur la diastase, les principaux produits de ses réactions et leurs applications aux arts industriels. *Ann. de chim. et phys.* LIII. (1833). 73.

PEKELHARING (274). *Virchow's Festschrift* 1891. Bd. 1.

PEKELHARING (271, 274). Untersuchungen über das Fibrinferment. *Verhand. d. Kon. Akad. v. Wet. Amsterdam* 1892.

PEKELHARING (271, 415, 433). Ueber die Beziehung des Fibrinferments des Blutserums zum Nucleoproteid des Blutplasmas. *Du Bois-Reymond's Archiv* 1895. 213 and *Centr. Physiol.* 9 (1895). 102.

PEKELHARING (189). Ueber eine neue Bereitungsweise des Pepsins. *Zeit. physiol. Chem.* 22 (1896). 233.

PEKELHARING (278). Ueber das Vorhandensein eines Nucleoproteïds in Muskeln. *Zeit. physiol. Chem.* 22 (1896). 245.

PERDRIX (347). Sur les fermentations produites par un microbe anaérobic (*Bacillus amylozyme*). *Ann. de l'Institut Pasteur* 5 (1891). 287.

PETERS. Untersuchungen über das Lab und die labähnlichen Fermente. *Preisschrift.* Rostock 1894.

PETIT (189). Études sur la pepsine. Paris 1881.

PETIT (28, 74). Variations des matières sucrées pendant la germination de l'orge. *Compt. Rend.* 120 (1895). 687.

PETIT et LABOURASSE. Sur la solubilisation des matières azotées du malt. *Compt. Rend.* 131 (1900). 349.

PFEFFER. *Sitzber. d. König. Sächs. Ges. d. Wiss.* 1893. 1192.

PFEFFER (32, 73, 398). Ueber die regulatorische Bildung von Diastase. *Ber. d. math.-phys. Classe d. Kön. Sächs. Gesell. der Wissensch. zu Leipzig* (1896). 513.

PFEFFER. *Bied. Centr.* 1897. 26 400 from *Ber. math.-phys. Klasse Kgl. Sächs. Gesell. Wissen. Leipzig.*

PICK. Ein neues Verfahren zur Trennung von Albumosen und Peptonen. *Zeit. f. physiol. Chem.* 24 (1897). 246.

PICK. Zur Kenntniss der peptischen Spaltungsprodukte des Fibrins. *Zeit. physiol. Chem.* 28 (1899). 219.

PIÉRI et PORTIER (317). Sur la présence d'une oxydase dans les branchies, les palpes et le sang des Acéphales. *Compt. Rend.* 123 (1896). 1314.

PIRIA. Recherches sur la salicine. *Ann. de Chem. et de Phys.* (3) 14 (1845). 257.

v. PLANTA (115). Ueber den Futtersaft der Bienen. *Zeit. physiol. Chem.* 12 (1888). 327.

POHL. Zur Kenntniss des oxydativen Fermentes. *Arch. f. exper. Path. u. Pharm.* 38 (1897). 65.

POPOFF. Ueber die Einwirkung von eiweissverdauenden Fermenten auf die Nucleïnstoffe. *Zeit. f. physiol. Chem.* 18 (1894). 533.

PORTES (148). Recherches sur les amandes amères. *Journ. de Pharm. et Chim.* 26 (1877). 410.

POTTER. On a bacterial disease of the Turnip. *Proc. Roy. Soc.* 67 (1901). 442.

POTTEVIN (168). La tamase. Diastase dédoublant l'acid gallo-tannique. *Compt. Rend.* 131 (1900). 1215.

POZZI-ESCOT. Les diastases et leurs applications. *Paris: Masson et Cie.* 1900.

PREGL (34, 134). Ueber Gewinnung, Eigenschaften, und Wirkungen des Darmsaftes vom Schafe. *Pflüger's Archiv* 61 (1895). 359.

PROCTER (166). Observations on the volatile oil of *Betula lenta*. *Amer. Journ. of Pharmacy* 15 (1844). 241.

PRUNET (22). Sur le mécanisme de la dissolution de l'amidon dans la plante. *Compt. Rend.* 115 (1892). 751.

PURIEWITSCH (30). Ueber die selbstthätige Entleerung der Reservestoff-behälter. *Ber. deut. bot. Gesell.* 14 (1896). p. 207.

PURIEWITSCH. Physiologische Untersuchungen über die Entleerung der Reservestoffbehälter. *Pringsheim's Jahrb.* Vol. 31 1897. p. 1.

QUEVENNE (4). Étude microscopique et chimique du Ferment, suivie d'expériences sur la fermentation alcoolique. *Journ. de Pharm.* (2) t. XXIV. (1838). 265.

RACHFORD (239). The influence of bile on the fat-splitting properties of pancreatic juice. *Journ. of Physiology* XII. (1891). 72.

RACIBORSKI. Ein Inhaltskörper des Leptoms. *Ber. d. deut. bot. Gesell.* 16 (1898). 52, 119.

RASETTI. *L'Orosi* 21 (1898). 289.

REINITZER (109, 419). Ueber die wahre Natur des Gummifermentes. *Zeit. physiol. Chem.* 14 (1890). 453.

REINITZER. Ueber das zellwandlösende Enzym der Gerste. *Zeit. physiol. Chem.* 23 (1897). 175.

REISS (104). Ueber in den Samen als Reservestoff abgelagerte Cellulose und eine daraus erhaltene neue Zuckerart, die "Seminose." *Ber. d. deut. chem. Gesell.* 22 (1889). 610.

REISS (104). Ueber die Natur der Reservecellulose, und über ihre Auflö-sungsweise bei der Keimung der Samen. *Ber. d. deut. bot. Gesell.* VII. (1889). 322.

REYCHLER (415, 426). La saccharification diastasique. *Bull. Soc. chim. Paris* 3e sér. t. 1 No. 5 286, also *Ber. d. deut. chem. Gesell.* 22 (1889). 414.

REY-PAILHARDE (313). Rôles respectifs du philothion et de la laccase dans les graines en germination. *Compt. Rend.* 121 (1895). 1162.

RICHET (368). De la fermentation lactique du sucre de lait. *Compt. Rend.* 86 (1878). 550.

RICHET. De quelques conditions de la fermentation lactique. *Compt. Rend.* 88 (1879). 750.

RICHET. De la diastase ureopoïétique. *Compt. Rend. Soc. Biol.* S. x. 1 (1894). 525.

RIDEAL (218). Some notes on papaïn digestion. *Pharm. Journ.* 3 ser. 24 (1893–4). 845.

RIDEAL (218). The conditions of papaïn digestion. *Pharm. Journ.* 3 ser. 25 (1894–5). 183.

RIESS und WILL. Einige Bemerkungen über "fleischessende" Pflanzen. *Bot. Zeit.* (1875). 713.

RINGER (273). The influence of certain salts upon the act of clotting. *Journ. of Physiol.* XI. (1890). 369.

RINGER (256). Regarding the action of lime salts on Casein and on Milk. *Journ. of Physiol.* XI. (1890). 464.

RINGER (256). Further observations on the behaviour of Caseinogen. *Journ. of Physiol.* XII. (1891). 164.

ROBERTS (253). Note on the existence of a milk-curdling ferment in the Pancreas. *Proc. Roy. Soc.* 29 (1879). 157.

ROBERTS. The digestive ferments. *Lumleian Lectures* 1880.

ROBERTS (65, 201). On the estimation of the amylolytic and proteolytic activity of pancreatic extracts. *Proc. Roy. Soc.* 32 (1881). 145.

ROBIQUET et BOUTRON (147). Nouvelles expériences sur les amandes amères et sur l'huile volatile qu'elles fournissent. *Ann. de chim. et phys.* 44 (1830). 352.

ROBIQUET und BOUTRON (147). Geschichtliche Darstellung der Arbeiten über die bitteren Mandeln, nebst einigen Betrachtungen über die in den Annal. der Pharmac. Bd. XXII. S. 1 enthaltene Abhandlung von Wöhler und Liebig. *Journ. de Pharm.* 23 (1837). 589 and *Liebig's Annal.* 25 (1838). 175.

RÖHMANN (34). Zur Kenntniss des diastatischen Ferments der Lymphe. *Pflüger's Archiv* 52 (1892). 157.

RÖHMANN (132). Zur Kenntniss der Glucase. *Ber. d. deut. chem. Gesell.* 27 (1894). 3251.

RÖHMANN. Zur Kenntniss der bei der Trypsinverdauung aus dem Casein entstehenden Produkte. *Ber. d. deut. chem. Gesell.* 31 (1898). 2188.

RÖHMANN und LAPPE (144). Ueber die Lactase des Dünndarms. *Ber. d. deut. chem. Gesell.* 28 (1895). 2506.

ROLLESTON. Forms of animal life. 2nd edn. 1888.

ROUX (327). Sur une oxydase productrice de pigment sécrété par le coli-bacille. *Compt. Rend.* 128 (1899). 693.

DU SABLON, Leclerc (247). Sur la germination des graines oléagineuses. *Rev. gén. de Botanique* 1895. 145.

DU SABLON, Leclerc. Sur la digestion de l'amidon dans les plantes. *Compt. Rend.* 127 (1898). 968.

SACHS (242). Ueber das Auftreten der Stärke bei der Keimung ölhaltiger Saamen. *Bot. Zeit.* 1859. 178 et seq.

SACHS (100). Zur Keimungsgeschichte der Dattel. *Bot. Zeit.* 1862. 241.

SACHS. Ein Beitrag zur Kenntniss der Ernährungsthätigkeit der Blätter. *Arbeit. d. Bot. Inst. in Würzburg* 3 1884. 1.

SACHS. Lectures on the Physiology of Plants. Eng. Trans. by Marshall Ward. Oxford 1887.

ST JENTYS (21). *Verhandl. Akad. Wiss. Krakau* 1893.

SALKOWSKI. Ueber das eiweisslösende Ferment der Fäulnissbakterien und seine Einwirkung auf Fibrin. *Zeit. Biol.* 25 (1889). 92.

SALKOWSKI (41). Notiz über das diastatische Ferment der Leber. *Pflüger's Archiv* 56 (1894). 351.

SALKOWSKI und KATSUSABURO YAMAGUWA (330). Zur Kenntniss des Oxydationsfermeunts der Gewebe. *Virchow's Archiv* 147 (1897). 1.

SALTER (404). Contributions to a fuller knowledge of starch grains. *Thesis for degree of D.Sc.* London 1898.

SARTHOW (327). Sur une oxydase retirée du Schinus Molle. *Journ. de Pharm. et Chim.* (VI) 11 (1900). 482.

DE SAUSSURE (3). Nouvelles observations sur la composition de l'alcool et de l'éther sulphurique. *Ann. de chimie* 89 (1814). 273.

SCHÄFER (280). Experiments on the conditions of coagulation of fibrinogen. *Proc. Physiol. Soc. Journ. of Physiol.* XVII. (1895). p. xviii.

SCHEIBLER. Vorläufige Mittheilung über die Metapectinsäure aus Zuckerrüben. *Ber. d. deut. chem. Gesell.* 1 (1868). 58.

SCHEIBLER. *Zeitsch. f. Rübenzucker Industrie* 1874. 309.

SCHIFF (205). Leçons sur la Physiologie de la digestion. Paris 1867.

SCHIFFERER. Ueber die nicht krystallisierbaren Produkte der Einwirkung der Diastase auf Stärke. *Inaug. Diss.* Kiel 1892.

SCHIMPER. Ueber die Bildung und Wanderung der Kohlehydrate in den Laubblättern. *Bot. Zeit.* 1885. 738.

SCHLEICHERT. Die diastat. Fermente d. Pflanzen. 1893. p. 85.

SCHLESINGER. Zur Kenntniss der diastatischen Wirkung des menschlichen Speichels, nebst Abriss der Geschichte dieses Gegenstandes. *Virchow's Archiv* 125 (1891). 146 and 340.

SCHLOESING et MÜNTZ (380). Recherches sur la nitrification par les ferments organisés. *Compt. Rend.* 84 (1877). 301 ; 85 (1877). 1018 ; 86 (1878). 892.

SCHLOESING et MÜNTZ. Recherches sur la nitrification. *Compt. Rend.* 89 (1879). 891, 1074.

SCHMIDT (335). Handwörterbuch der Chemie. 1848.

SCHMIDT, Al. (267). Weiteres über den Faserstoff und die Ursachen seiner Gerinnung. *Arch. f. Anat. u. Phys.* 1862. 428, 533.

SCHMIDT, Al. (267). Neue Untersuchungen über die Faserstoffgerinnung. *Pflüger's Archiv* 6 (1872). 445, 457.

498 BIBLIOGRAPHY.

SCHMIDT, R. H. Ueber Aufnahme und Verarbeitung von fetten Oelen durch Pflanzen. *Flora* 74 (1891). 304.

SCHMIDT-MÜLHEIM. Beiträge zur Kenntniss des Peptons und seiner physiologischen Bedeutung. *Du Bois-Reymond's Arch. Phys.* 1880. 33.

SCHMIEDEBURG, O. (307). Ueber Spaltungen und Synthesen im Thierkörper. *Arch. Exp. Path. Pharm.* 14 (1881). 379, Histozyme 382.

SCHNEEGANS (166). Betulase, ein in *Betula lenta* enthaltenes Ferment. *Pharm. Centr.* 38, 27, 1896.

SCHNEEGANS. Zur Kenntniss der ungeformten Fermente. *Journ. de Pharm. von Elsass Lothringen* 1896. 17.

SCHNEEGANS und GEROCK (166). Ueber Gaultherin, ein neues Glykosid aus *Betula lenta* L. *Arch. der Pharmacie* 1894. 437.

SCHŒNBEIN (316, 420). Ueber Ozon und Ozonwirkungen in Pilzen. *Philosoph. Mag.* XI. No. 70, p. 137.

SCHRÖTTER. Beiträge zur Kenntniss der Albumosen. *Monatsch. f. Chem. Wien* 16 (1895). 609 and 17 (1896). 199.

SCHÜTZENBERGER (5, 243). Les Fermentations. 6th edn. Bib. Sci. Int. Paris 1896.

SCHÜTZENBERGER (184). Recherches sur la constitution chimique des peptones. *Compt. Rend.* 115 (1892). 208, 764.

SCHULZ (367). *Journ. f. prakt. Chem.* 87 (1862).

SCHULZE, E. Zur Chemie der pflanzlichen Zellmembranen. *Zeit. physiol. Chem.* 16 (1892). 387.

SCHULZE. Ueber die Zellwandbestandtheile der Cotyledonen &c. *Ber. d. deut. bot. Gesell.* 14 (1896). 66.

SCHULZE, E. Ueber den Umzatz der Eiweissstoffe in der lebenden Pflanze. *Zeit. physiol. Chem.* 24 (1897). 18.

SCHULZE und BARBIERI. Zur Bestimmung der Eiweissstoffe in den Pflanzen. *Landw. Versuchsstat.* 26 (1881). 213.

SCHULZE, STEIGER und MAXWELL. Zur Chemie der Pflanzenmembranen. *Zeit. physiol. Chem.* 14 (1890). 227; 19 (1894). 38.

SCHUMBERG. Ueber das Vorkommen des Labferments im Magen des Menschen. *Virchow's Archiv* 97 (1884). 260.

SCHUNCK (162). On Rubian and its products of decomposition. *Phil. Trans.* 1851. 433.

SCHUNCK (162). Erythrozyme. *Phil. Trans.* 1853. 74.

SCHWANN (4, 438). Vorläufige Mittheilung betreffend Versuche über die Weingährung und Fäulniss. *Poggendorff's Ann.* 41 (1837). 184.

SCHWANN (und MÜLLER) (6). Versuche über die künstliche Verdauung des geronnenen Eiweisses. *Müller's Archiv* 1836. 66.

SCHWANN (et MÜLLER) (6). Expériences sur la digestion artificielle. *Ann. Sci. Nat. Zool.* (2) 7 (1837). 313.

SCHWEDER, C. G. (193). Zur Kenntniss der Glutinverdauung. *Dissert.* Berlin 1867. *Virchow's Jahresbericht* 1867. Bd. 1 152.

SCHWIENING (42). Ueber fermentative Prozesse in den Organen. *Virchow's Archiv* 136 (1894). 444.

SEEGEN (63). Ueber die Umwandlung von Glycogen durch Speichel- und Pancreasferment. *Pflüger's Archiv* 19 (1879). 106.

SEEGEN (328). On the glycolytic action of blood. *Centr. Physiol.* 5, 821, 869.

SEEGEN und KRATSCHMER. Zur Kenntniss der saccharificirenden Fermente. *Pflüger's Archiv* 14, 593.

SEEGEN und KRATSCHMER (39). Die Natur des Leberzuckers. *Pflüger's Archiv* 22 (1880). 206.

SEYFFERT. Untersuchungen über Gerste und Malz Diastase. *Zeit. f. d. ges. Brauwesen* 1898.

SHARP, Gordon (217). Papaïn digestion. *Pharmaceutical Journal* 3 ser. 24 (1893–4). 633.

SHARP, Gordon. The action of papaïn on egg- and serum-albumin. *Pharm. Journ.* 3 ser. 24 (1893–4). 757.

SHARP, Gordon. Papainverdauung : Völlige Abwesenheit von Pepton. Ueber Papaïn. *Chem. Centr.* 1894. 1, 512.

SHORE, L. E. On the effect of peptone on the clotting of blood and lymph. *Journ. of Physiol.* 11 (1890). 561.

SIEGFRIED und BALKE (184). Ueber Antipepton. *Zeit. f. physiol. Chem.* 27 (1899). 335.

SIGMUND (171, 245). Ueber fettspaltende Fermente im Pflanzenreiche. *Monats. f. Chem. Wien* 11 (1890). 272, also *Sitzungsber. der k. Akad. der Wiss. in Wien* 99 (1890). 407 and 100 (1891). 328.

SIGMUND. Beziehungen zwischen fettspaltenden und glycosidspaltenden Fermenten. *Sitzber. d. k. Akad. d. Wiss. in Wien* 101 (1892). 549.

SMITH, F. Analysis of Saliva of horse. *Veterinary Journal* 26 June 1888.

SOXHLET (258). Beiträge zur physiologischen Chemie der Milch. *Journ. f. prakt. Chem.* (new ser.) 6 (1872). 1.

SOXHLET (257). *Milch-Zeitung* 1877. 573.

SOXHLET. Human Milk and Cow's Milk. *Münchener Medicinische Wochenschrift* 1893, trans. in *Pharm. Journ.* 3 ser. 23 (1892–3). 785.

SPATZIER. Ueber das Auftreten und die physiologische Bedeutung des Myrosins in der Pflanze. *Pringsheim's Jahrb.* 25 (1893). 39.

SPITZER (330). Die zuckerzerstörende Kraft des Blutes und der Gewebe. *Pflüger's Arch.* 60 (1895). 303.

SPITZER (330, 434). Die Bedeutung gewisser Nucleoproteïde für die oxydative Leistung der Zelle. *Pflüger's Archiv* 67 (1897). 615.

SPITZER (330). Weitere Beobachtungen über die oxydativen Leistungen thierischer Gewebe. *Pflüger's Archiv* 71 (1898). 596.

STADELMANN. Ueber Fermente im normalen Harne. *Zeit. f. Biol.* 24 (1888). 226.

STADELMANN (200, 208). Bildung von Ammoniak bei Pankreas-Verdauung von Fibrin. *Zeit. f. Biol.* 24 (1888). 261.

STADELMANN (200). Ueber das beim liefen Zerfall der Eiweisskörper entstehende Proteinochromogen. *Zeit. f. Biol.* 26 (1889). 491.

STONE (24). Action of enzymes on starch of different origins. *Office of Expt. Stat. U. S. Agricultural Department* 34 (1896). 29–44.

SUNDBERG (189). Ein Beitrag zur Kenntniss des Pepsins. *Zeit. f. physiol. Chem.* 9 (1885). 319.

SZILAGYI. Ueber Diastase. *Zeit. f. g. Brauwesen* 1891. 258.

TAIT, Lawson (226). Insectivorous plants. *Nature* 12 (1875). 251.

TAMMANN. Die Reactionen der ungeformten Fermente. *Zeit. f. physiol. Chem.* 16 (1892). 271.

TAMMANN. Zur Wirkung ungeformter Fermente. *Zeit. f physik. Chem.* 28 (1895). 426.

TANRET, C. et G. (165). Sur la rhaminase et la xanthorhamnine. *Bull. Soc. Chim.* (III) 21 (1899). 1073.

TANRET, C. et G. (165). Sur le rhaminose. *Compt. Rend.* 129 (1899). 725.

TAPPEINER. Ueber Celluloseverdauung. *Ber. d. deut. chem. Gesell.* 15 (1882). 999.

TAPPEINER. Ueber Cellulosegährungen. *Ber. d. deut. chem. Gesell.* 16 (1883). 1734.

TAPPEINER. Ueber die Sumpfgasgährung im Schlamme der Teiche, Sümpfe, und Kloaken. *Ber. d. deut. chem. Gesell.* 16 (1883). 1740.

TAPPEINER. Untersuchungen über die Gährung der Cellulose insbesondere über deren Lösung im Darmkanale. *Zeit. f. Biol.* 20 (1884). 52.

TAPPEINER. Untersuchungen über die Eiweissfäulniss im Darmkanale der Pflanzenfresser. *Zeit. f. Biol.* 20 (1884). 215.

TAPPEINER. Nachträge zu den Untersuchungen über die Gährung der Cellulose. *Zeit. f. Biol.* 24 (1888). 105.

TEBB, Miss (56, 134). On the transformation of Maltose to Glucose. *Journal of Physiology* 15 (1894). 421.

TEBB, Miss (61). Hydrolysis of Glycogen. *Journ. of Physiology* 22 (1898). 423.

THÉNARD (3). Sur la fermentation vineuse. *Ann. de Chimie* t. 46 (1803). 224.

THOMÉ (148). Ueber das Vorkommen des Amygdalins und des Emulsins in den bittern Mandeln. *Bot. Zeit.* 1865. 240.

THOMSON und RICHARDSON. Ueber die Zersetzung des Amygdalins durch Emulsin. *Ann. de Pharmacie* 29 (1839). 180.

TIEDEMANN und GMELIN (193). Die Verdauung nach Versuchen. Heidelberg 1826. Bd. I. 171.

TIEGEL. Ueber eine Fermentwirkung des Blutes. *Pflüger's Archiv* 6 (1872). 249.

TISCHUTKIN (226). Die Rolle der Bacterien bei der Veränderung der Eiweissstoffe auf den Blättern von *Pinguicula. Ber. d. deut. bot. Gesell.* 7 (1889). 346.

TISCHUTKIN (226). Ueber die Rolle der Mikroorganismen bei der Ernährung der insektenfressenden Pflanzen. *Arbeit. d. St Petersburg naturf. Gesell.* 1891. Abs. in *Bot. Centr.* 50 (1892). 304 and 53 (1893). 322.

TOLLENS. Ueber die Constitution der Pectinstoffe. *Liebig's Annalen* 286 (1895). 292.

TOLLENS und GANZ. Ueber Quellen und Salepschleim. *Liebig's Annalen* 249 (1888). 245.

TOLOMEI (325). *Real Acad. Linc.* 1896. 5, 1, 52.

TRAUBE, M. (444). Zur Theorie der Gährungs- und Verwesungs-Erscheinungen, wie der Fermentwirkungen überhaupt. *Poggend. Annal.* 103 (1858). 331.

TRAUBE, M. Zur Theorie der Fermentwirkungen. *Ber. d. deut. chem. Gesell.* 7 (1874). 115.

TRAUBE, M. Die chemische Theorie der Fermentwirkungen und der Chemismus der Respiration. *Ber. d. deut. chem. Gesell.* 10 (1877). 1984.

TREUB (171). Sur la localisation, le transport et le rôle de l'acide cyanhydrique dans le *Pangium edule. Ann. du Jardin bot. de Buitenzorg* 13 (1895). 1.

TUBBY and MANNING. Human succus entericus. *Guy's Hospital Rep.* 48. 271.

TURPIN (4). Sur la cause et les effets de la fermentation alcoolique et acéteuse. *Compt. Rend.* 7 (1838). 369.

ULRICH (56). *Chem. Zeit.* 1895. 19, 1523.

VADAM (327). Ferments oxydants de l'hellébore fétide. *Journ. Pharm. et Chim.* (6) 9 (1899). 515.

VAN RIJN. Die Glykoside. Berlin 1900.

VAN TIEGHEM (302). Sur la fermentation ammoniacale. *Compt. Rend.* 58 (1864). 533.

VAN TIEGHEM. Recherches physiologiques sur la germination. *Ann. des Sc. Nat.* (5) 17 (1873). 205.

VAN TIEGHEM (120). Inversion du sucre de canne par le pollen. *Bull. Soc. Bot. de France* 33 (1886). 216.

VASSILIEF (232). *Archives de l'Institut de médecine expérimentale de St Pétersbourg* t. III. 1893.

VAYLESTEKE. Contribution à l'étude de la diastase. *Bull. Acad. Roy. Belg.* (3) 24, 577.

VIGNAL (31, 103, 232, 264). Contribution à l'étude des bactériacées. *Thèse pour le doctorat-ès-sciences naturelles.* Paris.

VINES (226, 414).   On the digestive ferment of Nepenthes.   *Journ. Linn. Soc. (Bot.)* 15 (1877). 427.

VINES (19).   On the presence of a diastatic ferment in green leaves. *Annals of Botany* 1891. 409.

VINES (227).   The proteolytic enzyme of Nepenthes.   *Annals of Botany* 11 (1897). 563.

WARD, Marshall (91).   A lily disease.   *Ann. of Botany* 2 (1888). 319.

WARD, Marshall (342).   The Ginger-beer plant and the organisms composing it.   *Phil. Trans. Roy. Soc.* B. 1892. 125.

WARD (Marshall) and DUNLOP (165).   On some points in the histology and physiology of the Fruits and Seeds in Rhamnus.   *Ann. of Botany* 1 (1887). 1.

WARD (Marshall) and GREEN (Reynolds) (344).   A sugar bacterium.   *Proc. Roy. Soc.* Mar. 9, 1899.

WARRINGTON (381).   On Nitrification.   *Journ. Chem. Soc.* 1878 p. 44, 1879 p. 429, 1884 p. 637, 1891 p. 484.

WASSERZUG (121).   Sur la production de l'invertine chez quelques champignons.   *Ann. de l'Institut Pasteur* 1 (1887). 525.

WATSON.   Notes on the effect of alcohol on Saliva and on the Chemistry of digestion.   *Journ. Chem. Soc. Trans.* 1879. 539.

WEHMER (372).   Préparation d'acide citrique de synthèse par la fermentation du glucose.   *Compt. Rend.* 117 (1893). 332.

WEINLAND (144).   Beiträge zur Frage nach dem Verhalten der Milchzuckers im Körper, besonders im Darm.   *Zeit. Biol.* 38 (1899). 16.

WEINLAND (145).   Ueber die Lactase des Pankreas.   *Zeit. Biol.* 38 (1899). 607.

WENT.   On the influence of nutrition on the secretion of enzymes of *Monilia sitophila* (Mont.) Sacc.   *Koninklijke Akademie van Wetenschappen te Amsterdam* 1901 p. 489.

WENT en PRINSEN GEERLIGS (346).   Over suiker- en alcoholverming door organismen in verband met de verwerking der naproducten in de rietsuikerfabrieken.   *Bot. Zeit.* 1895. 2ª Abt. 143.

WENZ.   Ueber das Verhalten der Eiweissstoffe bei der Darmverdauung. *Zeit. f. Biol.* 22 (1886). 1.

WEYL und BISCHOF.   Ueber den Kleber.   *Ber. d. deut. chem. Gesell.* 13 (1880). 367.

WIESKE, SCHULZE und FLECHSIG.   Kommt der Cellulose eiweisssparende Wirkung bei der Ernährung der Herbivoren zu?   *Zeit. f. Biol.* 22 (1886). 373.

WIESNER (109, 419).   Ueber das Gummiferment ; ein neues diastatisches Ferment.   *Sitzber. d. M. N. der K. Akad. der Wiss. in Wien* 92 (1885). 40, and *Ber. d. deut. chem. Gesell.* (1885). 619.

WIGAND.   Das Protoplasma als Ferment-organismus.   Marburg 1888.

WIJSMAN (60). Diastase considered as a mixture of Maltase and Dextrinase. *Rec. Trav. Chem.* 9. 1–13.

WILLIS (437). De fermentatione. 1659.

WILSING. Ueber die Mengen der vom Wiederkäuer in den Entleerungen ausgeschiedenen flüchtigen Säuren. *Zeit. f. Biol.* 21 (1885). 625.

WINDISCH und SCHELLHORN. Sur la diastase digestive des albuminoïdes de l'orge germée. Abstracted in L'offre et la demande en brasserie 4 (1901). 552 from *Wochenschrift für Brauweri* June 15 1900.

WINOGRADSKY (381, 382). Recherches sur les organismes de la nitrification. *Ann. de l'Inst. Pasteur* 4 (1890). 213, 257, 760; 5 (1891). 92, 577.

v. WITTICH (46). Ueber eine neue Methode zur Darstellung künstlicher Verdauungsflüssigkeiten. *Pflüger's Archiv* 2 (1869). 193.

v. WITTICH. Weitere Mittheilungen über Verdauungsfermente. *Pflüger's Archiv* 3 (1870). 339.

v. WITTICH. Weitere Mittheilungen über Verdauungsfermente des Pepsins und seine Wirkung auf Blutfibrin. *Pflüger's Archiv* 5 (1872). 435.

v. WITTICH (39). Ueber das Leberferment. *Pflüger's Archiv* 7 (1873). 28.

v. WITTICH. Zur Statik des Leberglycogens. *Centralb. f. d. Med. Wiss.* 19 (1875). 113.

WOOD (232, 264). *Laboratory Reports, Roy. Coll. Phys. Edinburgh* vol. II.

WOODS (327). The destruction of Chlorophyll by oxidising enzymes. *Cent. f. Bakt.* 2 Abt. 5 (1899). 745.

WOOLDRIDGE. On the coagulation of the blood. *Journ. of Physiology* IV. (1883). 226, 367.

WOOLDRIDGE. Blood plasma as Protoplasma. *Arris and Gale lectures. Royal College of Surgeons,* June 1886.

WOOLDRIDGE (279). Beiträge zur Lehre von der Gerinnung. *Du Bois-Reymond's Archiv* 1888. 174.

WORTMANN (24, 75). Untersuchungen über das diastatische Ferment der Bacterien. *Zeit. physiol. Chem.* 6 (1882). 287.

WORTMANN (19). Ueber den Nachweis, das Vorkommen und die Bedeutung des diastatischen Enzyms in den Pflanzen. *Bot. Zeit.* 1890. 37, 582, 598, 617, 634, 659.

WROBLEWSKI (191). Zur Kenntniss des Pepsins. *Zeit. f. physiol. Chem.* 21 (1895). 1–18.

WROBLEWSKI (426). Ueber die chemische Beschaffenheit der Diastase und über das Vorkommen eines Arabans in den Diastasepräparaten. *Ber. d. deut. chem. Gesell.* 30 (1897). 2289.

WROBLEWSKI. Zur Classification der Proteïnstoffe. *Ber. d. deut. chem. Gesell.* 30 (1897). 3045.

WROBLEWSKI. Ueber die chemische Beschaffenheit der Diastase und über die Bestimmung ihrer Wirksamkeit unter Benutzung von löslicher Stärke. *Zeit. f. physiol. Chem.* 24 (1898). 173.

WROBLEWSKI. Was ist Osborn'sche Diastase ? *Ber. d. deut. chem. Gesell.*
31 (1898). 1127.
WROBLEWSKI. Ueber die chemische Beschaffenheit der amylolytischen
Fermente. *Ber. d. deut. chem. Gesell.* 31 (1898). 1130.
WÜRTZ (214). Sur la papaïne ; contribution à l'histoire des ferments
solubles. *Compt. Rend.* 90 (1880). 1379.
WÜRTZ. Nouvelle contribution à l'histoire des ferments solubles. *Compt.
Rend.* 91 (1880). 787.
WÜRTZ et BOUCHUT (426). Sur le ferment digestif du *Carica papaya.*
Compt. Rend. 89 (1879). 425.
WÜRTZ et BOUCHUT (214). Le Papaïn. Paris 1879.

YOSHIDA (309). Chemistry of lacquer. *Journ. Chem. Soc.* 43 (1883). 472.

ZOPF (230). Die Pilze. Breslau 1890.
ZUNZ. Die fractionirte Abscheidung der peptischen Verdauungsprodukte
mittelst zinksulfat. *Zeit. f. physiol. Chem.* 27 (1899). 219.

INDEX.

Printed in the United States
By Bookmasters